Fractal Approaches for Modeling Financial Assets and Predicting Crises

Inna Nekrasova
Southern Federal University, Russia

Oxana Karnaukhova
Southern Federal University, Russia

Bryan Christiansen
PryMarke, LLC, USA

A volume in the Advances in Finance, Accounting, and Economics (AFAE) Book Series

IGI Global
DISSEMINATOR OF KNOWLEDGE

Published in the United States of America by
IGI Global
Business Science Reference (an imprint of IGI Global)
701 E. Chocolate Avenue
Hershey PA, USA 17033
Tel: 717-533-8845
Fax: 717-533-8661
E-mail: cust@igi-global.com
Web site: http://www.igi-global.com

Library of Congress Cataloging-in-Publication Data

Names: Nekrasova, Inna, 1962- editor. | Karnaukhova, Oxana, editor. | Christiansen, Bryan, editor.
Title: Fractal approaches for modeling financial assets and predicting crises / Inna Nekrasova, Oxana Karnaukhova and Bryan Christiansen, editors.
Description: Hershey : Business Science Reference, [2017]
Identifiers: LCCN 2017022431| ISBN 9781522537670 (hardcover) | ISBN 9781522537687 (ebook)
Subjects: LCSH: Finance--Mathematical models. | Fractals.
Classification: LCC HG106 .F715 2017 | DDC 332.01/51472--dc23 LC record available at https://lccn.loc.gov/2017022431

This book is published in the IGI Global book series Advances in Finance, Accounting, and Economics (AFAE) (ISSN: 2327-5677; eISSN: 2327-5685)

British Cataloguing in Publication Data
A Cataloguing in Publication record for this book is available from the British Library.

All work contributed to this book is new, previously-unpublished material. The views expressed in this book are those of the authors, but not necessarily of the publisher.

For electronic access to this publication, please contact: eresources@igi-global.com.

Advances in Finance, Accounting, and Economics (AFAE) Book Series

Ahmed Driouchi
Al Akhawayn University, Morocco

ISSN:2327-5677
EISSN:2327-5685

Mission

In our changing economic and business environment, it is important to consider the financial changes occurring internationally as well as within individual organizations and business environments. Understanding these changes as well as the factors that influence them is crucial in preparing for our financial future and ensuring economic sustainability and growth.

The **Advances in Finance, Accounting, and Economics (AFAE)** book series aims to publish comprehensive and informative titles in all areas of economics and economic theory, finance, and accounting to assist in advancing the available knowledge and providing for further research development in these dynamic fields.

Coverage

- International Trade
- Field Research
- International Economics
- Microeconomics
- Public Finance
- Accounting information systems
- Investments and Derivatives
- Managerial Accounting
- Theoretical Issues in Economics, Finance, and Accounting
- Health Economics

IGI Global is currently accepting manuscripts for publication within this series. To submit a proposal for a volume in this series, please contact our Acquisition Editors at Acquisitions@igi-global.com or visit: http://www.igi-global.com/publish/.

The Advances in Finance, Accounting, and Economics (AFAE) Book Series (ISSN 2327-5677) is published by IGI Global, 701 E. Chocolate Avenue, Hershey, PA 17033-1240, USA, www.igi-global.com. This series is composed of titles available for purchase individually; each title is edited to be contextually exclusive from any other title within the series. For pricing and ordering information please visit http://www.igi-global.com/book-series/advances-finance-accounting-economics/73685. Postmaster: Send all address changes to above address.

Titles in this Series

For a list of additional titles in this series, please visit: www.igi-global.com/book-series

Regulation and Structure in Economic Virtualization Emerging Research and Opportunities
Denis Ushakov (Suan Sunandha Rajabhat University, Thailand)
Business Science Reference • copyright 2018 • 238pp • H/C (ISBN: 9781522549666) • US $145.00 (our price)

Accountancy and the Changing Landscape of Integrated Reporting
Ioana Dragu (Babes-Bolyai University, Romania)
Business Science Reference • copyright 2018 • 303pp • H/C (ISBN: 9781522536222) • US $195.00 (our price)

Foreign Direct Investments (FDIs) and Opportunities for Developing Economies in the World Market
Venkataramanaiah Malepati (University of Gondar, Ethiopia) and C. Mangala Gowri (University of Gondar, Ethiopia)
Business Science Reference • copyright 2018 • 280pp • H/C (ISBN: 9781522530268) • US $215.00 (our price)

Economic Growth in Latin America and the Impact of the Global Financial Crisis
Mauricio Garita (Universidad del Valle de Guatemala, Guatemala) and Celso Fernando Cerezo Bregni (Universidad del Valle de Guatemala, Guatemala)
Business Science Reference • copyright 2018 • 193pp • H/C (ISBN: 9781522549819) • US $180.00 (our price)

Uncertainties and Risk Assessment in Trade Relations
Marianne Ojo (George Mason University, USA)
Business Science Reference • copyright 2018 • 385pp • H/C (ISBN: 9781522541318) • US $225.00 (our price)

Economic Reforms for Global Competitiveness
Denis Ushakov (International College Suan Sunandha Rajabhat University, Thailand & Russian Academy of National Economy and Public Administration–South, Russia)
Business Science Reference • copyright 2018 • 363pp • H/C (ISBN: 9781522538561) • US $235.00 (our price)

Alternative Decision-Making Models for Financial Portfolio Management Emerging Research and Opportunities
Narela Spaseski (International University of Sarajevo, Bosnia and Herzegovina)
Business Science Reference • copyright 2018 • 333pp • H/C (ISBN: 9781522532590) • US $175.00 (our price)

Ranking Economic Performance and Efficiency in the Global Market Emerging Research and Opportunities
Benjamen Franklen Gussen (University of Southern Queensland, Australia & Australian Law and Economics Association, Australia)
Business Science Reference • copyright 2018 • 372pp • H/C (ISBN: 9781522527565) • US $195.00 (our price)

IGI Global
DISSEMINATOR OF KNOWLEDGE

701 East Chocolate Avenue, Hershey, PA 17033, USA
Tel: 717-533-8845 x100 • Fax: 717-533-8661
E-Mail: cust@igi-global.com • www.igi-global.com

Editorial Advisory Board

Table of Contents

Detailed Table of Contents

Econophysicists have begun to make progress in answering significant questions. In particular, these collaborations have the potential to change the paradigm for understanding fluctuations. New theoretical approaches to predict complex markets may be proposed, by the captivating formulation of the stock market concerning statistical correlation to be given, where some simple (non-differential, non-fractal) expressions are also suggested as general stock price formulae in a closed form that can generate a variety of possible price movements in time. A given attribute of mechanics may be submitted as a likely option to cover the price movements regarding traditional concepts where utilising stock mechanics to grow the portfolios in real markets may be realised. The ideas prove useful in risk evaluation, extreme value statistics, critical limit theorems for sums of independent variables with power law distribution, random walks, fractals, and multfractal formalisms, etc.

The chapter is aimed at identification of criteria to select financial assets for investment; observing price fluctuations at small time intervals (up to one week) as possible predictors of the future of a significant increase in the price fluctuations amplitude; determining a fractal dimension of the financial markets on the basis of R/S-analysis; constructing a fractal index indicator to identify a bifurcation point, which gives birth to a possibility of crisis phenomena in economy. Therefore, the practical significance of the chapter lies in the idea of equipping academics and practitioners with new methods and tools for analysis and forecasting future development and dynamics of the financial markets.

This chapter uses the relationship between behavioral factors and the creditworthiness of small-scale enterprises to increase access of SMEs to credit facilities. The inability of several small businesses to secure loans cannot be overemphasized. Heuristics affecting entrepreneurs are explained in this chapter, and a regression model showing the dependence of creditworthiness on behavioral factors is proposed. If banks consider using psychometric tools in testing for creditworthiness of small-scale entrepreneurs, access to credit facilities will be significantly increased and businesses will flourish. Regressesion models such as the one explained in this chapter may be imbedded in psychometric tools to enhance creditworthiness testing and improve the quality of loans that banks give.

The purpose of this chapter is twofold: to assess the relationship of the nonlinear dynamics of pension systems and economic cycles, and to develop a descriptive evolutionary model of the stages of pension systems. Hodrick-Prescott filter is used to identify cycles in the pension and economic dynamics. The study proved empirically that the evolutionary dynamics of pension systems depends on the cyclicity of national and world economies. In addition, the bifurcation points associate with big Kondratiev cycles, and the fluctuations of the indicators of pension systems correlate with the medium-term Juglar cycles. The crisis starts pension reforms. The results of this study indicate that public pension spending growth is countercyclical and coincident indicator relative to the global business cycle in 13 countries from 21 of the OECD countries studied. The amount and volatility of public pension spending depends on the basic pension model and has higher values in Bismarck-model countries.

With the growing tension in the global community and the instability of the overall situation, problems with financial centers and regions have arisen, and as a result, the necessity of reforms have appeared in the global governance system and in the financial and economic system. BRICS countries are very interested in solving problems and developing relations within the bloc. Each country pursues its own interest, but the emerging dialogue of countries allows them to find a compromise in complex issues. The current economic situation in the world, where Western countries speak the language of sanctions, makes BRICS countries strengthen their interaction. It concerns the provision of financial and economic security. Increased financial risks in the capital market as well as the dominance of the dollar and the euro level, push BRICS countries to strengthen the financial mechanisms. BRICS countries support the creation of a global network of financial instruments.

Tax authorities often have to make decisions under conditions of incomplete information about the actual situation. The incompleteness and uncertainty of information may arise from the actions of the two sides with distinct interests. One side is the tax authority (local, regional, or national level), and the other is the taxpayers (organizations or individuals). This largely explains the interest of tax authorities in the creation of software capable of providing quantitative and qualitative analysis of the declarations submitted by taxpayers to identify suspicious declarations for their additional inspection. The purpose of this chapter is to carry out a comparative pre-forecast analysis of the time series enterprise performance indicators and time series of its tax payments.

The purpose of this chapter is to develop an analytical system for forecasting prices of financial assets with the use of artificial neural networks technology. Proposed by the authors, the analytical system consists of several neural networks, each of which makes the forecast of financial assets prices. The system includes recurrence (with feedback) neural networks with sigmoidal activation formula. This allows the networks to "remember" a sequence of reactions to the same stimulus. The learning process of neural networks is performed using an algorithm of back propagation of error. The key parameters of forecast for this analytical system are the indicators presented by the terminal MetaTrader 4-broker Forex Club: Average Directional и Movement Index; Bollinger Bands; Envelopes; Ichimoku Kinko Hyo; Moving Average; Parabolic SAR; Standard Deviation; Average True Range; and others.

Speaking of the preliminary findings of this chapter and considering empirical data observed in the course of stock exchange trading, it is possible to forge a connection between the actual behavior of the investors and the NLP meta-program classes. In particular, there is a palpable correspondence between the 13 distinct meta-program categories affecting workplace motivation and performance, commonly known as the language and behaviour profile or LAB Profile, of Shelle Rose Charvet, and the behavior patterns of investors. This may represent a development of the idea of "the conventional wisdom" formulated in his time by John Kenneth Galbraith. However, in its new incarnation of collective conscious and collective subconscious, it may affect decision making processes around the globe.

Preface

In the context of the recent financial crisis that has engulfed the most influential and economically developed countries, there is an increasing interest in research methods that can not only characterize the ongoing processes, but also with a high probability predict the waves of socio-economic development. Over the past 40 years, a language has emerged that could discuss what happens in seemingly random natural systems-for example, weather or financial markets-and can be forecasted. The classical mathematical models used throughout much of the twentieth century usually predict that extreme events with rapid changes could never occur. The cornerstone for calculating the behavior of financial markets and macroeconomic analysis is the modern theory of the portfolio (Markowitz, 1952), which tries to maximize profitability for a given level of risk. The mathematics underlying the theory of the portfolio copes with extreme situations, considering major market shifts that are too unlikely. Indeed, the theory of the portfolio can explain what happens in 95 percent of the cases. But its economic picture of the world does not reflect reality, if we assume that global events are part of the remaining five percent.

Another paradigm of economic and finance theory preoccupied academic and practitioners' world for at least twenty years was the efficient market hypothesis (EMH). After the financial crisis of 2008, additional empirical data analysis on financial markets this paradigm was over (Fox, 2009) and removed by the more effective fractal market hypothesis (FMH). There are phenomena observed in real data collected from trends in concentration of economic activities, financial markets, investment horizons, industrial dynamic etc. that cannot be explained by the traditional economic and finance theories (Farjoun & Levin, 2011; Andronache et al., 2016). Local heterogeneous trends and randomness together with global determination of the economic dynamic fortify the fractal approach positioning.

Since its introduction in 1963, the fractal approach to the analysis of financial markets, in particular, and global economic development as a whole, is still an innovative approach to the study of financial phenomena, the use of chaos theory and quantum theory to characterize and predict market predictability (Mandelbrot, 1963). The Mandelbrot set is a fractal, a term invented by Mandelbrot himself, which has a complex definition, but, in fact, refers to coarse forms that look the same if viewed from a long-range or close-up perspective. This phenomenon is common in nature: for example, broccoli has a fractal shape, mountain ranges, tree branches, and clouds. In each case, a small fragment, when it is enlarged, resembles a larger whole. With the invention of the fractal geometry theory, Mandelbrot expanded the possibilities of mathematical analysis so that it could be applied to complex and simple objects.

The concept of fractals was expanded to all areas of science and technology; however, the concept was born in the field of economics when Mandelbrot studied the changes in prices in the open market.

He empirically discovered that the chart of changes in market prices for cotton looks like another chart with a different time resolution. Such a large-scale invariance can be the key to describing many complex phenomena around us.

More than 40 years have passed since the initial discovery of fractals, and the economy is becoming one of the hottest topics in applying the fractal approach. Fractals, such as the Mandelbrot set, are usually associated with a study of complexity, also known as the fashionable nickname of chaos theory. Fractal analysis of the market reveals and characterizes the order hidden behind seeming random financial events, and determines the probability of future fluctuations.

The defining characteristic of a fractal picture is that it has a repeating pattern in all time frames; that is, smaller components have the same basic shape and structure as larger components. Markets are fractal and look the same. Thus, it is impossible to view the unmarked price schedule and determine whether it is an hourly, monthly or even a 5-minute trading schedule. The main advantage of knowledge of fractal patterns and projections is that they determine specific critical equilibrium points, where the potential energy can be directed to one or another route. It is interesting to note that in the field of fractal analysis, as practiced in financial markets or economic dynamic in general, there are few publications, especially of applied nature (Brenner et al., 1991; Mantegna et al., 1995; Inaoka et al., 2004). This circumstance makes the proposed volume highly relevant.

As a comprehensive collection of cases, the publication represents efforts of experts in applied methods in economics. Authors of the volume cover a variety of issues in the fractal approach implementation, illustrating them with examples of macro-, meso- and micro- level. The aim of the volume is to provide an introduction to the use of fractal approach to describe the logic and trends in several cases within in comparative perspective and to combine discussion of theoretical framework with empirical findings. It gives a balanced and insightful analysis of the scope and limits of the fractal approach.

The chapters provide provoking and comprehensive overview of academic thinking and practical cases. Therefore, some chapters would be interesting for researchers working in the relatively new areas of econophysics, neuro-linguistic programming, behavioral economics and neural networking. Some researchers and practitioners would find interesting case-studies with comparative inter-state or in-state perspectives. Finally, there is a group of specifically practice-oriented chapters focusing on development of economic phenomena, like taxpaying, pension system, creditworthiness, banking, and so forth.

The volume is unique not only by its wide scope but also its balance between theory and practice. It has an added-value bringing an innovative theory with relevant empirical evidences coming from the real-world experience. Authors write from the grassroots' perspective to reflect current realities of global and country-based economic development. The volume covers different and diverse topics related to challenges and opportunities of the fractal analysis implications to complex social and economic phenomena.

In Chapter 1, Cem Cagri-Donmez introduces the econophysics approach to analysis of uncertainty in dynamics of complex market structural models. As it comes from the idea of predictability of extrapolative fluctuations in the capital markets, the chapter presents a specific methodological approach design. Since the results of experimental work could be validated through comparative studies of historic data, the new and alternative indicators could be added to the ordinary technical analysis in Econophysics. Market indicators could be exploited within the context action-reaction mechanisms reliable and open to experimentation, in which laws of energy are also applicable, and wherein indicators changes (other than those pertaining to data numbers) can occur from one country to another, from one market to another, and even from one financial instrument to another within the same market. The dynamic phenomena of the econophysical architecture presented by modeling form the degree of its applicability to the markets and provide empirical data collected from high-frequency futures data corresponding to different futures assets. It demonstrates the effect of a trade order on the underlying price, which depends on the duration of mutual trade by examining the volatility and clustering of stock fluctuations. The author, focusing on the temporary progress of financial assets, is considering the possibility of parallel forecasts of the proposed model with a relatively rich ensemble of historical implementations. The presentation of financial time series in the clustering of volatility can be modeled and evaluated using a model for a particular market with a given risk factor. Although it seems that prices are formed under external influences, this fact indicates that internal market events determine prices, and external dynamics are indicated as a cause only after events have occurred.

In Chapter 2 as it is explained by Inna Nekrasova, Oxana Karnaukhova and Oleg Sviridov the index of fractality is used for the traditional analysis in the fractal structure of time series. With the same amount of data, the index (μ) is determined several times more accurately than the Hearst index (H). In this case, the minimum scale necessary for determining μ with a reasonable accuracy is several tens of points, whereas for the determination of H, several thousand data are needed. This fact makes it possible to use μ as a local fractal index in the analysis of time series. Financial time series demonstrate the behavior in which fluctuations in the fractal structure lead to a change in the type of behavior of the series. The research demonstrates that for the time series of prices in financial assets the value of μ is related to the stability of the series. In this case, the larger the value of μ is, the more stable the series behavior.

Chapter 3 attempts to investigate the impact of the long memory property on the market risk across two sub-periods based on four country-specific events (Tunisian revolution; Egyptian revolution; assassination of Prime Minister Rafik El Hariri and series of suicide bomber in Morocco). Salim Ben Sassi and Azza Bejaoui use the modified R/S test and the GPH test to demonstrate long-range dependence behavior, which changes over time. The focus is on the process, where the long memory property in returns and volatility tend to be significant in the pre-crisis period for all emerging markets. The study reveals that the long memory property in the post-crisis period seems to be as significant as in the pre-crisis period. As authors state, "market thinness and nonsynchronous trading, among others, as well as the country-specific circumstances can be considered as factors responsible for such behavior of long memory property before and after political events". The empirical results also show that long memory models maybe useful for depicting cyclical fluctuations in index returns, namely: the information from "today" is not instantaneously absorbed by the price in the stock market and also investors react with delay to information. So this leads to re-examining the concept of informational efficiency in the Middle East and North Africa (MENA) markets during the specific-country events.

Predicting financial crisis is also the focus of the Chapter 4 with an application to the case of Turkey. Dr. Asli Yuksel Mermod, Ülkü Yüksel and Catherine Sutton-Brady consider several incidents to highlight fundamental causes of financial crises, such as The Great Depression of 1929 - early 1940s, the 1997 Asian Financial Crises, the 1998 Russian Financial Crisis, and the Liquidity Crisis of 2008, and make a comparison among them and their various outcomes. This case study specifically analyses the Turkish Banking System that was restructured after the enormous financial crises in Turkey in 2001, which caused many Turkish banks to collapse. However, the precautions taken in the aftermath of the financial turmoil allowed them to survive the liquidity crises in 2008. The study specifies the cues that emerge in the financial system that may help governments predict upcoming financial crises through early warning signals.

In the aftermath of the global financial crisis, the author of the Chapter 5, Andrea Quintiliani aims to shed light on the determinants of the financial distress costs between Italian and Germany small and medium enterprises (SMEs). The author proposes an innovative formulation of the expected costs arising from the financial crisis, expressed as a result of the expected likelihood of a financial disaster multiplied by the total financial cost of the disaster, if such a situation occurs. The model is estimated using the panel data methodology for samples from Italy and Germany as two European countries. The obtained results show that the amount of expenses related to posts depends on the use of derivative financial instruments, the use of intangible assets and relations with local banks (small local banks, rather than large banking groups). It is emphasized that current work limits the scope of its research to several variables without completely affecting other elements of uncertainty that could adversely affect the likelihood of financial shocks and the value of SMEs.

In Chapter 6, Dmitry Shevchenko and Ellah Igoche Godwin demonstrate the effects of behavioral factors on the creditworthiness of small scale enterprises to increase access of SMEs to credit facilities. Authors remark that the inability of several small businesses to secure loans cannot be overemphasized. Heuristics affecting entrepreneurs are explained in this paper and a regression model showing the dependence of creditworthiness on behavioral factors is proposed. If banks consider using psychometric tools in testing for creditworthiness of small scale entrepreneurs, access to credit facilities will be significantly increased and businesses will flourish. Regression models may be imbedded in psychometric tools to enhance creditworthiness testing and improve the quality of loans that banks give.

The task of Chapter 7 by Larisa Yakimova is to assess relationship of nonlinear dynamics in pension systems and economic cycles, and to develop a descriptive evolutionary model of pension systems. The author applies the Hodrick-Prescott filter to determine cycles in the pension system and economic dynamics. An empirical study shows that the evolution of pension systems depends on cyclicality in the national and global economy. The point of bifurcation is associated with the large Kondratieff cycles, while the fluctuations in indicators in pension systems correlate with the medium-term Yoglar cycles. A comparative study of the OECD countries indicates the countercyclicity of growth in public spending on pension provision and the relativity of the coincident indicators in the global business cycle in 13 countries from 21 analyzed. The size and volatility of public pension spending depends on the basic pension model and have higher values in the countries of the Bismarck model.

Summing up the potential results of the measures in financial and economic security of BRICS countries examined in Chapter 8, Elena V. Fomicheva and Marina S. Shikun state that they reinforce the claims of five states for their wider representation in international financial institutions and, in general, in the management of the world economy. The creation of a joint financial infrastructure within the BRICS group will give an important impetus to the real sector of the economy. The use of national currencies in mutual settlements and lending will reduce the association's dependence on the conjuncture in traditional centers of the world economy. The development of measures to manage financial globalization in the face of increasing threats and challenges is necessary to ensure the financial and economic security of the BRICS group. Looking at the near future, there are serious reasons to discuss the trend of further strengthening and stability of all BRICS currencies. They are associated with strengthening the positions of countries within the group's closer economic alliance, creating their own "financial security cushion" in the form of a banking and a pool (in the future, it is possible to create an analogue of SWIFT and a joint aggregate currency) and a more intensive shift in world trade in CHN and rubles. The forecast for the future of currencies has good reasons.

Chapter 9 by Galina Artemenko describes the use of fractal analysis in assessing the reliability of taxpayer's information. The analysis concerns economic and mathematical tools aimed at creating information systems for decision support in complex interactions between tax authorities and taxpayers. For the purposes of strategic planning (forecasting) of economic indicators in the behavior of taxpayers, the author recommends a model of preliminary forecasting. This model came from the R / S analysis algorithm and phase analysis. Comparative analysis of the series of the dynamics of tax revenues and the corresponding series of the tax base dynamics reveals regularities and similar trends in the dynamics of time series of the taxation and indicators of accounting reporting of taxpayers. The practical significance of the study lies in the substantial applicability of automated information systems in any department and satisfaction of the need for tax control with a reduction in the resource costs carried by governments.

Chapter 10 written by Dmitry Averchenko and Artem Aldyrev is based on the results of applying neural networks for modeling of financial assets. According to the authors, neural networks as a distinguished statistical method of multidimensional classification is based on the parallel processing of information and possesses the ability to conduct self-learning. This means the neural networks allow to obtain a valid result based on data that was not encountered in the learning process. The ability to model nonlinear processes, adaptability and work with noisy data could apply neural networks to solve a wide class of financial problems.

Chapter 11 written by Mamed Babaev and Oxana Savenko applies neuro-linguistic programming based modeling to stock markets. Authors note that "in real life, many decisions do not entirely correspond to the rational part of the human intelligence and therefore LAB profiling may play the role of a rectifier that makes it possible to "weaponise" the concept of the Nash equilibrium within the framework of stock trading". It is recognised that not all the NLP modeling techniques are fruitful due to their intrinsic peculiarities, therefore the results of NLP implications are highly experimental and debatable. And yet, the astonishing success of the proponent of the reflexivity theory George Soros has already demonstrated an innovative character and possible implications of NLP to financial market analysis and forecasting.

The volatility spillover during crises is the main focus of Chapter 12 by Rıfat Karakuş, Şeyma Yılmaz Küçük, and İbrahim Bozkurt. It compares developed and developing markets in the situation of crises and questions possible changes in volatility spillover. The analysis demonstrates that volatility spillover from the developed markets to the developing markets is not valid in crisis periods, while commencing again in the post-crisis period. Moreover, the volatility spillover from developing to developed markets being limited in the crisis period increases in the post-crisis period. The promising idea come from the study is that despite the prejudice the developing markets constitute an important alternative for investors during the crisis.

Chapter 13 on Riesz Potential in Generalized Hölder Spaces by Boris Vakulov, Galina Kostetskaya, and Yuri Drobotov provides an overview of the advanced studies of the multidimensional Riesz potential operator in the generalized Hölder spaces. While the interest is represented by mathematical modeling in economics, theoretical physics and other fields of knowledge, the Riesz potential plays a significant role in the analysis of fractal sets, and this aspect is briefly described. The generalized Holder spaces provide convenient terminology for the formalization of the concept of "smoothness", which is described here. Operators of potentials of constant and variable order are considered, including a two-pole spherical potential. As a sphere, in a sense, a convenient set for analysis, there are two results, proved in detail: conditions for spherical fractional integration of a variable order for being bounded in generalized Holder spaces that the local continuity of the module occupies the dominant that can vary from point to point, and the ones, for a constant order of two-pole spherical potential type of operator.

Financial markets and even economic systems should be studied as chaotic, non-linear systems, but only some study them. But over the next few years, this will certainly change. We hope these chapters will generate ideas for future research and for the development of relevant policies that will influence global as well as national prosperity and will help to predict future crises. Fractal analysis pushes the still unexplored border of economic thought, but we believe that in 10 years the fractal analysis will assume its well-defined place.

Oxana Karnaukhova
Southern Federal University, Russia

Inna Nekrasova
Southern Federal University, Russia

Bryan Christiansen
PryMarke, LLC, USA

REFERENCES

Andronache, I. C., Peptenatu, D., Ciobotaru, A.-M., Gruia, A. K., & Gropoşilă, N. M. (2016). Using Fractal Analysis in Modeling Trends in the National Economy. *Procedia Environmental Sciences*, *32*, 344–351. doi:10.1016/j.proenv.2016.03.040

Brenner, Y. S., Kaelble, H., & Thomas, M. (1991). *Income Distribution in Historical Perspective*. Cambridge, UK: Cambridge University Press.

Farjoun, M., & Levin, M. (2011). A Fractal Approach to Industry Dynamism. *Organization Studies*, *32*(6), 825–851. doi:10.1177/0170840611410817

Fox, J. (2009). *Myth of the Rational Market*. Harper Business.

Inaoka, H., Ninomiya, T., Taniguchi, K., Shimizu, T., & Takayasu, H. (2004). *Fractal Network derived from banking transaction – An analysis of network structures formed by financial institutions*. Bank of Japan Working Paper. Retrieved in January 06, 2018, from: http://www.boj.or.jp/en/ ronbun/04/data/wp04e04.pdf

Mandelbrot, B. B. (1963). The variation of certain speculative prices. *The Journal of Business*, *36*(4), 394–419. doi:10.1086/294632

Mantegna, R. N., & Stanley, H. E. (1995). Scaling behavior in the dynamics of an economic index. *Nature*, *376*(6535), 46–49. doi:10.1038/376046a0

Markowitz, H. M. (1952). Portfolio Selection. *The Journal of Finance*, *7*(1), 77–91.

Chapter 1
An Econophysics Approach to Introduction Uncertainty in Dynamics of Complex Market Structural Models

Cem Cagri Donmez
Marmara University, Turkey

ABSTRACT

Econophysicists have begun to make progress in answering significant questions. In particular, these collaborations have the potential to change the paradigm for understanding fluctuations. New theoretical approaches to predict complex markets may be proposed, by the captivating formulation of the stock market concerning statistical correlation to be given, where some simple (non-differential, non-fractal) expressions are also suggested as general stock price formulae in a closed form that can generate a variety of possible price movements in time. A given attribute of mechanics may be submitted as a likely option to cover the price movements regarding traditional concepts where utilising stock mechanics to grow the portfolios in real markets may be realised. The ideas prove useful in risk evaluation, extreme value statistics, critical limit theorems for sums of independent variables with power law distribution, random walks, fractals, and multfractal formalisms, etc.

INTRODUCTION

In this chapter, we will discuss financial data analysis base on Econophysical approach focus on the context as a complex market, and we consider some key features of the relationships that exist between economics and physics. Enable to understand and analyse financial systems (that means Complex Systems such as stock markets, exchange markets...etc.) and applied for Econophysical phenomena. The Complex Markets have become progressively more globalised; many buyers and sellers now increasingly have the entire world economic stage as their planning horizon. International transactions are increasingly important to companies and financial institutions. However, such internationalisation has been forced

DOI: 10.4018/978-1-5225-3767-0.ch001

to cope with increased risk arising from deregulated financial markets. Exchange rates and interest rates are now much more volatile compared to earlier decades; they are now more like traditionally volatile outcomes such as share prices. Financial derivative products are means of avoiding or minimising this risk and uncertainty. In the multivariate systems (Complex Market), fractality is between the stochastic and deterministic approaches or stochastic rather than deterministic. Originates as multiplicative interaction results. The models of Volatility diffusion with multiple stochastic factors can generate the structure of fractality. In a few cases, such as exchange rates, the underlying structural equation will also give increase to fractality. Fractals are the common name for complex geometric shapes that have the property of self-similarity. Fractal principles can be used to develop predicting algorithms. Financial predicting Modelling exhibits high degrees of nonlinear variability and frequently has fractal properties. When the fractal dimension of a base on financial time series is non-integer, this associated with two features: (1) inhomogeneity: extreme fluctuations at irregular intervals, and (2) scaling symmetries: proportionality relationships between variations over different separation distances.

New theoretical approaches to predict complex markets may be proposed, by the captivating formulation of the stock market concerning statistical correlation to be given, where some simple (non-differential, non-fractal) expressions are also suggested as general stock price formula in a closed form that can generate a variety of possible price movements in time. A given attributes of mechanics may be submitted as a likely option to cover the price movements regarding traditional concepts where utilising stock mechanics is to grow the portfolios in real markets may be realised. The ideas that have proved useful in risk evaluation, extreme value statistics, critical limit theorems for sums of independent variables with power law distribution, random walks, fractals and multifractal formalisms, etc. We are discussed in an immediate and direct way so as to provide ready-to-use tools for analysing and representing power law behaviour in natural phenomena.

Contrariwise, fractal markets hypothesis (FMH) (Peters, 1994) has been constructed based on the most general characteristics of the markets. In its core, it builds on a notion altogether omitted in efficient markets hypothesis (EMH) liquidity. In according with FMH, liquidity provides smooth pricing process in the market, making it stable. If liquidity ceases, the market becomes unstable and extreme movements occur. In the literature, FMH usually connected with detection of fractality or multifractality of the price processes of financial assets (Onali et al., 2011). However, it has not put to the test on its predictions about causes and implications of critical events in the financial markets. Currently under development of microscopic models of financial markets is to reproduce the observed statistical features of market movements.

This chapter divided into 11 main sections. The second part Conceptual Literature Review, The next part dedicates to the methodological interaction between The Fields of Physics, Finance and Economy; in the third focuses on the meaning of Econophysics with Perspective of Complex Market Variability; Section four deals with the Existing Financial Variability Structure, particularly regarding the five section the Indeterminateness of Chaotic Structures; following section is Micro Structure of Markets; subsequent to section seven Random Walk Model; afterwards section eight Heterogeneous and Fractal Market Hypotheses; in the ninth focuses on The Relationship Between Chaos & Physics; the conclusion follow.

CONCEPTUAL LITERATURE REVIEW

The field's roots are dated back to a 1987 conference of physicists and economists at the Santa Fe Institute. The Boston physicist H. Eugene Stanley described the term "Econophysics" in 1995. In this research, after providing a definition and a brief historical background of Econophysics, a summary of some applications of models and methods of statistical physics to economic problems are introduced. Econophysicists have begun to make progress in answering significant questions. In particular, these collaborations have the potential to change the paradigm for understanding fluctuations. Moreover, then a new word, econophysics, has entered the scientific realm. The first explicit use of the latter term we are aware of dates back to a "Workshop on Econophysics" held in Budapest in July 1997, while the first book containing the word econophysics in its title (Mantegna et al., 2000). The origins of the word, which combines the names of the two disciplines economics and physics, are the result of physicists starting to apply statistical physics to finance: high-frequency finance, financial risk and correlations, and complexity in finance are parts of the titles of this book (Dacorogna et al., 2001).

As we may see, every action, every form or every activity has its own unique formulations in financial complex systems. If we accept it, then we have also to agree that index graphics cannot express by a single formula. The reason for that is that these graphics generated in the course of time (every minute, every hour, and every day) by many activities (and Econophysics is certainly interested in the reasons for those daily fluctuations). Therefore, we need another formula for every action, and in this context, we will jump right into some physics' topics in this section and then we will express how the physics' information applied to the econophysics.

Nevertheless, the precise relationship between economics and physics dates back at least to the 18th and 19th centuries, when the Marginalises began to make extensive use of mathematics, borrowing their tools from physics. Other social sciences have also tried to borrow the mathematical methods of physics, but economics has been the more devout. Thus, the debate on the use of mathematics in economics has been going on since the 1870s (Debreu, 1991). A large asymmetry that lasted until only a few years ago must be stressed, i.e. the fact that only economists contributed to the debate of the methodological interaction between physics and economics. A notable exception is that of the theoretical physicist Ettore Majorana who wrote an article in the 1930s on the meaning of statistics in physics and social sciences (Majorana, 1942). As we are interested in the interaction between physics and economics, a first section has to consider the method/methods of scientific inquiry briefly.

The primary objective of activity in Econophysics has been the financial markets and having to accept the stock market as a complex network of natural events that to some degree follow the rules of economics, which provides challenges for physicists by navigating through its terabytes of well-defined time series data. Econophysicists were to apply to social phenomena the methodology that proved so successful in physics and based on the territory of interest and analyse the socioeconomic phenomena through methodologies typically used in statistical and theoretical physics. One of the most important objectives of statistical mechanics is to find some of the macro-states which contain a huge number of microstates and which would most appropriately represent the thermodynamic equilibrium conditions and Investigate the connection between econophysicists and economists. It is an Investigative study of the probability of distribution and price changing.

Econophysics has already made some significant empirical contributions to our understanding of the social and economic world. These fall mainly into the areas of finance and industrial economics, where in each case there is a significant amount of reasonably clear data. More recently, Econophysics has also begun to tackle other areas of economics where data is in short supply and much less reliable. Also, econophysicists have attempted to apply the theoretical approach of statistical physics to try to understand empirical findings. Here the motivations behind the Econophysical perspective will be laid out and assessed (including the flaws in mainstream economic theory). We will question whether it is necessary (conceptually and practically) to shift to this new approach. Econophysics is an evolving enterprise; whatever it is and whoever is doing it is not going to stop shortly, whatever its longer-term evolution. As ecological pressures force physicists out of physics and into complex systems.

It has been a long-proven fact that physics and money have many common characteristics. After all, managing money and physics experiments both include dealing with equations and numbers. However, recent studies show that this similarity goes way beyond this simple fact. Physics is based on developing mathematical models for specific physical events in the world. From the movements of molecules to the dynamics of stars in a galaxy, physics is trying to solve some complex issues. Some scientists have been working recently on how the models developed for physics can be applied to the "financial world", and they are focused on ways to adapt mathematical models obtained in the "molecular world" to financial literature. The banking sector took a particularly strong interest in this subject. Many models have been developed using processes creating random results. Furthermore, specific research has been done on financial instrument pricing, future movements of money, capital markets and trends of the market. This research has attempted to provide instruments for investors to reach profitable investment strategies by using the results of these studies.

Amongst its aims, is to apply a more precise empirical methodology, which will not include the biases that are perceived in many classical economic theories. The major methodological tools from physics that are most useful in this interdisciplinary field are statistical physical methods of investigating systems that include a high degree of uncertainty, such as nonlinear dynamics and chaos theory. Econophysics' reasoning is as follows: complex price movements occur because many investors, who are creating these movements, are interrelated to each other in a system depending on specific rules, just as in particle physics. According to econophysicists, the price behaviour of stocks results from the interactions among the speculators and corrects itself accordingly, as interactions among molecules are creating gas pressure. Physics theories are models that reflect the structure and the functioning of the universe. Modern science, or rather classical physics developed by Newton and his contemporaries, imitates our world in which enormous complexity is nested in a perfect working order to an accurately working mechanistic model. According to this model, every component of the universe contributes mathematically determined forces to each other; and these components, under the influence of these forces, make the predetermined moves depending completely on these forces and their initial states.

There are many interesting and valuable open questions about market dynamics that interest researchers' econophysicists. Raising questions about how to properly measure and explain the important properties of market dynamics, about the stability of markets, and about what the differences are in the behaviour of variants of markets. There is an interest in how the dynamics of markets impacts society. Econophysics, also known as the physics of finance was started in the mid-1990s by several physicists working in the subfield of Statistical mechanics. They decided to tackle the complex problems posed by economics, especially by financial markets.

In general, scientists search for patterns and call the events which do not conform to their conceptual framework as anomalies since these do not match their conceptual framework based on symmetry. However, most of the entities in our physical world do not obey Euclidean geometry at all and exhibit different symmetries. Through the application of Euclidean geometry to a drawing, we can only create an approximation of a tree. In the real world, trees consist of a network of branches which are very similar to the overall shape of a tree but each branch is different (Peters, 1994).

Furthermore, there are other branches on branches on smaller scales (Successive Generation of Branches). The individual branch level, each branch has a different size but share certain common properties. This is call "Self-Similar Property" is one of the features of Fractal geometry. That means the actual structure of a tree includes both local randomness and deterministic point of view.

The other way around, could be fluid heated from below. Near the source, the fluid is heated by the way of convection. Then the fluid is going to reach an equilibrium state in which maximum entropy occurs. During this heating period, all fluid molecules move independently. When the temperature passes a critical level, molecules which move independently start to behave coherently, that means heat flows by means of convection. In that case, the convection result is known by scientists but direction of roles of molecules is unknown. That means local randomness and global determinism coexist together. From this point of view stock markets are also driven both by microeconomic considerations such as profit levels of firms and macroeconomic considerations such as employment and manufacturing data. The former can be compared to the individual molecules, the latter to the mass action of the molecules.

In this respect, the science of chaos theory and fractals are the places that chance (randomness) and determinism seem together. Economic systems also exhibit complicated dynamic (chaotic) evidences by large amplitude and periodic fluctuation in economic indices, for instance, stock market prices, currency prices, Gross Domestic Product (Peters, 1994). The classical approach to economic anabolisms is the Newtonian one in which economic fluctuations are evaluated as linear perturbations near the equilibrium. However, large fluctuations in economic indicators show that economic systems are driven from the equilibrium points such that nonlinearity takes a role and gives the clues of chaotic complex systems.

THE FIELDS OF PHYSICS, FINANCE, AND ECONOMY

Econophysics is a field of study in which physicists make use of the theories available in physics and apply it to financial complex systems. The common of work in this field has been done on the probability distributions of returns of financial data and this is mostly due to the availability of models in physics, in particular statistical mechanics.

Traditionally for predicting the financial markets historical data on past performance and market fluidity have been essential. With the establishment of the New York Stock Exchange`s (NYSE) the important data has begun to populate the trader's archives. In 1897 French-born scientist Pareto, suggested that the economy and market indices could be explicated through the similar structure to that of physical science. This could be exemplified as gas pressure resulting from intermolecular interactions; prices are determined by an interaction between the investors.

In 1900's a French Professor Paul Levy, presented the stock market price distribution function. In his research, the calculations behaviour of the independent random variable, "Levy Distribution" has been proposed as models for several types of physical and economical systems.

The probability density function of the Levy distribution over the territory $x \geq \mu$ is

$$f(x,\mu,c)=\sqrt{\frac{c}{2\pi}}\frac{e^{-\frac{c}{2(x-\mu)}}}{(x-\mu)^{3/2}}$$

μ = Location Parameter,
c = Scale Parameter.

A stochastic process (S_t) is a Lévy process if it has independent and stationary increments and has a stochastically continuous sample path, i.e. for

$$\varepsilon>0,\ Lim_{h\downarrow 0}P\left(\left|S_{t+h}-S_t\right|>\varepsilon\right)\rightarrow 0.$$

Levy processes provide a natural simplification of the calculation of independent distributed random variables.

In the Levy process model for stock prices where the stock price process (S_t) is assumed to have the form

$$S_t = S_0\exp(rt)\exp(L_t)$$

r = Infinitely Divisible Distribution Constant,
t = Time

where in Levy process (L) satisfies some integrability condition. This class comprises all models mentioned above, except for the Mandelbrot model, which suffers from a lack of integrability (Raible, 2000).

In the 1900's Bachelier seems to be the first to have provided an analytical valuation for stock options. His work is rather remarkable because, by addressing the problem of option pricing, Bachelier derived most of the theory of diffusion processes. The first theoretical description of a random walk in the natural sciences was performed in 1905 by Einstein (Einstein, 1905) in his famous paper dealing with the determination of the Avogadro number. In subsequent years, the mathematics of the random walk was made more rigorous by Wiener (Wiener, 1923). Nowadays the random walk concept has spread across almost all research areas in the natural sciences (Mantegna et al., 1999). Actually, the first formalisation of a random walk was not in a publication by Einstein, but in a doctoral thesis by Bachelier (Bachelier, 1900). Bachelier investigated price changes, whereas economists are mainly dealing with changes in the logarithm of the price. The distribution function for the drift free case was on the effect of an absorbing barrier on Brownian motion, this was addressed by Bachelier in 1901.

In the 1960s, Exponential law, it is "fractals" (geometrical expressions of exponential distributions) begin to expand. After Alan K. Wilson in 1982 awarded the Nobel Prize the physicists start to use his idea for their research. According to Wilson "Statistical mechanics and field theory; in some circumstances, macroscopic properties are independent of the microscopic interactions".

In 1963 Mandelbrot's discovery of scaling behaviour of cotton prices: The "statistical physics connection" also underpins much of this earlier work: the idea is that one can ignore microscopic detail (the individual social entities) in favour of the coarse-grained macro-description (the groups of individuals bound together by social forces), with the aim of deriving macro-level laws (Mandelbrot, 1963).

In the 1970s, F. Black, M. S. Scholes and C. Merton assumed that the model developed within the random movement of the geometric distribution. The Black and Scholes model is often regarded as either the end or the beginning of the option valuation history. Using two different approaches for the valuation of European style options, they present a general equilibrium solution that is a function of "observable" variables only, therefore the model subject to direct empirical tests. In 1997 Black and Scholes received a Nobel Prize (Black and Scholes, 1973).

In fact, the most revolutionary development in the theory of speculative prices since Bachelier's initial work (Cootner, 1964) is Mandelbrot's hypothesis that price changes follow a Lévy stable distribution (Mandelbrot B.B., 1963). Lévy stable processes are stochastic processes obeying a generalized central limit theorem. By obeying a generalized form of the central limit theorem, they have a number of interesting properties. They are stable (as are the more common Gaussian processes) – i.e., the sum of two independent stochastic processes x_1 and x_2 characterized by the same Lévy distribution of index α is itself a stochastic process characterized by a Lévy distribution of the same index. The shape of the distribution is maintained (is stable) by summing up independent identically distributed Lévy stable random variables (Mantegna et al., 1999).

On Black Monday, 19 October 1987, stock markets around the world crashed. The crash began in Hong Kong, spread west to Europe, then hit the United States, causing the largest daily percentage loss in the history of the Dow Jones Industrial Average, -22.61%. In Fact, "The starting point which gives an idea of physicists in 1987 on Black Monday crash."

The history of Econophysics was further developed by leading scientists in the field of physics and finance, where in the mid-1990s physical journals began to publish economics studies. The term "Econophysics" was first introduced by the important theoretical physicist Eugene Stanley in 1995 at the conference Dynamics of Complex Systems, which was held in India (Kolkata) as a satellite meeting to the STATPHYS-19 conference in China (Chakrabarti et al., 2005). The term appeared in print for the first time in the paper by Eugene Stanley (Stanley et al., 1996), in the proceedings of the Calcutta conference. The paper presented a platform of the new field, arguing that "behaviour of large numbers of humans (as measured, e. g., by economic indices) might conform to the analogy of the scaling laws that have proved useful in describing systems composed of large numbers of inanimate objects". After that, the first Econophysics conferences were organised: "International Workshop on Econophysics", It took place in Budapest in 1997 and in the next year in Palermo (1998) "International Workshop on Econophysics and Statistical Finance". At the end of these organizations first was published book is "Introduction to Econophysics" (Mantegna et al., 1999).

As scientists may use the term "Phynance" as an acronym of the terms "Physics" and "Finance" (Derman, 2012), we could criticise actually for this not to be true; this does not mean the implementation of physics to economics. Also the "phy" of finance and physics because of the similar abbreviation that makes no sense does not mean anything, others have adopted the terminology "Financial Physics"; however, but the Econophysics word is seated, in a clear analogy with terms such as Biophysics, Geophysics and Astrophysics.

Current researchers have developed many computers aided simulation models by carefully working on previously developed methods and researches. Especially, Prof. D. Sornette from Nice University has developed a very successful model that is based on learning from within the current sources and managed to earn serious income out of this. In the 2000`s as well as some other applications, Sornette`s "Exponential law applications" gave successful results. Same equations are also being used for Earthquake timing predictions. Dietrich Stauffer also states that Statistical Mechanics may provide more accurate and successful results than non-traditional computer applications. In 1989 Mirowski presented a table that indicates Economics Events as Physical Events. Mirowski indicated that there were similarities and connections between Economical and Physical concepts and, by determining the equivalents of physical units for economical unit; it was possible to formulate identical physical expression as for economical equations.

This topic has immensely expanded with the discovery of the potential use of scientific and physical prediction feature over financial markets can also be applicable for various other areas with the realisation of the common similarities. This development has peaked in popularity by the accurate prediction of the latest financial crisis which is similar to the 2000 financial crisis after D. Sornette and his team`s "scaling invariance & power law" method in "Statistical Mechanics" has started using in all worldwide stock markets especially for the ones like Dow Jones Index.

ECONOPHYSICS WITH PERSPECTIVE OF COMPLEX MARKET VARIABILITY

Scientific discipline that explains economic causes and affect, by using physical causes and effects in order to analyse the economy solidly? Most of the time, a market behaves like a liquid or a gas. In fact,

Figure 1. Attributes of econophysics in context and structure

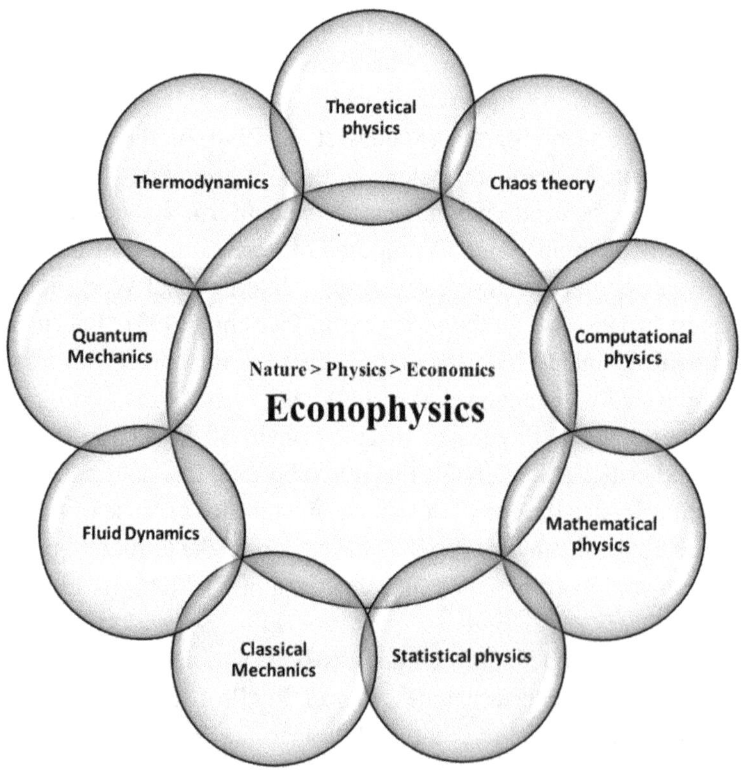

the cooperation of physics and finance had started a long time ago. At the beginning of the 20th century, while working on price fluctuations in the Paris Stock Exchange, Louis Bachelier contributed to the founding literature of the field with the term "random walk", which is the best mixture of physics and mathematics, which Einstein later developed and used to explain the distribution of the particles in a gas. The first attempt of using the random walk theory in a study was in 1900, completed by Louis Bachelier, who was working on commodity prices and concluded that the current price of a commodity can be used as an estimator for its future prices. However, Bachelier did not use the term "random walk" in his studies. After Bachelier's work, studies on stock prices have paused until computers began to be used effectively in research.

During the 1960's and 70's, three Americans, physicist Fisher Black, economist Myron Scholes and engineer Robert Merton, had worked on a mathematical model that includes variables such as time, price, interest rate and volatility, and can predict price changes. Their research had given a jump-start for the basis of a rather new and big financial industry: derivative instruments[1].

EXISTING FINANCIAL VARIABILITY STRUCTURE

The complex structure of financial prediction problems allows mathematical and statistical solving techniques be applied to solve them, yet these techniques are not sufficient alone. It is possible to linearize complex, nonlinear movements. In this way, analytical solutions can be achieved. This also forces the use of models that can be solved with computers. These solutions can be simplified by assuming fixed proportions among different values of the variables. The feature of these models is the concept that a marginal change in the independent variables causes a small change in the dependent variables. In this case, marginal measurement errors do not create huge discrepancies in the estimation. Modelling an irregular pattern includes many difficulties. All simulation models follow the sequences encountered in mathematical models. Modelling does not only depend on mathematical equations, but also on efforts to understand the processes and experiments.

Traditional analysis has begun to explain irregularities and the complexities, while previously it had focused on prediction and control issues. In order to cope with the rapid change and uncertainties, contemporary scientists develop new methods to define complex systems using natural events. Within the new theories of complex systems, there are many cases of interesting parallels. In the physical system, complex behaviour has been partially understood, meaning it is possible to predict general attitudes by measuring basic principles.

Table 1. Paired comparison of economics and statistical physics

Field	Economics	Statistical Mechanics
System	Agents	Particles
Conserved quantity	Money, m	Energy, ε
Probability distribution	P (m) - ?	Gibbs law,
Dynamics	Stock price movements	Price Fluctuation
Organization	Stock Markets	Energy container, Band of electrons

At the heart of investment are the purchase and sale of financial assets and their price valuations, as well as managing inherent risk. Risk is optimally distributed between the person and the institution who are exchanging the asset. At this point, scientists who are studying physics step in, and they perform calculations about possible future interest rates and prices. They test their models as if they are doing observations and experiments in their physics laboratories, and with the computer simulations that have been developed, they compare the difference between the expected and realised results. One important battlefield in this matter is "Volatility", in other words, the amount fluctuations of the market (Zumbach & Lynch, 2001). A powerful volatility model that can predict potential risks in different situations is critical for financial analysts. There has yet to be such tool that may predict a financial crisis as of now. A way to construct such a model could be robust patterns designed by numbers. Market dynamics actually consist of nonlinear movements, meaning that changes are not directly related to the time between the purchase and the sale, and thus the results demonstrate randomness. Therefore, physicists who are involved in academic studies started to develop tools that provide useful knowledge for the seemingly irregular patterns. One technique they are using in this end is "fractals" along with "chaos theory". Fractals are the common name for complex geometric shapes that have the property of self-similarity. Fractals are different from classical, or Euclidian, geometric shapes, such as the square, circle, or sphere. Fractal shapes have the capability to describe irregular objects and phenomena that cannot be described by Euclidian geometry. All fractals resemble to itself, if not completely, at least at some parts. In a self-similar object, the parts that constitutes the whole, i.e. the components, resembles to the complete substance. Irregular details or patterns repeats themselves at the diminishing scales, and for the abstract objects, this process can go to infinity such that when every small part of every part is magnified, the part of the part still resembles to the whole. Fractal phenomenon can be easily observed in snowflakes or in tree barks. All the natural fractals like these, as well as some other mathematical forms of fractals are random, therefore, they can only be scaled statistically. Another important feature of fractals is a mathematical parameter called "Fractal Dimension". This is a property of such fractal objects that no matter how much it is zoomed in, or how much the angle of view is changed, they stay the same. As opposed to the Euclidian dimension, fractal dimension usually represented with a non-integer number, or with a fraction.

Another important feature of financial physics is the vast amount of research on a model constructed on the principles of physics. When basic variables fail to represent the market, the requirements of that model should be modified for different structures. At this point, specific skills of physicists set in. In fact, while researchers in financial markets perform trend analyses, physicists deal with creating new pricing models or develop existing ones in our competitive world. These skills can be applied to the fluctuation of stock prices, as well as enabling us to predict potential changes in the market.

As mentioned earlier, Econophysics is the process of constructing and developing microscopic models of the financial markets. In fact, the kinetic theory of gases, and modelling the effects of the collision of gas particles, attempt to show the behaviour of gases under heat, pressure and volume, by using equations.

Consequently, by considering individual investors as microscopic particles, a universal financial model that would include all the particles and their behaviours' can be constructed. The molecular mass of a substance can explain the molecular behaviour of gases. This approach is known as the Kinetic-Molecular Theory of Gases.

During the mid-1800s Ludwig Boltzmann, James Clerk Maxwell and R.J.E Clausius developed this theory (Kennard, 1938). It is possible assume the following for the theory: gases are composed of particles called molecules that are completely identical for a particular gas; molecules of different gases have different masses and sizes; molecules in a container continuously and randomly move depending on the temperature; and the collisions of gas molecules, both with each other and with the container, are absolutely flexible. The collision of gas molecules with the container creates what is known as pressure. Pressure is the force that the molecules place on the unit surface. The pressure of gas molecules on the container under a fixed temperature does not change in time, and molecules do not encounter the force of friction while colliding. Therefore, their kinetic energies do not diminish. The kinetic energy of a molecule is proportionate to the absolute temperature of the environment. Under low pressure, the distance between gas molecules compared to their diameter is immense. Therefore, molecular gravitation, which can be defined as the inverse of the square of the distance between the molecules, is negligible. The volume of the gas molecules, compared to the volume of the container, is very small and therefore can be neglected in the calculations.

The volatility of the market, which shows similarities to the temperature of gas in some aspects, depends on the actions of all investors in the market. It is impossible to make an approximation of how investors would behave because profit occurs when behaviour contradicts that of all others under normal conditions. In a sense, this can be described as a technique of the minority against the majority. In addition, investors, unlike gas particles, have the ability to recall information, in order to continuously adapt their actions. It is hereby attempted to model new physical formulae that would provide the opportunities for closing this gap. Recently, researchers have applied every possible concept, from magnetism to game theory, neural networks, other learning theories, and pattern recognition. The synergy that was born from the cooperation with finance led to a new field in physics. Even though it seems that this new field is currently under the thumb of finance, in time, it will be applied to other scientific disciplines, such as information technology or biochemical systems, with the help of new models.

THE İNDETERMINATENESS OF CHAOTIC STRUCTURES

Chaotic indeterminateness carries a particular meaning. According to it, when a measurement is performed relating to a system, no matter how precisely that measurement is conducted, there will be errors made, and indeed, errors will be made; in reality, Heisenberg's uncertainty principle tells us that a measurement will never be of an infinite degree of precision, and that there will always have to be an amount of imprecision. This unavoidable "imprecision" that emerges in every real measurement states that even the most perfect measurement device that could be imagined, having been designed and being utilized in the most perfect way, in order to be able to record its measurements with a finite precision, would need to be able to show infinite number of digits, which is not possible and for which no realism exists. In terms of dynamic sciences, the existence of imprecision for each measurement implies that each measurement has a certain sensitivity and dependency on "initial conditions". That is, if this dependency on initial conditions reaches an inordinate level, this is called Dynamic Instability and accordingly, a system that displays high dynamic instability, where initial errors may have excessive effects on the predictability of a future state of the system, is called a "Chaotic System". So the definition is formed as follows: chaos is "dynamic instability"; and a "chaotic system" is a "dynamically instable system". According to this definition, when a measurement is performed with a chaotic system, it will not be possible to obtain very sound and very dependable long-term results based on these measurements. The principle of determinism, on the other hand, had accepted as one of the fundamental principles of classical physics the axiom of precise measurements, while not always possible in practice, at least in theory. According to this, if such an error-free measurement is performed, it is possible to know, looking forward, the entire future, and looking back, the entire history, of a system "in full and without any omission". Dynamic instability observed in certain physical systems as a time-dependent special behaviour was first discovered in 1900 by Henri Poincaré. Poincaré, who was interested in the movements of planets around the sun, realized that the results obtained from observing and measuring planet movements did not concur with the determinism proposed by Newtonian mechanics. Certain celestial systems did not conform to the principle that reducing imprecision in a starting condition would in turn reduce the imprecision in the results, and an extreme level of sensitivity to initial conditions excessively affected the results, making them indeterminate, a situation which later came to be referred as "dynamic instability", or "chaos" for short.

Therefore, the existence of chaotic systems in the nature seems to place limits on our ability to apply deterministic laws of physics to forecast movements with a certain amount of accuracy. That is, in order to perform forecasts with a certain level of accuracy, an infinite number of measurements are necessary. In time it will become a known fact of chaotic systems that in a system with two separate initial conditions, the most minutely imaginable difference will lead to huge variations, both in the time following and in the time preceding.

MICRO STRUCTURE OF MARKETS

The micro structure of a market deals with the purchase and sale of financial assets such as stocks or bonds and the functioning of these markets. The micro structure investigates the short-term price effects of trading in stock markets The investors who make up the supply and demand, the intermediaries who perform the trading activities (dealers, brokers) and the institutions in which the stock transactions are carried out (the stock exchange) are the constituents of the market.

Perspective of the micro structure, information is the most important element that ensures the efficiency of the market. The market efficiency implies that there are specific information regarding securities in addition to those already known by all investors, and that the investors make investments using both information (Galluccio et al., 1997). The purpose of this structure that investigates micro structure relations is to gain a better understanding of the methods through which information that ensures market efficiency into is integrated into prices. Micro structures are divided into two groups, which are the competitive models and the strategic models (Jones et al, 1994).

In competitive models, asymmetric information leads to the purchase and sale of securities. In these types of models, the trade volume is positively correlated with the information quality of the informed investor. As the informed investor knows the real value of the stock, he/she compares this specific information with market prices and trades accordingly. In case the specific information obtained by the informed investor is of a nature that would lead to significant price changes in the company's stocks, the informed investor will perform a large volume stock purchase or sale transaction. For this reason, the trade volume leads to the "reverse selection issue" in security trading. This is because informed investors prefer to engage in large-volume trade at any price, while certain investors might interpret the information differently and prefer to withdraw from the market or to avoid purchase and sale transactions.

In strategic models, security purchase and sale are also performed according to asymmetric information. However, an investor who is knowledgeable regarding strategic models performs a large number of small volume transactions, instead of a single large volume transaction. With numerous small volume transactions, the investor attempts to conceal from the other investors the transactions he\she performs based on the specific information. In this way, transaction frequency rather than transaction volume provides more information to market participants.

RANDOM WALK MODEL

Studies on efficient capital markets were initially based on the random walk model. Two more models appeared later on. After, these two models in addition to the random walk model. One of these models is the fair game model, while the other one is the martingale and sub-martingale models. Among these models, the random walk model is more limiting than the fair game or martingale models. In case the random walk model can successfully determine stock prices, it can be said that the market is efficient in the weak-form; however, if a market is efficient in the weak-form, it will not be considered as being in accordance with the random walk model (Abrosimova et al., 2002).

Analyses used within this context argue that the past price behaviour of a stock is a strong indicator of its future behaviour. Fama discussed the random walk model theory in detail and tested the empirical accuracy of the model. According to the random walk theory, the future course of a stock price can only be estimated as much as the course of the cumulative random number series (Fama, 1965). Statistically, this purports that consecutive price changes conform to an independent and series distribution.

The random walk theory states that the expected return distribution of a stock is formed by the investors. The equation below describes the Random Walk model with the abovementioned characteristics:

Figure 2. Example of random motion

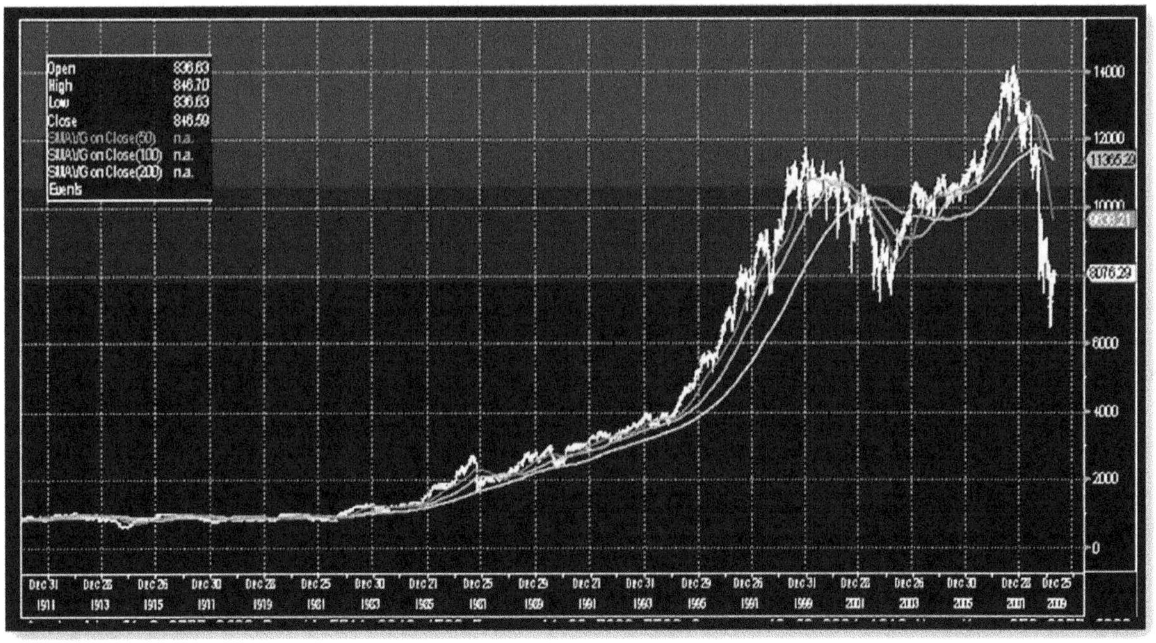

$$f\left(r_{i,t+1} \mid r_{i,t-1} \ldots, r_{i,t-n}\right) = f\left(r_{i,t}\right)$$

in which i is the probability distribution of stock returns from $f\left(r_{i,t}\right) = t - n$ to t.

If stock prices follow the Random Walk model, the return will remain the same over time, and the serial correlation tests will provide the zero correlation coefficient.

HETEROGENEOUS AND FRACTAL MARKET HYPOTHESES

The market is fractal with heterogeneous trading behaviour and this was due to the successful fractal approach which gave rise to the hypothesis. The investors are generic and the FMH (The Fractal Market Hypothesis) affirms the impact of the information and investment horizons on the behaviour of investors. Investors want to sell or buy or even hold a security because of the available information known as "Generic Item". Investors always desire to maximize their return and are knowledgeable on how to value the current information. This is sometimes considered as a rational price-ticker. With this generic approach where the information and investors are general cases, this indicates that all kinds of information have an impact on all the investors equally. However, this is where it fails and as below the assumptions were introduced for FHM (Peters, 1994):

1. Individuals who make up the market overall with a huge number of different investment horizons;
2. There will not be a long term trading, liquidly, short –term information domination if a security has no connection to the economic lifecycle;
3. Prices reflecting a mix of long-term fundamental valuation and short term technical trading;
4. Various investment horizons are impacted differently by information;
5. Balancing of supply and demand; liquidly is present when the market is made of many investors with number of different investment horizons.

FMH gives a model which shows the investor behaviour and market price movements which is fit for observations. Without a doubt, EMH works well during the times when the market is stable although similar singularities in physics, this cracks and model breaks down during the panic and stampedes. But this is not unknown to anyone as EMH is an equilibrium model and it cannot cope the transition to turbulence. FMH on the other hand states that the information is valued accordingly to the investment horizon of the investor. The underlining factor is that with FMH model, market undoubtedly is stable when it has no characteristic time scale or investment horizon. This is where the unbalancing occurs; when the market loses its structure and makes an assumption of a fairly uniform investment horizon.

By analysing t time intervals of varying values, we can perform a statistical investigation of financial data based on a fractal perspective. According to this approach, the relation between the volatility size and the time intervals can be determined through the scaling law described by Müller (Müller et al., 1990):

$$v\left(t\right) = c\left(\Delta t\right)^{D},$$

in which the relation between volatility $v(t)$ and time Δt is described. In this context, is the empirical constant while D represents the empirical drift exponent. In this context, a behavior that differs from the Gaussian hypothesis will be observed. This is because the abovementioned formula holds true when Gaussian hypothesis when D is set to 0.5. On the other, empirical values differ considerably for D.

Such observations and characteristics regarding empirical time series have lead to the development of Heterogeneous Market Hypothesis (HMH), which provides a model that differs from the Homogenous Market. The difference centres around the assumption, held with homogenous markets, that news are interpreted in a similar fashion in the markets, and that the reactions to this news are also similar. In the Heterogeneous Market, on the other hand, the following interpretations apply according to the HMH:

1. Heterogeneity is also observed with regards to regional aspects (location of the actors, geographical factors, etc.);
2. The market has short-term, medium-term and long-term components, which leads the time horizons to assume fractal structures;
3. Different actors will contribute to volatility by trading at different prices and market situations;
4. All actors have different time horizons and dealing frequencies;

In addition to the points mentioned above, actors in a Heterogeneous Market can also differ from one another with regards to risk aversion, regulatory limitations and the costs involved in trading (Guillaume et al., 1997).

The analysis of Guillaume, evaluated several aspects regarding the intra-daily FX (foreign Exchange) market. Demonstrating a number of complex structures within this market, they categorized the observed empirical regularities according to three main groups: (I) Observations regarding prices changes and their distribution, (II) Observations on the processing of price information, (III) Observations on the heterogeneity of market structure.

Separately from the study conducted by Gallucio (Galluccio et al., 1997) have analysed the same market aspects concerning FX rates, and determined that the hidden interrelations and patterns within the observed data contributed to a more complicated structure (more complex than the one observed with random walk). These studies further elucidate concepts regarding the functioning of Fractal and Heterogeneous Markets by providing a theoretical context in which the relevant financial process can be analysed and discussed. In addition, the variations observed in market return levels are also relevant and significant within the scope of economic and financial analyses.

THE RELATIONSHIP BETWEEN CHAOS AND PHYSICS

The "Chaos Approach", which is based on the fundamental laws of physics, is also applicable in financial approaches. As is the case in many basic fields of science, it is argued by certain theoreticians that theories can describe and clarify the nature of the non-linear relations in the deterministic conditions of financial analyses. Conceptually, the chaos theory or disorder model is not a physics theory or a mathematical induction; instead it is a method that serves to describe, as a whole, the tendencies of the components of physical reality. As in many other aspects of real life, uniform behaviour should not be expected in economy; in other words, linear dynamic, liner and non-chaotic dynamic, and chaotic dynamic and noise behaviour should be expected. Chaos is actually formed by the combination of several concepts which, initially, may seem very disparate, but which are in fact inseparable from one another:

- Chaotic systems are extremely dependent on their initial conditions. (In other words, even the slightest change in initial conditions will deeply alter the system's trajectory)
- The system's evolution over time cannot be predicted. (For example, let us consider that the system began its evolution with a certain μ. Based on the observations performed at time t_1, the system's condition at time t_2 cannot be known for certain within the bounds of possibility, and only a small portion of the system's conditions can be estimated).
- As a result of the exertion of an external force onto the system, a marginally constant limit can transition to a chaotic mode due to the periodic folding of the cycle (such as speculation).
- When an oscillating system is subject to an external force (such as the business cycle)
- Fractal geometry (The new approach that we mention has also engendered a new understanding in geometry. When we look at nature, there are no straight lines and border; and regardless of how much we reduce the scale, a self-repeating structure is constantly observed. Fractal geometry was propounded by Benoit Mandelbrot.
- Mathematical Chaos (Due to the chaos theorem in mathematics, an underlying order has been discovered in many systems. Many others are on the verge of being discovered).

The behaviour of non-linear systems can be categorized under three different groups as (I) systems moving towards equilibrium or stability, (II) systems displaying periodic behaviour or stable oscillation, and (III) systems displaying chaotic behaviour. According to Mandelbrot, one of pioneers of the chaos theory, complexity is an element of daily life that cannot be controlled. In phenomena where others saw only turmoil, Mandelbrot was able to identify a fractal order. With fractals, the parts imitate the whole according to a definite and measurable formula when the scale is increased or decreased. As the scale of simple fractals vary in a similar fashion in both directions, they are described as having a self-similar pattern (Mandelbrot and Hudson, 2006).

Nature is constituted entirely of dynamic systems. When the behaviours of dynamic systems are controlled, their evolution over time can be observed. The dynamic systems in nature have different oscillations. By observing the period and frequency of the oscillation, it is possible to decide whether a system is chaotic of periodic. First of all, for a system to be chaotic, it must be non-linear. This is because linear systems cannot generate chaos. Furthermore, as another condition for the system to be chaotic, it is necessary for the system to be fading. In order to analyse the identity of a system, the main distinctions that we have mentioned are utilized. In this case, the behaviour of the systems can be described as periodic, random, aperiodic and chaotic. Whether a system's long-term behaviour becomes unpredictable is dependent on the presence or absence of chaos. This is because even a minor variation in the initial conditions of a system increases at an exponential rate over time, thus leading to an increase in uncertainty. Non-linear behaviour is a requirement for observing chaos in a system; however, it is not a sufficient condition on its own. Chaos and order are generally observed together in non-linear dynamic systems. It is possible to linearize non-linear complex movements, and to thereby generate analytical solutions. These solutions can be simplified by assuming a fixed proportionality between the different values of the variables. For this reason, "the transition from chaos to order and from order to chaos is always possible." Actually, the index and stock prices are the single-dimensional outputs of a complex and multi-dimensional system that is affected by many different macro and micro-level factors. One situation that needs to be taken into consideration is that many phenomenon in the real world actually "have a chaotic structure, and that all chaotic structures have an order in themselves." For example, the rising of cigarette smoke by forming a set of irregular spirals, the fluttering of a flag in the wind, or the water initially dripping from a tap at regular intervals but then becoming irregular are all events that lead to chaos. These forms of behaviour conform to the laws of this new field of science.

These processes in question cannot be described with the context of the notion of causality, and are designated as dynamic systems. It is argued that, with these processes, the chaos approach takes the place of determinism and brings a new breath to science which runs into problems when dealing with such systems. At the same time, this approach uses to a limited extent the methods of modern science, and focuses on the behaviour of non-linear complex systems. The classical modern science "founded" by Newton deals mainly with linear systems, and sometimes with closed system that are non-linear. The econophysical approach, on the other hand, describes an entirely non-linear system, and attempts to establish internal associations with other systems.

Many phenomenons in nature take place within a certain system and possess a chaotic structure. To describe non-linear complex systems, let us take the Stock Market (batch) as example. In this system, if we assume that there is no interaction between the stocks, when we take one of the stocks from the batch (in order words, when the sale or purchase of the stock takes place in the stock pool), the other remaining stocks will graphically assume a new shape in order to adjust to the changing conditions. To be able to predict this new structure, it will be necessary to first consider the position and configuration of each stock (paper) under the initial conditions, and to also consider the importance and characteristics of the "changing conditions" following the purchase of the stocks.

Chaos is noteworthy due to its ability to reproduce the characteristics of time series that are similar to the cases observed in financial models and theories. Its most important consequence is that it does not impose restrictions on physicists and mathematicians. We observe that the majority of studies are divided into two groups as stochastic and deterministic studies. While stochastic models describe many fluctuations in terms of external and random shocks, chaotic systems describe that these fluctuations are part of the deterministic order and that they are generated internally. In case the equations can be solved, it will be possible to rearrange the previous states, or to predict the future states. It is necessary to increase the represent ability of irregular or random events, and to expand the existing models. One of the underlying phenomenon's that form the basis of financial markets is the alternating "cycle of recession", growth and decline. Within this cycle, it is not possible to ascertain whether a decline is temporary or whether it is the beginning of a period of recession. However, it is possible to describe this chaotic structure by using physics and modelling methods; yet, the models that we form will not demonstrate for certain whether the predictions are correct. We can say that the fundamentals of these models can be described in terms of physics, and that these models are especially recognized and accepted in the financial markets. Thus, we can say that the chaos theory is a beginning point rather than an end.

CONCLUSION

The objective of this research was to design a methodological approach in which extrapolative fluctuations of the capital markets could be predicted. This work presents the introduction of Physics theories to predict the time progression of financial assets (market value), information fractal approach.

This approach essentially constitutes experimentation that could be validated through comparative studies of historic data. In this respect, it was apparent that the new and alternative indicators could be added to the usual technical analysis indicators of Econophysics. Even when looking at the currently used indicators, it is possible to observe the assertion that classical physics is affected by laws on force, energy and oscillation. Market indicators could thus be used within the context action-reaction mechanisms that are more reliable and open to experimentation, in which energy laws are also applicable, and wherein indicators changes (other than those pertaining to data numbers) can occur from one country to another, from one market to another, and even from one financial instrument to another within the same market.

By employing statistical physics as a nonlinear dynamic model, resulted in entrancing formulation that have provided plausible (non-differential, non-fractal) equations of stock price in systematic form. This generates a variability of possible price movements in time referred to as predicted future. To this end, general philosophies of statistical infrastructure are exposed and then central factor illustrated through simulation of boundary elements.

The dynamic phenomena of Econophysics architecture presented through simulation, forms the degree of its applicability for markets and provides empirical evidence assembled from high frequency futures data corresponding to various liquid futures assets. It illustrates the impact (as measured from the standard definition or using the return variance) of a trading order on the core price which depends on mutual-trade duration through investigating volatility and clustering of stocks fluctuation. The underlining time progress of financial assets has been addressed in a case where parallel predictions of the proposed model may be possible, with a relatively rich ensemble of history realizations. The representation of financial time series in volatility clustering can be simulated and estimated through the model for a given market with a given risk factor.

The impression that the prices make as if they are reacting to the external factors could affect all investors at the same time but what is important in a neutral environment. Through simulated model it was possible to expound how investors influence each other and whether they would have the same energy at a given time or not. This in turn, is a state that could be evaluated within the statistical infrastructure presented in context of Econophysics herein. Although it seems that the prices are formed under the external influences, the fact provided indicates formation through the internal dynamics of the market. Internal market events determine the prices and the external dynamics are pointed as the cause only after the events had taken place.

REFERENCES

Abrosimova, Dissanaike, & Linowski. (2002). *Testing Weak-Form Efficiency of the Russian Stock Market.* EFA 2002 Berlin Meetings Presented Paper.

Bachelier, L. (1900). *Theorie de la Speculation.* Paris: Gauthier-Villars.

Black, F., & Scholes, M. S. (1973). the Pricing of Options and Corporate Liabilities. *Journal of Political Economy*, *81*(3), 637–659. doi:10.1086/260062

Bouchaud, J. P., & Potters, M. (2000). *Theory of Financial Risks. From Statistical Physics to Risk Management.* Cambridge, UK: Cambridge University Press.

Carbone, A., Kaniadakis, G., & Scarfone, A. M. (2007). Where do we stand on econophysics? *Physica A*, *382*(1), xi–xiv. doi:10.1016/j.physa.2007.05.054

Cohen, J. B., Black, F., & Scholes, M. (2012). The valuation of option contracts and a test of market efficiency. *The Journal of Finance*, *27*(2), 399–417. doi:10.1111/j.1540-6261.1972.tb00969.x

Cont, R. (1997). *Beyond implied volatility: Extracting information from options prices. Econophysics*. Dordrecht: Kluwer.

Cootner, P. A. (1964). *The Random Character of Stock Market Prices*. Cambridge, MA: MIT Press.

Czarnecki, L., Grech, D., & Pamula, G. (2008). Comparison study of global and local approaches describing critical phenomena on the polish stock exchange market. *Physica A*, *387*(27), 6801–6811. doi:10.1016/j.physa.2008.08.019

Dacorogna, M. M., Gençay, R., & Olsen, R. B. (2001). An Introduction to High-Frequency Finance. London: Academic Press.

Debreu, G. (1991). The Mathematization of Economic Theory. *The American Economic Review*, *81*(1), 1–7.

Derman, E. (2012). *My life as a quant: Reflections on physics and finance*. Wiley.

Didier, S. (2004). A complex system view of why stock markets crash. *New Thesis, 1*(1), 5-18.

Didier, S., & Zhou, W.-X. (2004). Antibubble and prediction of China's stock market and real-estate. *Physica A*, *337*(1), 243–268.

Domino, K. (2011). The use of the Hurst exponent to predict changes in trends on the Warsaw Stock Exchange. *Physica A*, *390*(1), 98–109. doi:10.1016/j.physa.2010.04.015

Domino, K. (2012). The use of the Hurst exponent to investigate the global maximum of the Warsaw Stock Exchange WIG20 index. *Physica A*, *391*(1-2), 156–169. doi:10.1016/j.physa.2011.06.062

Einstein, A. (1905). On the Movement of Small Particles Suspended in a Stationary Liquid Demanded by the Molecular-Kinetic Theory of Heat. *Annalen der. Physik*, *17*, 549–560. doi:10.1002/andp.19053220806

Fama, E. F. (1965). The behavior of stock-market prices. *The Journal of Business*, *38*(1), 34–105. doi:10.1086/294743

Fama, E. F., & Malkiel, B. G. (2012). Efficient Capital Markets: A Review of Theory and Empirical Work. *The Journal of Finance*, *25*(2), 383–417. doi:10.2307/2325486

Financial Times. (2004). *Book review: Memoirs of a physicist on Wall Street, By Stefan Stern*. Financial Times.

Galluccio, S., Caldarelli, G., Marsili, M., & Zhang, Y. C. (1997). Scaling in currency exchange. *Physica A*, *245*(3), 423–436. doi:10.1016/S0378-4371(97)00316-6

Gligor, M., & Ignat, M. (2000). Dissipative Structures in the Stock Market Space. *Bulg. Journal of Physics*, *27*.

Grech, D., & Mazur, Z. (2004). Can one make any crash prediction in finance using the local hurst exponent idea? *Physica A, 336*(1-2), 133–145. doi:10.1016/j.physa.2004.01.018

Grech, D., & Pamula, G. (2008). The local hurst exponent of the financial time series in the vicinity of crashes on the polish stock exchange market. *Physica A, 387*(16-17), 4299–4308. doi:10.1016/j.physa.2008.02.007

Guillaume, D. M., Dacorogna, M. M., Davé, R. R., Müller, U. A., Olsen, R. B., & Pictet, O. V. (1997). From the bird's eye to the microscope: A survey of new stylized facts of the intra-daily foreign exchange markets. *Finance and Stochastics, 1*(2), 95–129. doi:10.1007007800050018

Jones, C. M., Kaul, G., & Lipson, M. L. (1994). Transactions, volume, and volatility. *Review of Financial Studies, 7*(4), 631–651. doi:10.1093/rfs/7.4.631

Kennard, E. H., & Kennard, E. H. (1938). *Kinetic theory of gases: with an introduction to statistical mechanics*. New York.

Kristoufek, L. (2010). Local scaling properties and market turning points at prague stock exchange. *Acta Physica Polonica B, 41*(6), 1223–1336.

Kristoufek, L. (2011). Multifractal height cross-correlation analysis: A new method for analysing long-range cross-correlations. *EPL, 95*(6), 68001. doi:10.1209/0295-5075/95/68001

Liu, Y., Cizeau, P., Gopikrishnan, P., Meyer, M., Peng, C. K., & Stanley, H. E. (1998). Volatility studies of the S&P 500 index. Center for Polymer Studies and Department of Physics, Boston University.

Majorana, E. (1942). Il valore delle leggi statistiche nella fisica e nelle scienze sociali. *Scientia, 71*, 58–66.

Mandelbrot, B., & Hudson, R. L. (2006). *The Misbehavior of Markets: A fractal view of financial turbulence*. Basic Books.

Mandelbrot, B. B. (1963). The variation of certain speculative prices. *The Journal of Business, 36*(4), 394–419. doi:10.1086/294632

Mantegna, R. N., & Stanley, H. E. (2000). An Introduction to Econophysics. Cambridge, UK: Cambridge University Press.

Mantegna, R. N., & Stanley, H. E. (1999). *Introduction to Econophysics: Correlations and Complexity in Finance*. Academic Press.

Merton, R., & Scholes, B. (1972). The Valuation of Options Contracts and a Test of Market Efficiency. *The Journal of Finance, 27*(2).

Morales, R., Di Matteo, T., Gramatica, R., & Aste, T. (2012). Dynamical generalized Hurst exponent as a tool to monitor unstable periods in financial time series. *Physica A, 391*(11), 3180–3189. doi:10.1016/j.physa.2012.01.004

Müller, U. A., Dacorogna, M. M., Olsen, R. B., Pictet, O. V., Schwarz, M., & Morgenegg, C. (1990). Statistical study of foreign exchange rates, empirical evidence of a price change scaling law, and intraday analysis. *Journal of Banking & Finance, 14*(6), 1189–1208. doi:10.1016/0378-4266(90)90009-Q

Onali, E., & Goddard, J. (2009). Unifractality and multifractality in the Italian stock market. *International Review of Financial Analysis, 18*(4), 154–16. doi:10.1016/j.irfa.2009.05.001

Onali, E., & Goddard, J. (2011). Are European equity markets efficient? new evidence from fractal analysis. *International Review of Financial Analysis, 20*(2), 59–67. doi:10.1016/j.irfa.2011.02.004

Peters, E. (1994). *Fractal Market Analysis – Applying Chaos Theory to Investment and Analysis.* New York: John Wiley & Sons, Inc.

Raible, S. (2000). *Lévy processes in finance: Theory, numerics, and empirical facts* (PhD dissertation). Universität Freiburg.

Sornette, Johansen, & Bouchaud. (1996). Stock market crashes, precursors and replicas. *Journal de Physique I France, 6*(1), 167–175.

Sornette, D. (2004). *Critical phenomena in natural sciences: chaos, fractals, selforganization, and disorder: concepts and tools.* Springer Verlag.

Sornette, D., & Johansen, A. (1997). Large financial crashes. *Physica A, 245*(3), 411–422. doi:10.1016/S0378-4371(97)00318-X

Stanley & Gabaix. (2006). *Economic Fluctuations and Statistical Physics: Quantifying extremely rare and less rare events in finance.* Academic Press.

Stanley, H. E., Afanasyev, V., Amaral, L. A. N., Buldyrev, S. V., Goldberger, A. L., Havlin, S., ... Viswanathan, G. M. (1996). Anomalous fluctuations in the dynamics of complex systems: From DNA and physiology to Econophysics. *Physica A, 224*(1-2), 302–321. doi:10.1016/0378-4371(95)00409-2

Stanley, H. E., Amaral, L. A. N., Canning, D., Gopikrishnan, P., Lee, Y., & Liu, Y. (1999). Econophysics: Can physicists contribute to the science of economics? *Physica A, 269*(1), 156–169. doi:10.1016/S0378-4371(99)00185-5

Stauffer, D. (2000). EconoPhysics-A New Area for Computational Statistical Physics? *International Journal of Modern Physics C, 11*(06), 1081–1087. doi:10.1142/S0129183100000948

Takayasu, H., Sato, A. H., & Takayasu, M. (1998). *Power law behaviours of dynamic numerical models of stock market prices.* Academic Press.

Vandewalle, N., & Ausloos, M. (1998). Multi-affine analysis of typical currency exchange rates. *J. Eur. Phys. B, 4*(2), 257–261. doi:10.1007100510050376

Weron, A., & Weron, R. (2000). Fractal market hypothesis and two power-laws. *Chaos, Solitons, and Fractals, 11*(1), 289–296. doi:10.1016/S0960-0779(98)00295-1

Wiener, N. (1923). Differential space. *Journal of Mathematical Physics, 2*(1-4), 131–174. doi:10.1002apm192321131

Zumbach, O., & Lynch, P. (2001). *Heterogeneous Volatility Cascade in Financial Markets.* University of Manchester Institute of Science.

ENDNOTE

1 The model that is usually referred as "Black-Scholes" equation, for which Scholes and Merton were awarded the Nobel Prize in Economics in 1997, actually included Black as well However, Black died 2 years prior to the prize being awarded. This model attracts attention because of its similarity to those in physics that investigate the diffusion of temperature (Merton & Scholes, 1972).

Chapter 2
Fractal Properties of Financial Assets and Forcasting Financial Crisis

Inna Nekrasova
Southern Federal University, Russia

Oxana Karnaukhova
Southern Federal University, Russia

Oleg Sviridov
Southern Federal University, Russia

ABSTRACT

The chapter is aimed at identification of criteria to select financial assets for investment; observing price fluctuations at small time intervals (up to one week) as possible predictors of the future of a significant increase in the price fluctuations amplitude; determining a fractal dimension of the financial markets on the basis of R/S-analysis; constructing a fractal index indicator to identify a bifurcation point, which gives birth to a possibility of crisis phenomena in economy. Therefore, the practical significance of the chapter lies in the idea of equipping academics and practitioners with new methods and tools for analysis and forecasting future development and dynamics of the financial markets.

INTRODUCTION

Despite the considerable research already conducted on measuring the market efficiency the problem is still under investigation. There is no common methodology, results of measuring differ one over another, some aspects are underestimated etc. For example, the behavior of the market efficiency during a crisis in general and from the position of countries' differences is not clear even nowadays.

The aim of this study is a synthesis of the Efficient Market Hypothesis (EMH) and Fractal Market Hypothesis (FMH) hypotheses to predict the financial crisis and investment decision-making. On the one hand, following the FMH, interpretation of price fluctuations based on the description of the mar-

DOI: 10.4018/978-1-5225-3767-0.ch002

ket agents' behavior can be changed significantly at different time intervals. For instance, for intraday transactions, where more than a half of the stock exchange transactions performed by robots (in the US markets), the behavior of agents is very close to the rational. On the other hand, for time intervals from few days to few months, social psyche plays a significant role and always contains an irrational element. Meanwhile, the unchanging nature of vibrations (according to the EMH) with the most common mode of a random walk takes place at all-time intervals, starting from the smallest. This suggests that the basis of these price oscillations is the common mechanism of delay. This delay mechanism is an important element of the decision-making for agents in the stock market.

However, despite the frequent lengthy local deviations, the prices tend to go back to the effective behavior, which describes the random walk model. Based on the foregoing, the following positions are formulated:

- Identifying criteria to select financial assets for investment;
- Observing price fluctuations at small time intervals (up to one week) as possible to predict the future of a significant increase in the price fluctuations amplitude;
- Determining a fractal dimension of the financial markets on the basis of R/S-analysis;
- Constructing a fractal index to identify a bifurcation point, which gives birth to a possibility of crisis phenomena in economy.

Therefore, the practical significance of the chapter lies in the idea to equip academics and practitioners with new methods and tools for analysis and forecasting future development and dynamic of the financial markets.

THEORETICAL FRAMEWORK

One of the most important objectives in finance is modeling and forecasting price fluctuations in risky assets. For analysts and policy makers volatility is a key variable for understanding market fluctuations. Analysts need accurate forecasts of volatility as an indispensable input for tasks such as risk management, portfolio allocation, value-at-risk assessment, and option and futures pricing. In financial markets volatility is a fluctuations measure of the financial instrument price over time. It cannot be directly observed, but has to be estimated via appropriate measures or as a component of a stochastic asset pricing model.

An important contribution in studying financial markets persistence has been done by Los (2003), who analyzed the existing methodology quite carefully. According to his research, one of the key indicators of the persistence level is the so called Hurst exponent. The Hurst exponent was created by hydrologist Hurst (1951). For current study there are quite important works of Kristoufek (2013), Roch (2011), Vacha et al. (2013), which who investigated signals of market crashes.

For a long time the study of financial time series has been developed in the form of two unrelated directions, and only recently a tendency towards their convergence appeared. The first direction, which can be called statistical, originates from the work of Bachelier (1900), where the researcher five years before Einstein proposed the first model of Brownian motion (random walk model) and applied it to describe fluctuations in share prices on the stock-market.

Strictly mathematically this model was justified in the early 1920s by Norbert Wiener and for a long time was developed exclusively in the academic environment. The latter is due to the fact that in accordance with this model it was impossible to obtain a profit exceeding the average of the market. However, this contradicted the experience of the real trade. Therefore, among investors the Brownian model got serious attention only in the early 1960s, when the concept of an "efficient market" was finally formed. Under efficient market it was understood a market where prices fully reflect all available information.

According to Fama (1970), for the existence of such market it is sufficient to assume that it has a large number of fully informed agents who instantly react to external events, while acting rationally and independently. Within the framework of this concept the theory of the optimal portfolio formation presented by William Sharpe (1970) has acquired the meaning and practical significance.

The core model of price behavior has become the model of Brownian motion. The highest achievement of this approach was based on this model and presented in works of Black and Scholes (1973), which allowed to accurately calculate "fair" option prices.

The classical model of Brownian motion was based on two postulates. First, increments of the process in any time interval have a normal (Gaussian) distribution with the zero mean, which results from the central limit theorem and gets obtained as the result of summing sufficiently large number of independent random factors. Second, the increments in disjoint intervals are statistically independent. However, observations of financial time series revealed a number of features that are not consistent with these postulates. The most important of them are related to the fact that strong changes of the time series occur much more often than one would expect on the assumption of Gaussian distribution (the problem of "thick tails"). Moreover, such changes are usually separated by fluctuations of relatively low intensity (clustering effect of volatility).

For this reason, in parallel with the Brownian model various generalizations of this model got developed. They were associated with the rejection of either the conditions for the normal distribution of increments (the first postulate) or conditions for their independence (the second postulate). In the first case one arrives to the Levi motion and, in particular, to stable Pareto distributions. In the second case processes with memory and fractional Brownian motion appeared (Osborne, 1964). Finally, if to abandon both postulates of the Brownian model, then one comes to ideas of autoregressive conditional heteroscedasticity, namely: dependence of the values of the time series on its previous values, when the variance is changed in time, or the ARCH - Engle model and its various generalizations.

The second direction, naturally called dynamic, was born and got development among practicing traders in early 1930s. This approach was called "technical analysis" as opposed to the fundamental analysis based on calculating the "fair" share price in the form of discounted value of future incomes. The ideological basis of this approach was the well-known position of Charles Dow, put forward by him at the end of the 19th century. Dow argued that the natural state of the price is the trend of directional movement up or down, which is the result of the joint crowd action and reflects the current social trend in the market. Therefore, this trend will continue until market changes the trend.

The aim of the technical analysis is to reveal internal regularities of the time series, so one can predict the transition from the trend to flat as relatively stable market conditions and back. Numerous forms of relatively stable behavior of the time series (figures of the technical analysis) were opened within this direction. The first study belongs to William Gann and Ralph Elliot. In 1950s almost all classical figures of the technical analysis, such as "triangle", "trapezium", "head and shoulders", etc., have been known. However, only in 1980s their systematization has been made in the well-known volumes of Murphy (1999) and Prehter (2007).

The most recent work coincided with the unexpected support made by the technical analysis from the theory of the dynamic chaos. From the general theory it was demonstrated that the time series, which outwardly looked like realization of a random process, could have been well generated by a nonlinear dynamical system of a small dimension. This means that it can be represented as a one-dimensional projection of the system trajectory in an extended phase space, which is described with a small number of ordinary differential equations.

According to the Dow concept, the behavior of stock prices is determined by the herd instinct, which (as convincingly evidence data in social psychology) is a subject to an extremely primitive mechanism. Therefore, it seems reasonable to hypothesize that such a mechanism is a dynamic system. In this case, using the Takens' theorem (1981), one can reconstruct the current value of the time series based on a sufficiently large number of historical data. Moreover, for such restoration it is not necessary at all to know a specific form and number of equations of the system. Essentially, this procedure turns the extrapolation problem of a one-dimensional series into the task of interpolation of some multidimensional function. The latter is a typical task for neural networks. Therefore, the theory of dynamic chaos can be considered as the ideological basis for the powerful introduction of neurotechnologies into business, which was observed everywhere in 1990s.

However, from the view of the stationary approach, an accurate forecast of the financial time series is impossible in principle. To resolve the contradiction between two approaches, one should turn to results of an experimental market study. As shown by similar studies, in the real market the whole aggregate of agents breaks up into clusters (reference groups); in each cluster agents imitate each other. Clusters can form quite a complex hierarchical links, or merge into larger ones and break down into smaller ones. It is clear that the concept of an effective market and the Dow concept describe two limiting cases. In the first case, there is a large number of approximately identical clusters in the market. The market is located in the most stable state and its evolution is mainly determined by external information, which has a random character. In the second case, there is one large cluster in the market, considerably exceeding all the others. The market is the closest to the collapse and its dynamics is a subject of exclusively internal factors. In connection with aforesaid, questions arise. What laws does the cluster structure evolve by and what is the reason for the formation of large clusters? An answer to these questions means the synthesis of these two concepts.

Thus, the main objective of this study is to combine statistical and dynamic approaches to the analysis of time series of financial assets to predict the dynamics of their changes in the future prices.

METHODOLOGY

According to the definition of Mandelbrot (1999), a fractal is a set, for which the Hausdorff' dimension D is strictly greater than its topological dimension D_t (Formula 1). For a compact set in an arbitrary metric space the fractal dimension is defined as follows (Hausdorff, 1919):

$$D = \lim_{\delta \to 0} \frac{\ln N\left(\delta\right)}{\ln N\left(1 \setminus \delta\right)} \tag{1}$$

where,

N (δ) = the minimal number of balls of radius δ covering the original set.

The main method of the fractal time series study is the R/S-analysis or the method of rescaled range. It was suggested by hydrologist Harold Edwin Hurst (1951), working on the Nile dam project in the mid-20th century. The task was to calculate the required volume of the dam reservoir, filling of which happened due to various natural sources: rainfalls, floods, etc. Usually, in such cases hydrologists started with an assumption that the water level in a river is a random series, where the value of the water level in following years do not depend on previous ones. But having read about floods in the last 800 years, Hurst discovered the following regularity: the year of a high water level is usually followed by another year with a high level, and the year with a low level of water by another year with low levels. It appeared like a cycle with an unpredictable period. A standard statistical analysis revealed no significant correlations between observations, so Hurst had to develop his own methodology.

There are at least two variations of the fractal dimension – D and A indicated in Table 1. The fractal dimension D (where D is the dimension of the time track – an assessment of the degree of affectation series) is defined due to Formula 2:

$$D = 2 - H \tag{2}$$

Mandelbrot and Van Ness (1968) have demonstrated that the fractal dimension is the reciprocal value of the Hurst exponent (H). For instance, if H = 0.5, the fractal dimension is equal to 2 (1/0.5), and if H = 0.8, the fractal dimension is equal to 1.25 (1/0.8). Therefore, the fractal dimension of Mandelbrot A (where A is the dimension of the probability space – estimation of the thickness of tails in the probability density function) is calculated due to Formula 3:

$$A = 1 / H \tag{3}$$

The Hurst exponent can be defined in the interval [0, 1], and is calculated within the following limits:

- **$0 \le H < 0,5$:** Data is fractal, the FMH is confirmed, the «heavy tails» of distribution, antipersistent series, negative correlation in instruments of value changes, pink noise with frequent changes in direction of price movement, trading in the market is more risky for an individual participant;
- **H = 0,5:** Data is random, the EMH is confirmed, movement of asset prices is an example of the random Brownian motion (Wiener process), time series are normally distributed, lack of correlation in changes in value of assets (memory of series), white noise of independent random process, traders cannot «beat» the market with any trading strategy;

Table 1. The values of the fractal dimension variations

	Straight Line	**Random Series**	**Infinite Linear Trend**
Hurst Index (H)	$H \approx 0$	H = 0,5	H = 1
Fractal Dimension D	$D \approx 2$	D = 1,5	D= 1
Fractal Dimension A	$A \to \infty$	A=2	A=1

- **0,5 < H ≤ 1:** Data is fractal, the FMH is confirmed, «heavy tails» of distribution, persistent series, positive correlation within changes in the value of assets, black noise, the trend is present in the market.

Hurst took the formula from the Einstein's work on Brownian motion of particles as a reference point per Formula 4:

$$R = \sqrt{T}, \tag{4}$$

where,

R = distance covered by a Brownian particle in time T;
T = time index.

According to this formula, a Brownian particle is moved by a distance equal to the square root of time spent on this movement. If H =0,5, a system runs in the time T the same distance as a Brownian particle. With the large values of H a system goes a considerable distance in the same time T in comparison with a Brownian particle. The Hurst exponent calculation can be carried out according to the following Formulas 5, 6:

$$R \setminus S = \left(aN\right)^{H}, \tag{5}$$

$$H = \frac{\log\left(R \setminus S\right)}{\log\left(aN\right)}, \tag{6}$$

where,

H = Hurst index;
S = mean-square deviation of an observations series x;
R = amplitude of the accumulated deviations Zu;
N = number of observation periods;
a = given constant, a positive number. Hurst has empirically calculated this constant for relatively short-term time series of natural phenomena. The constant is 0.5.

Many researchers in their studies use the Hurst index to evaluate the fractal structure of the time series (for instance, Nekrasova and Karnaukhova, 2016). However, the study showed that it has a number of imperfections. The first is the need for a large number of observations, and the second is a simple statement of the fractality presence without indicating the nature of further dynamics in the prices of financial assets. Therefore, the conclusion could be that the fractal index μ is more preferable for determining the fractal properties of financial assets.

The main advantage of the index μ in comparison with the Hurst (H) index is that the corresponding value $V_f(\delta)$ has a quick exit to the power asymptotic regime. This leads to the opportunity of using μ as of a local characteristic that determines the dynamics of the initial process. Such conditions include flats (periods of relative calm) and trends (periods of relatively long movement up or down).

Considering the main stages of determining a fractal index, let a function with at most a finite number of discontinuity points of the first kind be given on [a, b]: exactly these functions are usually considered as model functions. A uniform partition of the interval is introduced in Formula 7:

$$\omega_m = \left[a = t_0 < t_1 < \ldots < t_m = b\right], \tag{7}$$

where,

$$t_i - t_{i-1} = \delta = \frac{b-a}{m}, \left(i = \overline{1,m}\right).$$

Then the graph of this function is covered with rectangles in such a way that this covering is minimal in the area within the class of coverings by rectangles with base δ. The height of the rectangle on the segment $\left[t_{i-1}; t_i\right]$ will be equal to the amplitude $A_i(\delta)$, which is the difference between maximum and minimum value of the function $f(t)$ in this segment, as it is demonstrated in the Figure 1.

Let's introduce the quantity by the Formula 8:

Figure 1. The minimum cover of the graph f (t)

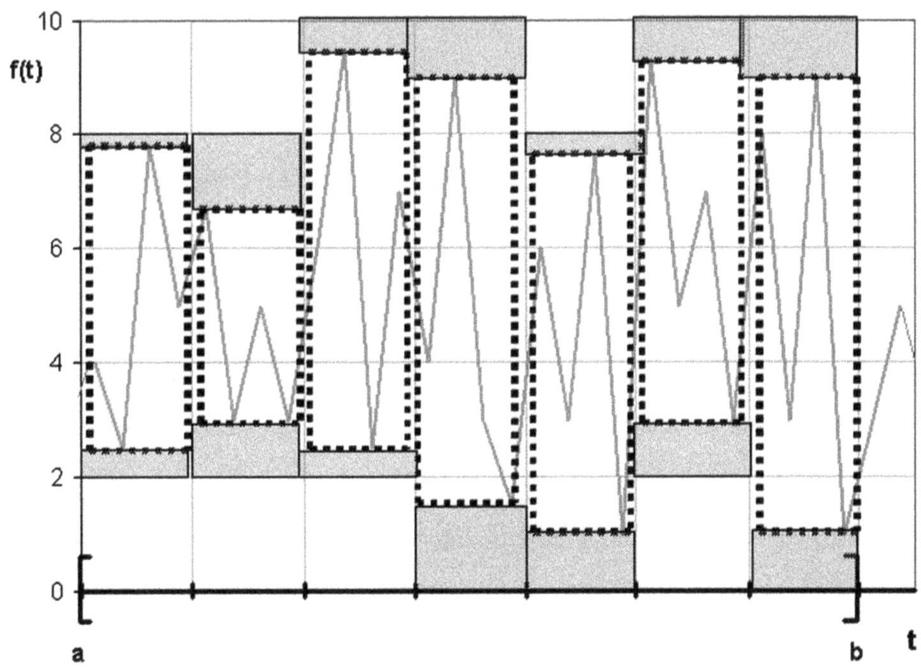

$$V_f(\delta) = \sum_{i=1}^{m} A_i(\delta) \tag{8}$$

Then the total area of the minimal covering can be written in the form, presented in the Formula 9:

$$S^{\mu}(\delta) = V_f(\delta) \cdot \delta \tag{9}$$

On the other hand, if to use the Hausdorff equation for the fractal dimension the result is calculated by the Formula 10:

$$N(\delta) \sim (1/\delta)^{D} \text{ at } \delta \to 0, \tag{10}$$

where,

$N(\delta)$ = minimal number of balls covering the original set.

Multiplying both sides of (10) by δ^2, the authors obtain the definition of dimension in the form of a power law for the area of cell coverings $S^c(\delta)$ as it is presented in the Formula 11:

$$S^c(\delta) \sim \delta^{2-D} \text{ at } \delta \to 0, \tag{11}$$

Because the $S^c(\delta) \sim S^{\mu}(\delta)$ at $\delta \to 0$ (Dubovikov et al., 2004), then in the Formula 12:

$$V_f(\delta) \sim \delta^{-\mu} \text{ at } \delta \to 0 \tag{12}$$

where, $\mu = D_{\mu} - 1$ for a one-dimensional function $D_T = 1$, then $\mu = D_{\mu} - D_T$.

Therefore, in this case the index μ is naturally called the fractality index, and the dimension D_{μ} by the dimension of the minimal covering. In accordance with these assumptions, it is possible to determine the coordinates of the regression equation for each δ how it is demonstrated in the Formula 13:

$$x = \ln\delta, \ y = \ln V_f(\delta) \tag{13}$$

Approximating a straight of the form $y = ax + b$ by least squares (OLS), assuming that $\mu = -a$, it can be determined in the slope of the regression line μ.

In order to relate the value μ to the behavior of the time series, it is natural to introduce the function μ(t) as the value of μ, determined at the minimum interval preceding t, where μ can still be calculated with acceptable accuracy. In the case of a continuous argument *t*, this would be an arbitrarily small interval, however, since in practice the time series always has a minimum scale δ_0 (in our case, δ_0 is 1 day), then μ_t has finite length (in this case the authors take μ_t = 32 days).

INTERRELATION THE FRACTAL STRUCTURE AND CRITICAL BEHAVIOR OF TIME SERIES OF FINANCIAL ASSETS

In many areas of research, the study of so-called critical or extreme events has a particular importance. First of all, this refers to natural disasters (earthquakes, hurricanes, tsunamis, floods, etc.) and human-made disasters (explosions, transport accidents, environmental pollution, technogenic climate change). However, a critical behavior is typical for significantly wider class of systems, which in the modern literature is customary to call complex systems.

Due to the wide spread of the EMH, the long-term crises and financial crashes were considered as statistical artifacts and were explained by a rare combination of external causes. However, the application by some researchers of the ideas within the theory of dynamic chaos to the behavior of market agents has demonstrated that the instability of financial markets can be caused by internal reasons related to the cooperative behavior of market participants. From the practical point of view traders are interested in similar price behavior, which can lead to sharp changes in the value of the investment portfolio. Commonly, critical behavior in finance is understood as a sharp drop in the value of most securities.

The most important are two types of such behavior. The first is a rapid and strong change in the amplitude of price fluctuations in large scales. This type of behavior corresponds to the transition of the market from flat to trend. The second type of critical behavior is a sharp change in the direction of the trend. In most cases in the market a change in the direction of the trend occurs through an intermediate stage - flat. This gives market participants time to analyze and make decisions. However, in some situations, an intermediate stage is absent.

Considering these types of critical price behavior for the price time series of three indices - FESX, MICEX, Dow Jones, Figure 2, Figure 3, and Figure 4 demonstrate a typical fragment of the price time series of these indices together with the calculated for this fragment of function $\mu(t)$. The study period is from June, 2003 to November, 2016 with the interval of the time series of one month. All prices are obtained with the help of the MetaTrader - the information and trading platform developed by the MetaQuotes Software Corp., designed for organizing a dealing service in the Forex, CFD and Futures markets. The index μ was defined as an average value consisting of 11 indices: AEX Index (Amsterdam Exchange Index 25), EURO STOXX 50, Nikkei 225 (Nikkei 225 Stock Average), DJI (Dow Jones Industrial Average 30), S&P 500 (SPX), FDAX (Germany 30), FTSE 100, IBEX 35 (Iberia Index), MIB (Italy 40), CAC 40, MICEX.

These graphs demonstrate the similar behavior of the prices within the FESX, MICEX, Dow Jones indices at these time intervals, as well as certain regularity between the dynamics of the fractality index μ and the time series of the index prices. Similar behavior of the price indexes FESX, MICEX, Dow Jones is the evidence of the globalization of financial markets. The synchronicity of price dynamics in the main indices of different countries is clearly observed over the long time intervals that are several years or more.

If the value of the fractality index $\mu < 0.5$ (as Figures 2, 3, 4 in the intervals from 06.2003 to 05.2008, from 05.2008 to 02. 2011), it is seen that the price series of stock indices are in trend. If the value of the fractality index $\mu > 0.5$ (Figures 2,3,4 in the interval from 02. 2011 to 05.2013), it is seen that the price series of stock indexes are in a flat, i.e. price fluctuations are insignificant and occur in the lateral price channel. When $\mu \approx 0.5$, it is clear that prices are in an intermediate state between trend and flat. Therefore, the initial series turns out to be the more stable; the larger is the value of μ. In this case, if

Figure 2. Graphs of daily values EURO STOXX 50 Index (FESX) and function $\mu(t)$ ($\mu(t)$ - red color, lower)

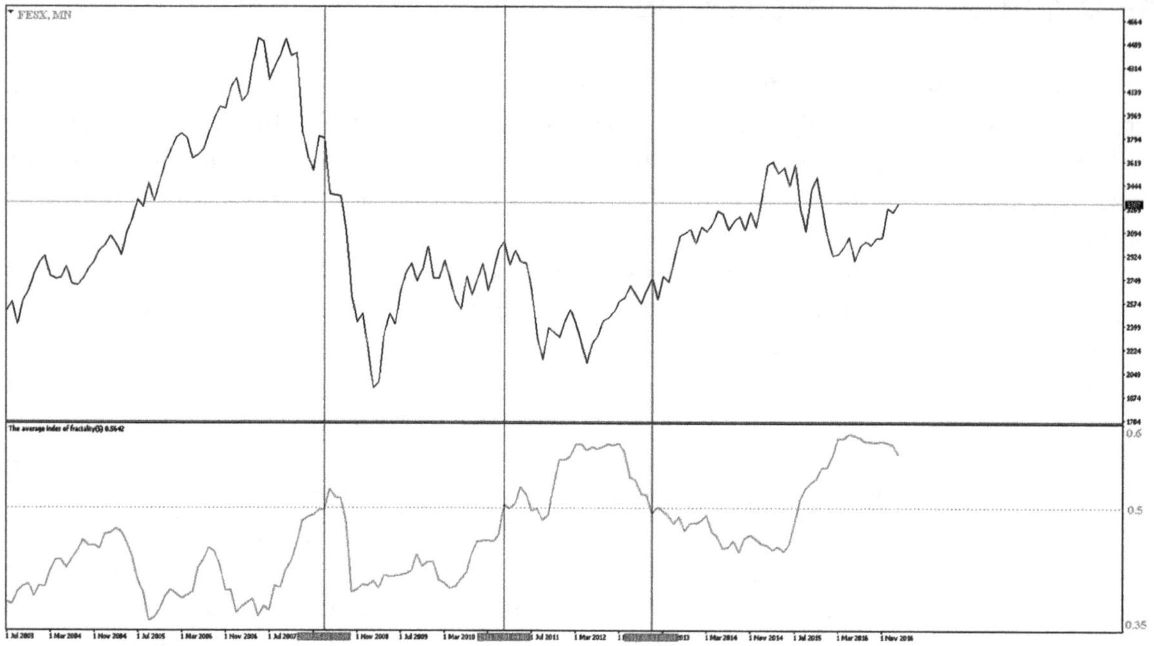

**For a more accurate representation see the electronic version.*

Figure 3. Graphs of the daily values of the MICEX Index and the function $\mu(t)$ ($\mu(t)$ - red color, lower)

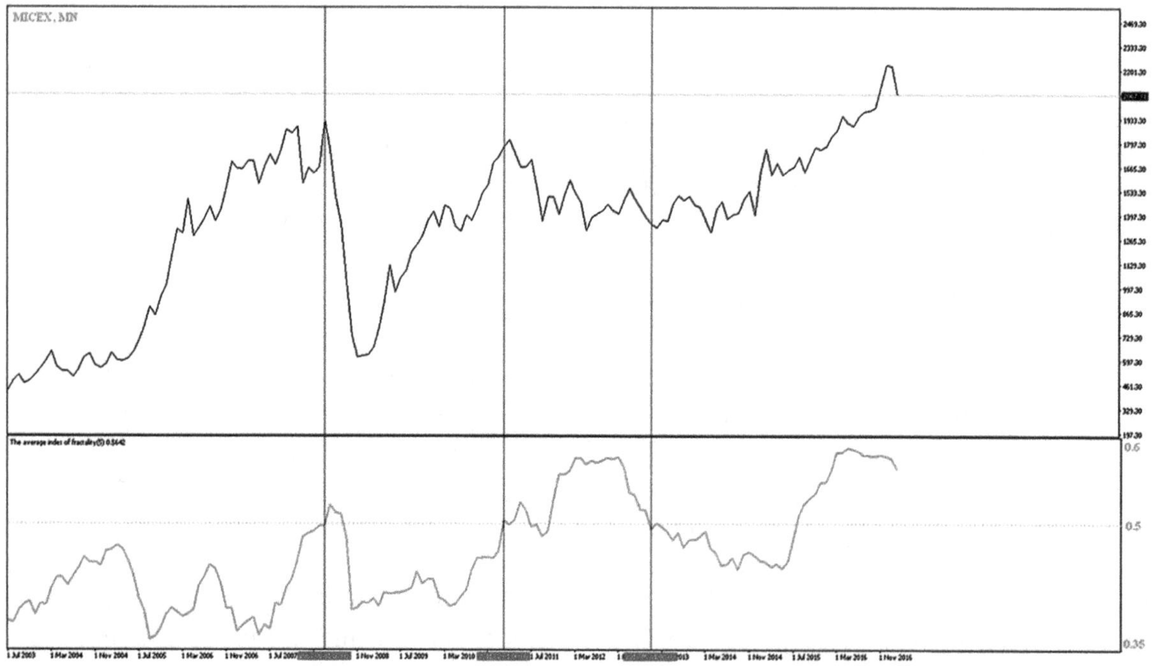

**For a more accurate representation see the electronic version.*

Figure 4. Graphs of the daily values of the Dow Jones Index and the $\mu(t)$ function ($\mu(t)$ - red color, lower)

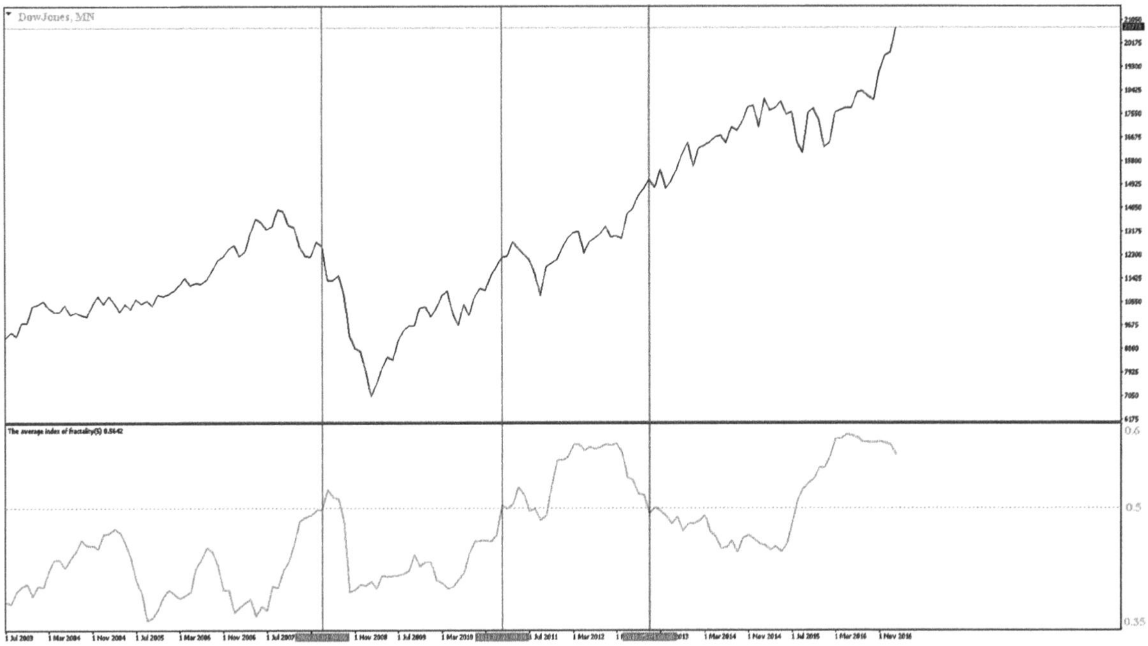

**For a more accurate representation see the electronic version.*

μ> 0.5, then a flat is observed, if μ <0.5, a trend is observed, and, finally, if μ ≈0.5, then the process is in an intermediate state.

In addition, in all three graphs another important regularity is observed. As soon as there is a decrease in the intensity of price fluctuations (in this case, μ ≈0.5), a radical change occurs in the subsequent price dynamics: either a change in trend direction, or a change of states from trend to flat or vice versa. For example, it can be seen in the graphs that at the point corresponding to the time period 05.2008 (μ ≈0.5), there is a change in the trend of price movement from growth to decline. At the point of 02.2011, the trend is replaced by a flat, and at the point 05.2013, the flat is replaced by a trend at μ ≈0.5. Thus, the financial time series do demonstrate a behavior, in which fluctuations in the fractal structure lead to a change in the type of behavior of the series. Indeed, for μ> 0 (the intensity of small-scale oscillations is larger than the intensity of large-scale oscillations), the amplitude of future large-scale oscillations decreases, i.e. time series seek to move towards to antipersistent behavior, i. e. flat. For μ <0, on average, the amplitude of future large-scale oscillations increases, i.e. time series seek to move towards to the persistent behavior (or trend).

Giving a theoretical acknowledgment of this dependence for the case of Gaussian random processes, it is important to begin with considering the Wiener model, which is the simplest case of such processes. The classical Wiener model of the Brownian motion is based on two postulates. First, the increments of the process over a certain time interval have a normal (Gaussian) distribution with zero mean, which follows from the central limit theorem and is obtained as a result of summing a sufficiently large number of independent (or weakly related) random variables with finite variance. Second, increments on non-inter-covering time intervals are statistically independent. The conclusion of these postulates could be expressed in the Formula 14:

$$\left[\left(X(t)-X(t_0)\right)^2\right]=\sigma^2\left|t-t_0\right|, \tag{14}$$

where,

The square brackets mean averaging,
$X(t)$ and $X(t_0)$ = values of the process, respectively, at times t and t_0,
σ_t and σ_{t_0} = the variances per unit time (in financial terms, the parameter σ is known as volatility).

From Formula14 it can be concluded that this process is converted into itself when the time scale is changed by a factor of b and the spatial scale is simultaneously changed $b^{1/2}$ times. The fractal dimension of the realization schedule for such a process is $D=1.5.\left(\mu=0.5\right)$.

Various generalizations of this model consist of the rejection either of conditions for independence of increments or of normality in their distributions. In the first case, processes with memory are derived and, in particular, a generalized Brownian motion. In the second, the Levy motion appears which has an infinite variance of displacement. Then the generalized Brownian motion model - $X_H(t)$ is considered.

As it is known, the generalized Brownian process has a zero mean increment and a variance of increments, as it is presented in the Formula 15.

$$\left[\left(X_H(t)-X_H(t_0)\right)^2\right]=\sigma^2\left|t-t_0\right|^{2H}, \tag{15}$$

where, *H* is the Hurst index associated with the fractal dimension in the graph of this process realization D (and, correspondingly, with the index μ) by the relation

$$H=2-D=1-\mu \ (0<H<1).$$

For convenience the system of units is used in which $\sigma^2=1$ and $X_H(0)=0$. Then, being based on the formula 15, the normalized correlation function of the past increments $-X_H(-t)$ with the future $X_H(t)$ is gained in the Formula 16.

$$C(t)=\frac{\left[-X_H(-t)X_H(t)\right]}{\left[X_H^2(t)\right]}=2^{2H-1}-1 \tag{16}$$

For *H>0.5* (*μ<0.5*), the correlation is positive, i.e. the tendency to increase (positive increment) in the past means, on average, a tendency to increase in the future, and vice versa, the downward trend (negative increment) in the past means a tendency to decrease in the future. This process is called persistent (preserving the trend) and it can be considered as a trend model.

For $H < 0.5(\mu > 0.5)$, the correlation is negative, i.e. in this case, an increase in the past means a likely decrease in the future, and a downward trend in the past means an increase in the future. Such a process is called antipersistent and it can be considered as a flat.

At $H = 0.5(\mu = 0.5)$ there is no correlation, and it is an intermediate state between the flat and the trend. Thus, the stability of the generalized Brownian motion is determined by the value of the index μ.

DETERMINATION OF THE RELATIONSHIP BETWEEN THE STABILITY OF FINANCIAL TIME SERIES AND THEIR FRACTAL STRUCTURE

Further, for study of the relationship between stability of financial time series and their fractal structure, empirical dependencies of various stability indicators on the value of the index μ were constructed. In total, five different indicators of stability were chosen and presented in the Formula 17:

$$F_1 = I_g\left(C_i / C_{i-32}\right), \tag{17}$$

where,

C_i = a current price closing price,
C_{i-32} = closing price 32 days ago.

The greater is the difference between the current closing price and the closing price 32 days ago, the greater is the value of F_1. Large modulo F_1 values mean a significant price movement over a period of 32 days, and small as a rule ones tend to be low price fluctuations, i.e. flat, as in the Formula 18.

$$F_2 = A_i / A_{i-32}, \tag{18}$$

where,

A_i = the amplitude of price fluctuations for the current period,
A_{i-32} = the amplitude of price fluctuations over the past 32-day period.
F_2 = the ratio of the amplitude of price fluctuations for the current and past 32-day periods. As a rule, if there is a trend - $F_2 > 1$, and if there is a flat - $F_2 < 1$.
$F_3 = \sigma\left[\log\left(C\right)\right]$ = standard deviation of closing prices for the last 32 days.

The greater is the value of F3, the clearer trend is in the behavior of prices.

F_4 = the value of the slope coefficient in the linear regression line, carried out for a number of prices. The larger is the value of F_4, the stronger is the slope of the regression line and, consequently, the clearer trend.

F_5 = the difference between the current price and the price of 32 days ago, normalized to the sum of the modules of daily increments in prices for 32 days, as in the Formula 19.

$$F_5 = \frac{C_i - C_{i-32}}{\sum_{j=i-31}^{i}\left(C_j - C_{j-1}\right)} \tag{19}$$

where,

$C_i - C_{i-32}$ = the difference between the current price and the price of 32 days ago.

In the case that for 32 days the price increment never changed sign (there was observed movement up or down) | F5 | = 1. If the flat on price chart can be observed, then the value of | F5 | is close to zero.

For all selected stability indicators, the following statement is true: the more stable is the behavior of the initial series (the oscillations occur near one level), the closer is F to zero and vice versa, the more pronounced is the trend, the more modulo is F.

For investigation of the Fi (μ) dependencies (i = 1..5), the original series were divided into intervals of 32 days, displaced relatively to each other by 1 day. For each source row, 8145 days in length, 8113 intervals were derived. For each of them the value of μ and the value of Fi were calculated. The obtained values were plotted on the dot diagrams, where the value of μ was plotted along the x axis, and the value of Fi was along the y axis. Typical diagrams derived are presented in Figure 5, Figure 6, and Figure 7 for a number of prices of the MICEX, Dow Jones and the EUR / USD currency pair.

Figure 5. Spot diagram of the relationship between F1 and μ for the MICEX index for the period from 20.05.2004 to 23.03.2017

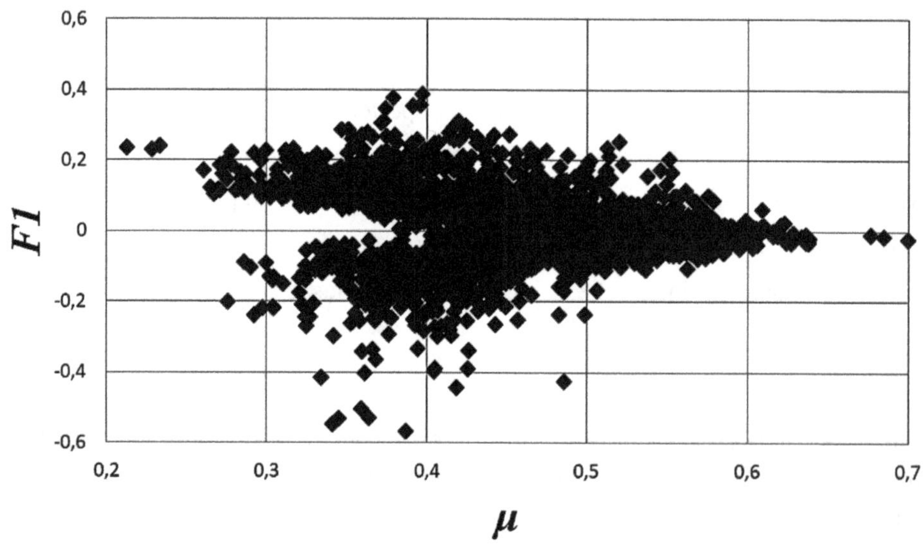

Figure 6. Spot diagram of the relationship between F4 and μ for the Dow Jones for the period from 20.05.2004 to 23.03.2017

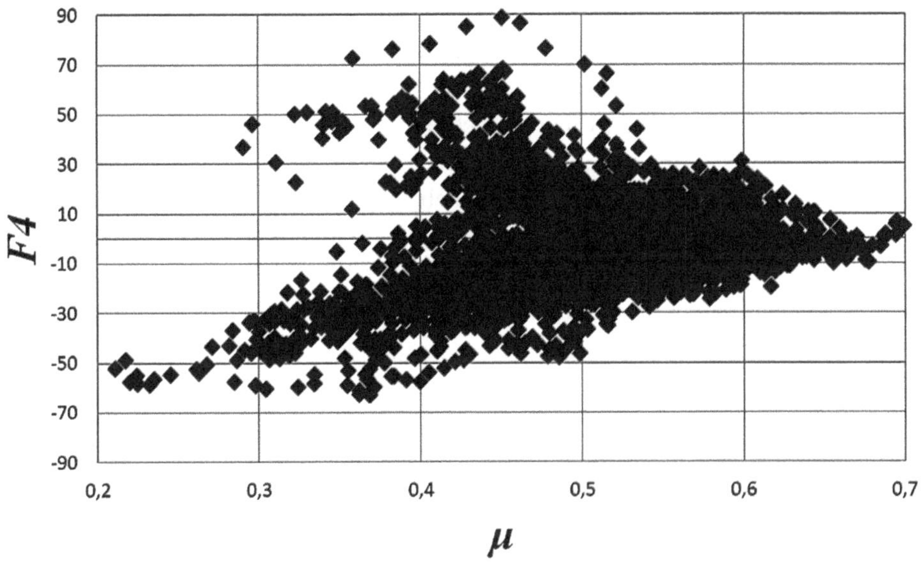

Figure 7. Spot diagram of the relationship between F5 and μ for the EUR / USD pair for the period from 20.05.2004 to 23.03.2017

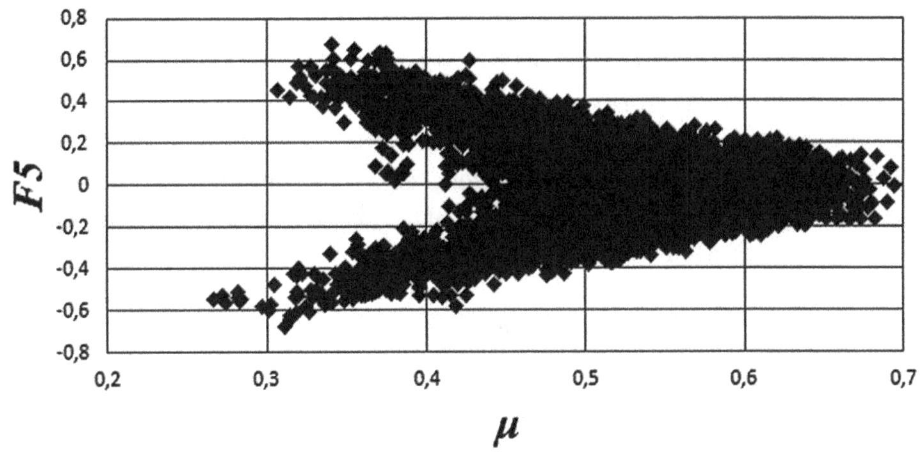

It can be seen from the derived type of diagrams that for the time series under the study the value of μ is related to the stability of the series. Moreover, the larger the value of μ is, the more stable is the behavior of the series, and, the smaller is the value of μ, the more expressed is the trend in the original series.

Now it's time to introduce the concept of the internal state of a stock at time t as the value of the stability index μ (t) and to understand the internal mechanism that determines the behavior of μ (t) and, consequently, the behavior of the share price. For this purpose the feedback mechanism is used, namely: any open system always acts in its environment and, accordingly, experiences the opposite effect. If with this reverse action the intensity of direct action increases, then the feedback is called positive (otherwise

- negative). Many abrupt and avalanche processes (including social and natural disasters) are the result of positive feedback. A negative feedback leads to the maintenance of a stable regime in all possible open systems. This mechanism is known in many fields of economic science, namely: on the one hand, in establishing equilibrium (negative connection), and on the other hand, with all possible transient processes (positive connection).

This mechanism also acts in the stock market, determining the nature of prices series deviation from a random walk. The principle of its work, in the authors' opinion, is based on the emergence of a certain relationship between the behavior of prices and the behavior of investors. For convenience, in addition to the index μ, the coefficient $\lambda = (0.5 - \mu)$ is used, which is called the feedback indicator. This option ($\mu(t) = 0.5$ and ($\lambda(t) = 0$) corresponds to a random walk of price (wiener process). In this case, investors behave independently and there is no pronounced tendency in price behavior. In this situation, it can be said that the price has "normal" stability. Since the price is absolutely passive with the respect to external influences, then there is no any "connection".

The case when $\mu(t) > 0.5$ and $\lambda(t) < 0$ corresponds to the situation of increased price stability relative to external influences, which may be due to increased investor confidence in the stability of a company and the lack of new information about the market. In this case, the stock prices fluctuate within a fairly narrow price range. With an increase in prices there is a sufficient number of sellers, and in case of a drop in prices, there are buyers who return prices to their original range by their actions. Obviously, the "connection" in this case is negative and its action "suppresses" the change in the share price, causing a stable price behavior.

The case when $\mu(t) < 0.5$ and $\lambda(t) > 0$ corresponds to a situation of low stability (or inertia) of the price. This is usually due to the emergence of new information and the reaction of investors to this information. Since in the first approximation it can be assumed that all market participants estimate the incoming information approximately similarly, a trend appears in the price movement, so there is a tendency corresponding to the information received. Under certain conditions, such a situation can lead to a sharp drop or increase in the share price, since in this case a positive feedback mechanism unfolds. In its pure form, the act in this way would be following: stock prices begin to rise (fall), but at the same time there are new people willing to buy (sell), that leads to even a greater growth (falling) of prices.

With free development, such a connection would ultimately lead to a collapse. But the modern financial market is stable as a whole. And this stability in the authors opinion is provided not only by the regulatory structures (central banks, commissions on securities, etc.), but also by the built-in stabilizer. It is represented by financial speculators, who with the slightest signs of slowing or slew reversing the existing trend, begin to sell or buy shares ("play" against this trend), thereby performing the function of a "silencer" of positive feedback. At the very beginning of the trend development, when the price is still relatively stable, the actions of the same speculators, who are striving to make quick profits on sharp price movements, intensify this trend. In other words, financial speculators "push" the internal state of the stock towards a random walk (i.e., $\mu(t) \approx 0.5$), where, as it is known, there are no arbitration opportunities. So in general terms, the mechanism that determines the nature of the price series behavior is observed.

Thus, knowledge of the numerical values of the series fractal indices can be used to predict the future behavior of time series. In finance such forecasts can be exploited both for trading and correct assessment of risks moments in preceding the emergence of crisis phenomena in the financial market.

SOLUTIONS AND RECOMMENDATIONS

The movement of prices has a fractal nature, because the actions and reactions of people in the market are repeated (the self-similarity property). The main indicator of self-similarity is the dimension D (or the exponent H). Based on a large amount of empirical data, it was proved that the index μ is an indicator of the stability of the financial assets time series. The index of fractality μ allows to solve, with acceptable accuracy, the most diverse practical tasks of the stock market professional participants: testing of financial assets time series, risk assessment, construction of indicators of strong changes in prices of financial assets. The authors of the chapter substantiated and confirmed the effect of increasing large-scale fluctuations upon suppression of small-scale ones. The well-known "lulling effect" (suppression of the high-frequency noise component) which usually precedes natural disasters (for example, earthquakes) is a particular manifestation of this effect.

CONCLUSION

Summarizing the main ideas, the following conclusions could be drowned. The index of fractality can be used for the traditional analysis in the fractal structure of time series. With the same amount of data, the index μ is determined several times more accurately than the Hearst index (H). In this case, the minimum scale necessary for determining μ with a reasonable accuracy is several tens of points, whereas for the determination of H, several thousand data are needed. This fact makes it possible to use μ as a local fractal index in the analysis of time series.

Financial time series demonstrate the behavior in which fluctuations in the fractal structure lead to a change in the type of behavior of the series. Indeed, for $\mu > 0$ (the intensity of small-scale oscillations is larger than the intensity of large-scale oscillations), the amplitude of future large-scale oscillations decreases, i.e. a number of strive to move to an antipersistent behavior or flat. For $\mu < 0$, on average, the amplitude of future large-scale oscillations increases; a number of attempts to shift to persistent behavior or trend.

The research has demonstrated that for the time series of prices in financial assets the value of μ is related to the stability of the series. In this case, the larger the value of μ is, the more stable the series behavior. Conversely, the smaller is the value of μ the more pronounced is the trend in the original series. If $\mu > 0.5$, then a flat is observed, if $\mu < 0.5$, a trend is observed, and finally, if $\mu \approx 0.5$, then the process is in an intermediate state between the trend and the flat. Therefore, it can be concluded that the fractality index can be used to predict the emergence of crisis events in financial markets, manifested in a sharp fall (or growth) in the prices of the financial assets with a significant decrease in the values of μ below 0.5.

Any open system always acts in its environment and, accordingly, experiences the opposite impact. If, with this reverse impact, the intensity of the direct action increases, then the feedback is called positive (otherwise negative). Many abrupt and avalanche processes (including social and natural disasters) are the result of positive feedback. Negative feedback leads to the maintenance of a stable regime of all possible open systems.

REFERENCES

Bachelier, L. (1964). *Theory of Speculation. In P. H. Cootner (Ed.), Random Character of Stock Market Prices*. Cambridge, MA: The MIT Press.

Black, F., & Scholes, M. (1973). The Pricing of Options and Corporate Liabilities. *Journal of Political Economy*, *3*(3), 637–659. doi:10.1086/260062

Dubovikov, M. M., Starchenko, N. S., & Dubovikov, M. S. (2004). Dimension of the minimal cover and fractal analysis of time series. *Physica A*, *339*(3-4), 591–608. doi:10.1016/j.physa.2004.03.025

Fama, E. F. (1970). Efficient Capital Markets: A Review of Theory and Empirical Work. *The Journal of Finance*, 47.

Gencay, R., Selcuk, F., & Whitcher, B. (2002). Scaling properties of foreign exchange volatility. *Physica A*, *289*(1-2), 249–266. doi:10.1016/S0378-4371(00)00456-8

Hausdorff, F. (1919). Dimesion und Ausseres Mass. *Mathematische Annalen*, *79*(1-2), 157–179. doi:10.1007/BF01457179

Hurst, H. E. (1951). Long-term Storage of Reservoirs: An Experimental Study. *Transactions of the American Society of Civil Engineers*, *116*, 770–799.

Kristoufek, L. (2013). Fractal Markets Hypothesis and the Global Financial Crisis: Scaling, Investment Horizons and Liquidity. *Scientific Reports*, *3*, 2857. doi:10.1038rep02857 PMID:24091386

Los, C. A. (2003). *Financial Market Risk: Measurement & Analysis*. London, UK: Taylor & Francis Books Ltd. doi:10.4324/9780203987636

Mandelbrot, B. B. (2002). *Gaussian Self-affinity and Fractals*. Springer.

Mandelbrot, B. B., & Van Ness, J. W. (1968). Fractional Brownian motions, fractional noises and applications. *SIAM Review*, *10*(4), 422–437. doi:10.1137/1010093

Merton, R. (1973). Theory of Rational Option Pricing. *The Bell Journal of Economics and Management Science*, *4*(1), 141–183. doi:10.2307/3003143

Murphy, J. (1999). *Technical Analysis of the Financial Markets*. Institute of Finance.

Nekrasova, I., & Karnaukhova, O. (2016). Plasticity and Memory in the Financial Markets. In Neuroeconomics and the Decision-Making Process. Hershey, PA: IGI Global.

Osborne, M. F. (1964). Brownian motion in the stock market. In P. H. Cootner (Ed.), *Random Character of Stock Market Prices*. Cambridge, MA: The MIT Press.

Prechter, R. R. Jr, & Parker, W. D. (2007). The Financial/Economic Dichotomy in Social Behavioral Dynamics: The Socionomic Perspective. *Journal of Behavioral Finance*, *8*(2), 84–108. doi:10.1080/15427560701381028

Roch, A. (2011). Liquidity risk, price impacts and the replication problem. *Finance and Stochastics*, *15*(3), 399–419. doi:10.100700780-011-0156-x

Sharp, W. F. (1970). *Portfolio Theory and Capital Markets*. McGraw-Hill.

Takens, F. (1981). *Dynamical Systems and Turbulence*. Berlin: Springer.

Vacha, L., Janda, K., Kristoufek, L., & Zilberman, D. (2013). Time-frequency dynamics of biofuel-fuel-food system. *Energy Econ*, *40*, 233–241. doi:10.1016/j.eneco.2013.06.015

Chapter 3

On the Impact of Long Memory on Market Risk:

Pre- and Post-Crisis Evidences - Long Memory and Stock Market Risk

Salim Ben Sassi
High Institute of Management of Tunis, Tunisia

Azza Bejaoui
High Institute of Management of Tunis, Tunisia

ABSTRACT

This chapter investigates the influence of the long memory behavior in returns and volatility on the market risk for four emerging stock markets during the pre- and post-crisis periods. In this respect, the authors consider four major political events (Tunisian revolution, Egyptian revolution, assassination of Prime Minister Rafik El Hariri, and a series of suicide bombings in Morocco). Using the modified R/S test and GPH test, they show the long memory property in returns and volatility over the two sub-periods. To explore the dual long memory property, the authors apply the joint ARFIMA–FIGARCH specification on the returns and volatility of the four emerging stock markets. The dual long memory property is prevalent in the returns and volatility of the emerging stock markets over the pre-crisis period. During the post-crisis period, the dual long memory process is only detected in the Moroccan market. The authors also display the dynamic behavior of VaR during the two sub-periods. In addition, based on the backtesting test, VaR performed better during the two sub-periods for all countries.

INTRODUCTION

Over the past decade, many researchers have correspondingly investigated the behavior of stock returns in different financial markets. Thereby, volatility clustering, asymmetric leverage effects, and long memory, among others, are considered as stylized facts in finance. In particular, whether stock returns long memory, they show significant autocorrelation among observations which are largely separated in

DOI: 10.4018/978-1-5225-3767-0.ch003

time (Barkoulas et al., 2000). In this respect, many researchers focus on investigating the long memory property in emerging markets. Even though the long memory property has become of paramount importance in empirical finance, the debate on existence of this issue in emerging stock returns still remain inconclusive. The recurring market microstructure change coupled with the alternate periods of financial turmoil can be impetus for analyzing the issue of long memory in such markets. Even though the rapid growth of emerging markets and their increasing importance attract the investors, these markets suffer from biases due to market severe thinness and nonsynchronous trading. As well, investors in emerging markets tend to react gradually and slowly to new information compared to developed markets (Barkoulas et al., 2000). Apart from the specific features of these markets, many emerging markets have experienced dramatic country-specific events such as the 1994 Mexican crisis, the 1997 East Asian crisis, the 2001 Argentine crisis and the 2011 Arab Spring. The revival of emerging financial markets, after the country-specific circumstances, represents a new challenge for many researchers. Notably, research on emerging markets focuses on the statistical nature of the market fluctuations and risk management, especially amid financial turmoil.

As aforementioned, emerging markets experienced great upheaval, involving that such markets collapsed and their volatility raised. A great deal of econometric research has been devoted to examine the influence of crises on the behavior of stock prices. In this respect, there is no precise and rigorous definition of the concepts of “stock market crisis”. These commonly used expressions are used to describe a brutal and widespread collapse of stock prices. The crash resulted in an almost continuous fall in stock market prices, following massive sales of securities held by investors. From empirical standpoint, the stock market crisis should not be confused with price volatility. The latter represents the amplitude of changes in stock prices over a given period (measured on the basis of stock market indices) and reflects the chronic instability of the stock market. The stock market crisis is generally accompanied by an increase in volatility. Thereby, the emerging markets exhibit an increased level of risk during the recurring turbulence in those markets. As a result, the use of standard measures can misevaluate the market risk. Under uncertainty in financial markets, measures of risk have a crucial role. Standard quantitative tools, such as Value-at-Risk (VaR) are used for measuring, managing and controlling risk. The Value-at-Risk (VaR) expresses the maximum loss of an asset or a portfolio in a given time line and confidence level. Several researches have focused on the evaluation of the VaR using different types of models. Despite the multitude of methods, there are no standard ones. The researches in this area have focused on the search for the most adequate method to the structure of financial returns especially in developed markets.

In this chapter, we focus our attention on the following questions: Does the long memory property in return and volatility change over time and among stock markets? If so, has the long memory property affected the market risk in the emerging markets? and how should the VaR be adjusted to these specific criteria?. For this end, we use stock markets indices for Tunisia, Egypt, Lebanon and Morocco for the different periods. Choosing such emerging markets is motivated by a comparable degree of financial markets and the country-specific circumstances. The sampled market data is divided into sub-periods based on the onset of political unrests.

The chapter is organized as follows. Section 1 presents the review literature on long memory and its definitions and measures. Section 2 reports the features of emerging markets. Section 3 presents the methodology and Section 4 reports data and descriptive statistics. Estimation results are shown in Section 5 and Section 6 concludes.

LONG MEMORY: DEFINITION, MEASURES, AND RELATED LITERATURE

The long memory depicts the higher order correlation structure of a time series. Especially, the existence of long memory in returns and volatility involves that temporal dependencies between distant (or widely separated) observations persists. Such series clearly display the hyperbolically decaying autocorrelations as well as low frequency distributions (Bhattacharya and Bhattacharya, 2012). Hiremath and Bandi (2010) define long memory in volatility as a slow hyperbolic decay in autocorrelation functions of the squared or log-squared returns. As is evident, the long memory issue implies the long-run effects of shocks compared to conventional approaches (Charfeddine and Ajmi, 2013). In other words, a shock at one point of time involves an increase in risk and uncertainty in the stock market, lasting a significant time when the long memory process is present (Bhattacharya and Bhattacharya, 2012). From a mathematical perspective, whether

$$\lambda_s = \text{cov}\left(y_t, y_{t-s}\right), s = 0, \pm 1, \pm 2, ...,$$

and there exists constants k and α, $\alpha \in (0,1)$ such as $\lim_{s \to \alpha} k\lambda_s s^{-1} = 1$ then y_t is a long memory process. The long memory, or long-range dependence, in stock returns therefore has key implications for investment decisions and capital market theory. Obviously, suspecting the existence of long memory in stock returns can lead to possibly predict the behavior of these series in a market setting.

A large volume of literature on econometrics and empirical finance focuses on modeling the temporal variation in financial markets. Many conditional volatility models have been developed to test for and capture the volatility persistence properties in conditional variance such as Autoregressive Conditional Heteroscedasticity (ARCH) and Generalized ARCH (GARCH) models pioneered by Engle (1982) and Bollerslev (1986), respectively. Nevertheless, such models fail to detect long memory in volatility given that the autocorrelation of stock returns seems to decay at a slower rate (Hiremath and Bandi, 2010). Granger and Joyeux (1980) and Hosking (1981) proposed the so-called Autoregressive Fractionally Integrated Moving Average (ARFIMA) model which is characterized by fractionally difference in the mean process. To account for the high persistence behavior, many models have been developed in the empirical literature such as the Integrated GARCH (IGARCH) model proposed by Engle and Bollerslev (1986) which suggest that a shock to the conditional variance seems to be crucial for the optimal forecasts of the variance for all future horizons. Similarly, Baillie et al. (1996) put forward the FIGARCH model, taking into account fractional difference in the conditional variance function.

The FIGARCH model provides high potential for modeling financial time series compared to IGARCH model (Charffedine and Ajmi, 2013). As Bollerslev and Mikkelsen (1996), a shock to the conditional variance in the FIGARCH model is transitory, "in the sense that the influence on the forecast of the future conditional variance dies out at a slow hyperbolic rate of decay". Mendes and Kolev (2006) suggest that the long memory in conditional variance overshadows the true dependence structure. Mandelbrot (1971) stipulates that perfect arbitrage is not possible if stock returns exhibit a long-range dependence.

Empirically, testing for the long memory in financial time series remains a key problem because of high autocorrelation of long memory processes which decay asymptotically over space and time (Bhattacharya and Bhattacharya, 2012). Nonetheless, some attempts have been made to capture and test for long memory properties in financial data series. For instance, the Rescaled-Range (R/S) method is a widely accepted method to capture long memory property in financial markets (Lin and Fei, 2013). According to Bhattacharya and Bhattacharya (2012), such measure is found on the assessment of Hurst's exponent of stationary time series and used by Mandelbrot (1971, 1972, and 1997) who claimed the outperformance of R/S analysis compared to the variance analysis. Nevertheless, Davies and Harte (1987), among others, display that R/S method could provide biased estimations and seems generally to confirm the long-range dependence hypothesis. In particular, in many cases, the R/S method is unable to discern the short memory from long memory in financial time series. Therefore, Lo (1991) proposed the so-called modified Rescaled-Range (R/S) analysis which is able to distinguish between long and short memory and overcome the drawback of Gaussian distributions and conditional heteroscedasticity.

Many studies search for the presence of long memory in financial markets in different financial markets. For instance, Harvey (1993) shows the existence of long memory in volatility of currency exchanges rates. Robinson (2003) displays the presence of long memory in financial time series of asset returns whereas Eldera and Serletis (2008) provide a support for long memory in future energy prices. Obviously, long memory in stock returns seems to be well-documented in different stock markets (e.g. Andersen & Bollerslev, 1997, 1998; Breidt et al., 1998). As reported by Charfeddine and Ajmi (2013), these findings spawned research into possible explanations and the development of alternate models such as FIGARCH model. Leafing through the literature on long memory shows that many studies have focused on this issue in developed markets and in particular the US market (e.g. Ding et al., 1993; Crato & Lima, 1994; Granger et al., 1997; Adreano, 2005; Gurgul & Wojtowicz, 2006). For instance, Greene and Fielitz (1997) report evidence of persistence in daily US stock returns series. Yet, Cheung and Lai (1995) find no evidence of persistence in many international stock returns series. As well, Crato (1994) reports similar evidence for the stock returns series of G7 countries.

The debate on existence of long memory in stock market returns still remain inconclusive, even in the context of emerging markets. For example, Koong et al. (1997) provide little support for long memory in Pacific Basin stock returns. Cavalcante and Assaf (2005) suggest a strong dependence in squared and absolute Brazilian stock returns over the period 1997-2002. Mendes and Kolev (2006) display strong presence of long memory in volatility in twelve emerging markets during the period 1995-2005. Korkmaz et al. (2009) show that the Turkish unfiltered index returns display strong evidence of long memory, but the results show weak evidence after treating structural breaks properly. Disario et al. (2008) establish long memory in Turkish return volatility whereas Kang et al. (2010) display similar evidence in China. Using stock returns for Korea, Malaysia, Singapore and New Zealand, Sadique and Silvapulle (2001) report evidence for long-term dependence in these markets. Jefferis and Thupayagale (2008) establish long memory in volatility for South Africa and Zimbabwe, while no such evidence was shown in Botswana. MacMillan and Thupayagale (2009) report evidence of long memory in volatility in seven of eleven studied African markets. Nath (2001) finds evidence of long memory property in Indian whereas Mukherjee et al. (2011) do not show long memory property in raw returns, but in absolute and squared returns in Indian stock markets. Lin and Fei (2013) examine the long memory property of Chinese stock markets by using the conditional ad actual volatility series, and report moderate positive degree of long memory in these markets.

EMERGING MARKETS: SPECIFICITIES, CRISES, AND MARKET RISK

Investigating of the statistical nature of the market fluctuations in emerging markets has become more important in recent years. Obviously, the rapid growth of emerging markets and its increasing importance in the global finance have attracted the attention of investors across the globe. As an example, Pownall and Koedjik (1999) argue that many Asian markets are characterized by highly volatility financial markets which when coupled with high returns can attract investors who search for diversifying their portfolios. Consequently, many researchers, investors and practitioners increasingly tend to understand the behavior of emerging markets. Despite emerging economies are considered as the new engines of global growth (Pitterle et al., 2015), many emerging markets are characterized by specific country-specific features. Noteworthily, the lack of institutional development and corporate governance, the thin-trading, high volatility and various frictions, as well as market microstructure generally characterize stock markets in emerging economies. As pointed out by Blavy (2002), the fundamental information in emerging markets is narrowed given the low requirements for disclosure and scare corporate information available for investors to approximate intrinsic values. Such environment combined with country-specific circumstance, the lack of sound regulatory and supervision framework, the lack of liquidity, high volatility, uncertainty and missing market makers increase speculative movements and unveil the development of these countries. In particular, emerging markets experienced upheaval, implying that many markets collapsed and their volatility increased. This is generally attributed to political shocks and political instability or social turmoil which are impediment for setting up country-specific reforms and policy measures to boost such economies. According to Pitterle et al. (2015), emerging economies have experienced a significant and widely unexpected growth slowdown since 2011.

By and large, the alternate episodes of financial turmoil have a severe impact on investors' exposure to the emerging markets, even leading investors to fled these markets. With this respect, combining the recurring turbulence and the increased expansion of trading activity in the emerging markets has seemingly propelled asset liquidity trading risk to the forefront of market risk management research (Al Janabi, 2011a).

Perceiving risk from investing in emerging markets can be much greater than the standard measure of risk (Pownall & Koedijk, 1999), implying the lack of diversification. Thereby, risk assessment and management have become of paramount importance in financial industry, not only for developed markets but also emerging markets. The depth and the synchronized nature of emerging market nature of the emerging market slowdown involve the large negative returns and the existence of additional downside risk, risk that seems more severe during periods of financial turmoil (Pownall & Koedijk, 1999). During these periods, using standard measures can underestimate the market risk. In this regard, Pownall and Koedijk (1999) argue that utilizing the variance of an asset's return distribution as the unique risk's measure can thus imply a serious underestimation of the true risk involved in holding such assets. Capturing and measuring the increased market risk require management techniques such Value-at-Risk (VaR). VaR provides an estimation of the worst expected loss for a time horizon and within a confidence level. Notably, it is able to assess possible losses in financial markets.

Many researchers adopt the GARCH-POT model with the aim to estimate VaR such as Allen et al. (2011), Bhattacharyya and Ritolia (2008), Bali and Neftçi (2003), and so on. Bali and Neftçi (2003) argued that the EVT-GARCH model gives accurate estimates than those of the GARCH-T model. Bhattacharyya and Ritolia (2008) employed the EVT-GARCH model to estimate the Indian stock market risk, and recommended its use to establish a system of adequate risk management. In Marimoutou et al. (2009), the EVT-GARCH model could capture the heteroscedasticity through GARCH model and model the tail-fatness of financial returns distribution via the EVT approach.

Other methods have been also proposed to estimate the downside risk. In this regard, the Markov-switching models have been widely applied in the VaR literature due to their aptitude to describe efficiently the time-varying volatility of financial returns[1]. Billio and Pelizzon (2000) extended the mixture of normal distribution method that was proposed by Zangari (1996) and Venkataraman (1997), by developing a multivariate MS model to estimate the daily VaR for ten Italian stocks. They found it able to capture the leptokurtic nature of financial returns and the discontinuity of changes in market risk that are generally irregular; the risk increases (resp. decreases) when the portfolio makes huge losses (resp. gains). In addition, it outperforms both RiskMetrics and GARCH models. Li and Lin (2004) used the SWARCH model to compute VaR. The SWARCH model could outperform the ARCH and GARCH models when dealing with tail-fatness, leptokurtosis, and skewness.

Kawata and Kijima (2007) develop the regime switching log-normal (RSLN) model to estimate VaR. They inserted adjustment factors for forecast probabilities when estimating their two-state RSLN model parameters using the simple EM algorithm. Assuming that the market switches through time between a stable low-volatility state and a more unstable high-volatility state, they employed a two-step procedure for estimating the returns based on closing prices of 200 stocks and returns based on closing prices of 200 stocks and TOPIX on the Tokyo Stock Exchange for the period from 26 September 1994 to 16 November 2004. First, they assumed that there exists a process ξ_t that alone can determine the market's regime. Given the data $y_t = \left(\xi_1, \xi_2, \ldots, \xi_T\right)$. They estimated all the RS model parameters. Second, using the estimated smoothed probabilities, they determined to which regime each particular date belonged. By taking into account the log-returns' conditional independence, they can estimate each regime's mean vector and the covariance matrix. Finally, all of the forecast probabilities and the portfolio VaR are then calculated. They found their model good enough for VaR estimation.

Samuel (2008) introduced some refinements in the traditional CEVT model. Explicitly, he proposed an EVT-SWARCH model to deal with the time-varying volatility nature of financial returns distributions and the tail behavior of the SWARCH residuals' distributions as well. His approach focuses on the whole distribution, instead of only the tail distribution (Byström, 2004). It starts with the use of SWARCH model to estimate the conditional mean and volatility. Next, he applies the POT method to model the residuals' distribution. The VaR is then calculated. Also, Samuel (2008) compared the VaR estimation performance of EVT-SWARCH with EVT, EVT-GARCH, Student-t, SWARCH and GARCH models. He concentrated on negative tails and forecasting performance of the models when they are applied to the daily stock return series of the S&P500, Hong Kong's Hang Seng Index and the Hang Seng China Enterprise Index. He found that both EVT-GARCH and EVT-SWARH models give much more accurate results for the three stock indices, mainly at the 99 and 99.5% confidence levels. The daily VaR estimates by the conditional EVT models adjusted closely with the volatilities' change. Specifically, the EVT-SWARCH model showed a great performance when capturing the extreme losses during the period of study.

METHODOLOGY

Long Memory in Returns

Modelling the stock returns is basically made using Autoregressive (AR) model). This model allows taking into account the long memory in stock returns. Unlike, we replace the AR model by the ARFIMA model, developed by Granger (1980), Granger and Joyeux (1980). Evidently, the ARFIMA model takes into consideration the long memory in returns.

Formally, the ARFIMA(p,d,q) model for returns is given as follows:

$$\phi(L)(1-L)^d(R_t - u_t) = \varphi(L)\,\varepsilon_t \tag{1}$$

where ε_t is a white noise sequence with zero mean and innovation variance σ_t^2. L denotes the backward-shift operator (Granger and Joyeux 1980). R_t are the stock returns. More precisely, the ARIMA model is written as follows:

$$R_t = u_t + \sigma_t Z_t \tag{2}$$

where

$$\begin{cases} \phi(L)(1-L)^d(R_t - u_t) = \varphi(L)\,\varepsilon_t & \textit{models the returns process} \\ \sigma_t^2 = \alpha_0 + \alpha_1 \varepsilon_{t-1}^2 + \beta\sigma_{t-1}^2 & \textit{models the volatility process} \end{cases}$$

Long Memory in Volatility

Modelling the volatility is generally made by GARCH model to take into account the long memory issue. In our case, however, we replace the GARCH model by the FIGARCH (Fractionally Integrated Generalized Autoregressive Conditional Heteroskedasticity) model proposed by Baillie et al. (1996). The FIGARCH model can measure the phenomenon of persistence of shocks and long memory in volatility and the effect of the propagation of shocks to volatility.

In the FIGARCH model, the volatility is modelled as follows:

$$\sigma_t^2 = \frac{a}{(1-\beta(L))} + \lambda(L)\varepsilon_t^2 \tag{3}$$

with

$$\lambda(L) = 1 - \frac{\phi(L)}{[1-\beta(L)]}(1-L)^d.$$

ϕ and β are two polynomials given by

$$\beta(L) = \beta_1 L + \beta_2 L^2 + \ldots + \beta_p L^p$$

and

$$\phi(L) = \phi_1 L + \phi_2 L^2 + ... + \phi_q L^q .$$

The parameter d captures the effect of shocks on the long-term volatility and $(1 - L)^d$ is the fractional differencing operator.

Estimation of Value-at-Risk

The VaR adjusted to the long memory in both returns and volatilities is estimated based on the Conditional Extreme Value Theory. Formally, the VaR is calculated as follows:

$$\hat{V}aR^t_{\alpha,pred} = -\hat{u}_{t+1} + \hat{\sigma}_{t+1}\hat{q}^Z_\alpha \tag{4}$$

with

$$\begin{cases} \hat{u}_{t+1} & \textit{is estimated by ARFIMA model} \\ \hat{\sigma}_{t+1} & \textit{is estimated by FIGARCH model} \\ \hat{q}^Z_\alpha & \begin{array}{l} \textit{is the extrem quantile of the quantile of the} \\ \textit{residuals } Z_t \textit{ estimated by the POT method} \end{array} \end{cases}$$

Testing for Long Memory Property

As aforementioned, the long memory refers to a significant dependence between distant observations in a time series. To detect long memory, Hurst (1951) applies R/S test on standard deviation of returns and volatilities. Geweke and Porter-Hudak (1983) proposed a semiparametric approach to the testing for long-memory. Nonetheless, the classical R/S analysis is sensitive to the presence of explicit short-range dependence structures. To overcome these shortcomings, Lo (1991) proposed a modified R/S statistic, given as follows:

The modified R/S test: H0: No long memory process vs. H1: Long memory process

$$Q_{n,l} = \frac{1}{\hat{\sigma}_n(l)}\left[\max_{1\le k\le n}\sum_{i=1}^{k}\left(R_i - \overline{R}_n\right) - \min_{1\le k\le n}\sum_{i=1}^{k}\left(\overline{R}_i - \overline{\overline{R}}_n\right)\right] \tag{5}$$

$Q_{n,l}\big/\sqrt{n}$ converge to V bridge brownien process.

with

$$\hat{\sigma}^2_n(l) = \frac{1}{n}\sum_{i=1}^{n}\left(R_i - \overline{R}_n\right)^2 + \frac{2}{n}\sum_{i=1}^{n} w_i(l)\left\{\sum_{j=i+1}^{n}(R_j - \overline{R}_n)(R_{j-i} - \overline{R}_n)\right\} \text{and } w_i(l) = 1 - i/(l+l).$$

Another test used in this chapter is GPH test. In this respect, Geweke and Porter-Hudak (1983) developed a semi-nonparametric approach to examine the issue of long memory. Particularly, the spectral density of the fractionally integrated process yt is expressed as follows:

$$f(w) = \left[4\sin^2\left(\frac{w}{2}\right)\right]^{-d} f_u(w) \tag{6}$$

where w is the Fourier frequency, and $f_u(w)$ is the spectral density which refers to u_t. Notice that the fractional difference parameter d can be estimated by the following regression:

$$\ln f(w_j) = \beta - d\ln\left[4\sin^2\left(\frac{w_j}{2}\right)\right] + e_j \tag{7}$$

For $j = 1,2,\ldots,n_f(T)$. Geweke and Porter-Hudak (1993) displayed utilizing a periodogram estimate of $f(w_j)$, the least squares estimate $\hat{d}$ employing the above regression is normally distributed in large samples if $n_f(T) = T^{\alpha}$ with $0 < \alpha < 1$:

$$\hat{d} \to N\left(d, \frac{\pi^2}{6\sum_{j=1}^{n_f}\left(U_j - \overline{U}\right)^2}\right) \tag{8}$$

where $U_j = \ln\left[4\sin^2\left(\frac{w_j}{2}\right)\right]$ and $\overline{U}$ is the sample mean of U_j, for $j = 1,2,\ldots,n_f$.

PRESENTATION OF SELECTED STOCK MARKETS, DATA, AND DATA DESCRIPTION

Presentation of Selected Stock Markets

The selected emerging stock markets were generally found around the mid-1970s. The Egyptian Exchange is one of the oldest stock markets in the Middle East and Africa (Alexandia Stock Exchange and Cairo Stock Exchange began in 1883 and 1903, respectively). In 1929, the Casablanca Stock Exchange was created in Morocco. Then, in 1969, The Tunis Stock Exchange was established in Tunisia and the Beirut Stock Exchange (Lebanon) in 1980.

Table 1 summarizes the main performance indicators for the four selected emerging stock markets. Obviously, the market capitalization varies widely among the four stock markets. In 2010, the Egyptian market seems to have the highest market capitalization ($ million 84,277) whereas the less important capital market is the Tunisian market. The ratio of the market capitalization of the listed companies to the Gross Domestic Product (GDP) is the highest for Morocco (74.2%) followed by Egypt (38.5%), Lebanon (33.4%) and Tunisia (24.2%) in 2010. As well, the number of transactions is greatest for the Morocco. The number of publicly listed firms in the selected emerging stock markets generally raised from 2010 and 2015 (except for Lebanon), implying the attractiveness of stock market financing for companies.

Data and Descriptive Statistics

In this chapter, the data employed comprise four emerging stock markets, i.e. Tunis Stock Exchange (TUNINDEX), Egyptian Stock Exchange (EGX30), Casablanca Stock Exchange (MASI), and Beirut Stock Exchange (BLOM). The stock markets under consideration were chosen due their comparable degree of financial markets. As well, these markets have experienced dramatic country-specific events. Table 2 reports these different events. The price data was provided from DataStream. From Table 1, the sample period is divided into two sub-periods based on the date of occurrence of the event. The continuously compounded daily returns for each stock market are expressed as: $R_t = Ln\left(\frac{P_t}{P_{t-1}}\right) * 100$; where P_t and P_{t-1} are the closing prices for each market at time t and t-1, respectively.

Figures 1 and 2 display daily indices and returns for our sample, respectively. At the first sight, the time series plots of the four stock indices (Figure 1) seem to exhibit cyclical movements over the sample period. Also, Figure 1 shows the emergence of several troughs with different amplitude. These troughs correspond to dates of occurrence of the dramatic country-specific events. For example, the Tunisian stock market exhibits the highest trough at the date of revolution (01/2011) in comparison with other trough due to events. The same in the case of Egypt, Morocco and Lebanon also shows a greater trough in different dates. From Figure 2, all return series tend to fluctuate over time. In this respect, all daily indices return appears to have volatility clustering behavior.

Table 3 displays the descriptive statistics of the data used in this study for the whole period as well as the two sub-periods. Descriptive statistics include the mean, standard deviation, variance, skewness, kurtosis and Jarque-Bera for normality test. All of these statistics display the asymmetric patterns in the evolution of different indices return, following the occurrence of the country-specific events. From Table 3, the descriptive statistics generally display the worsening of the market conditions following the onset of different events. For instance, in the Tunisian and Egyptian markets, the mean return has

Table 1. The performance indicators for the selected emerging markets-2005

Country	Number of Firms		Market Capitalization ($millions)		Market Capitalization (% of GDP)		Value of Shares Traded (% GDP)		Turnover Ratio (%)		Number of Transactions	Number of Day Traded
	2010	2015	2010	2015	2010	2015	2010	2015	2010	2015		
Egypt	227	250	84,277	55,192	38.5	16.7	17	4.5	44.2	26.7	-	-
Morocco	73	74	69,152	45,928	74.2	45.8	6.5	2.9	8.8	6.4	72331593	241
Tunisia	58	78	10,652	-	24.2	-	4.2	1.9	17.2	-	393095	235
Lebanon	10	10	12,697	-	33.4	-	4.9	1.4	14.8	-	9434	235

Source: World Bank and Arab Monetary Fund.

Table 2. Political events

Index	Pre-Crisis Period	Post-Crisis Period	Political Event
Tunindex	03/01/2000-13/01/2011	14/01/2011-25/09/2017	Tunisian Revolution
EGX30	20/06/2004-24/01/2011	25/01/2011-25/09/2017	Egyptian Revolution
BLOM	03/01/2000-14/02/2005	15/02/2005-25/09/2017	Assassination of Prime Minister Rafik El Hariri
MASI	05/01/2000-15/05/2003	16/05/2003-25/09/2017	Series of suicide bomber, 45 people were killed

Figure 1. Evolution of indices prices for the selected emerging markets

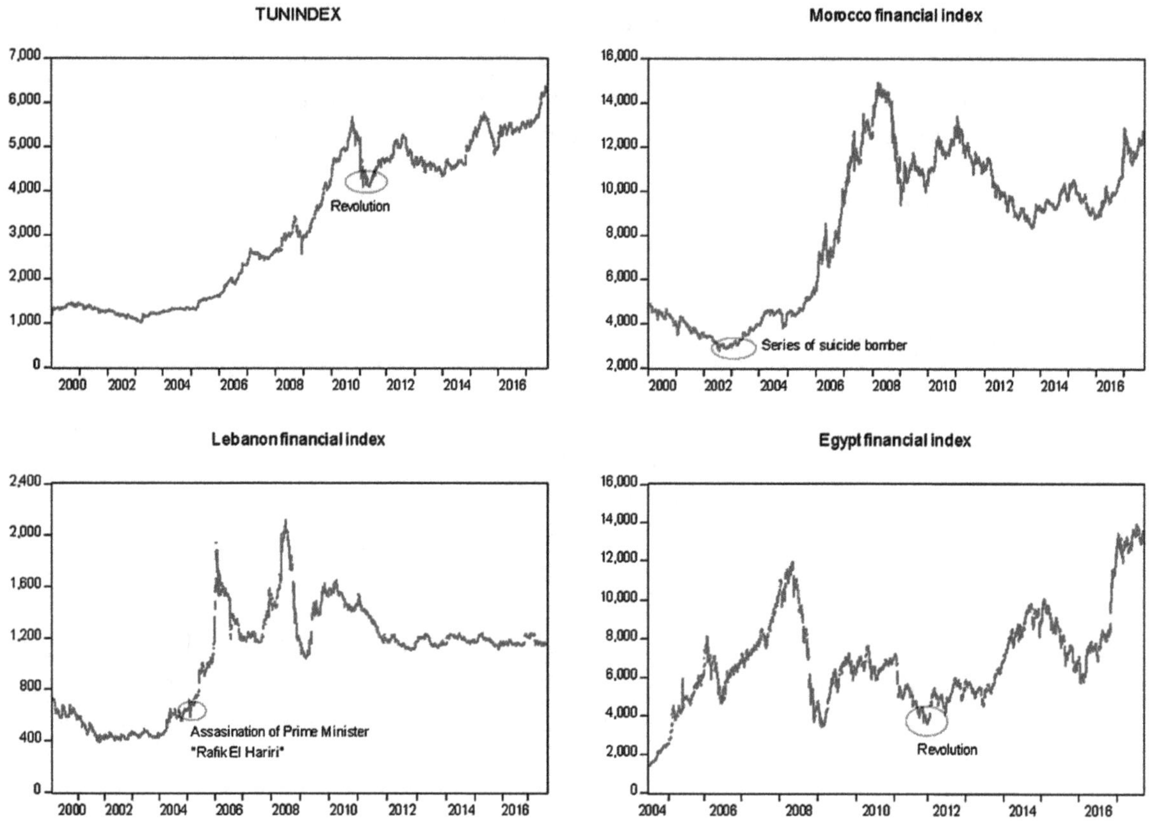

decreased from 0.0497% and 0.0969% over the pre-revolution period to 0.0171% and 0.0447% over the post-revolution period, respectively. As well, in the Moroccan case, the daily index return seems to be right skewed over the pre-crisis period then it becomes left skewed over the post-crisis period. This finding can approve the choice of event's date.

ESTIMATION RESULTS

Obviously, the presence of long memory in financial time series stipulates that future stock returns may be predicted from past stock returns. Thereby, as point out by Yajima (1985), the linear pricing models

Figure 2. Evolution of indices returns for the selected emerging markets

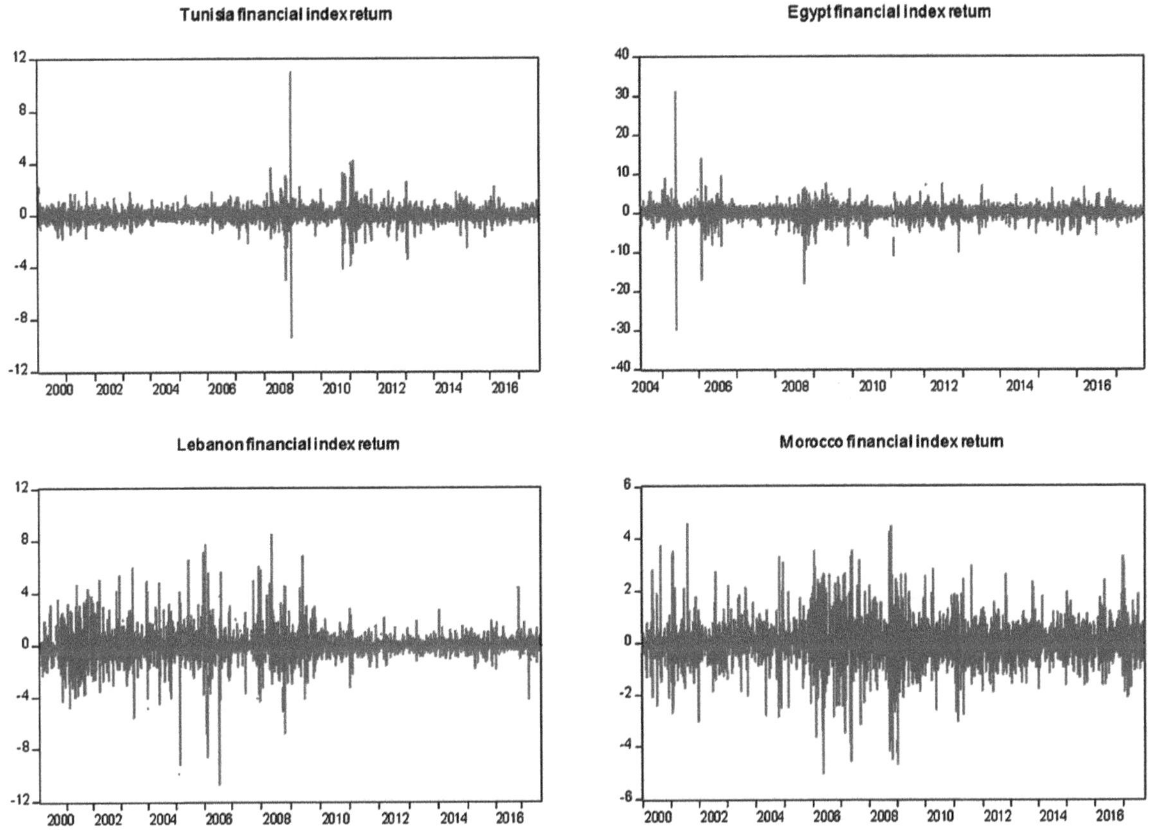

employed to predict the expected stock returns and statistical inferences about asset pricing models based on standard testing procedures can be inappropriate. In this study, to test for the presence of long memory property in returns and volatilities, we use both the modified R/S test and GPH test for the two sub-periods which are reported in Table 4. For each test, the test statistics and the corresponding significances are given. From Table 4, the indices return for all selected emerging markets clearly displays long memory properties in returns during the pre-crisis period for all emerging markets. Also, the different markets are characterized by long memory processes in volatility. Notably, under such conditions, the efficient market hypothesis is rejected in the four selected emerging markets due to the fact that the stock market prices do not follow a random walk. The aforementioned finding is somewhat not surprising given that the trading conditions and illiquidity in these markets could be primary determinant for the issue of long memory.

During the post-crisis period, the modified R/S test and GPH test reject the null hypothesis of no long memory at different levels. So, the results clearly indicate that the long memory property in the post-crisis period appears as significant as in the pre-crisis period. These emerging markets have experienced significant shocks and breaks associated with the dramatic country-specific events.

Kang and Yoon (2007), among others, profess that the long memory dynamics in different returns series are generally detected jointly in the conditional mean and variance. Overwhelmingly, it is worthwhile to investigate the dual long memory property in the conditional mean and variance using the

Table 3. Descriptive statistics of the selected MENA markets for different periods

Index Return	Mean	Variance	Skewness	Kurtosis	J.B
The Whole Period					
Tunindex	0.0374	0.3306	0.3043	55.845	513,200***
EGX30	0.0708	3.8396	-0.5059	45.846	244,300***
BLOM	0.0109	1.1859	0.1491	16.949	33,530***
MASI	0.0212	0.5741	-0.0797	8.6269	5,834.4***
The Pre-Crisis Period					
Tunindex	0.0497	0.3578	0.4167	70.863	527,780***
EGX30	0.0967	5.2103	-0.4754	47.713	133,760***
BLOM	-0.0036	1.5926	0.2718	5.7614	379.88***
MASI	-0.0466	0.4643	1.1886	11.2190	2540.8***
The Post-Crisis Period					
Tunindex	0.0171	0,2847	0.0095	14.459	9083***
EGX30	0.0447	2,4520	-0.5608	8.2861	1930.9***
BLOM	0.0166	1,0288	0.0628	25.892	65134***
MASI	0.0370	0,5983	-0.2947	8.2985	4246.7***

Notes: *: significant at 10% level, **: significant at 5% level, ***: significant at 1% level.

Table 4. The modified R/S test and GPH test of predictability over pre- and post-crisis period

	Modified R/S Test				GPH Test			
	Returns		Volatility		Returns		Volatility	
	Pre-Crisis	Post-Crisis	Pre-Crisis	Post-Crisis	Pre-Crisis	Post-Crisis	Pre-Crisis	Post-Crisis
Tunindex	4.6689***	1.4698	6.6669***	6.0007***	0.2639***	0.1363***	0.6770***	0.6619***
EGX30	3.1483***	1.3848	4.8521***	4.0744***	0.2266***	0.1501***	0.8006***	0.9742***
BLOM	3.6336***	4.2327***	4.7905***	11.9542***	0.4846***	0.3149***	0.6256 ***	0.64450***
MASI	1.9247**	4.5257***	3.1724***	10.3515 ***	0.4914***	0.3286***	0.5224***	0.6086***

Notes: *: significant at 10% level, **: significant at 5% level, ***: significant at 1% level.

ARFIMA-FIGARCH model. The ARFIMA-FIGARCH specifications allow to capture the issue of the long memory for all indices. Table 5 reports the estimation results of ARFIMA(1,d,0)–FIGARCH(1,d,1) model. From Table 5, both the long memory parameters (d-ARFIMA and d-FIGARCH) are significantly different from zero during the pre-crisis period, implying that the dual long memory property is prevalent in the returns and volatility of the emerging stock markets. However, during the post-crisis period, the dual long memory process is only detected in the Moroccan market.

The aforementioned empirical findings then lead to estimate the VaR based on the conditional EVT. Figure 3 highlights different behaviors of downside risk measure according to the stock cycle fluctuations during the two sub-periods (the pre-crisis period in the left-hand and the post-crisis period in the right-hand) for the four emerging markets. At the first sight, we note that the VaR value evolves over time by following the chronicles of stock returns. During the periods of high volatility, the VaR value tends

Table 5. Estimation results of the ARFIMA(1,d,0)–FIGARCH(1,d,1) model

	TUNINDEX		EGX30		BLOM		MASI	
	Pre	Post	Pre	Post	Pre	Post	Pre	Post
d-ARFIMA	0.1129	0.0546	0.0536	0.0061	0.0747	0.0283	0.1854	0.0595
t-Stat.	4.628	1.370	1.878	0.1919	1.815	1.466	2.723	2.425
P-Value	0.0000	0.1709	0.0606	0.8478	0.0698	0.1428	0.0066	0.0154
AR(1)	0.1092	0.1143	0.0585	0.1996	-0.0155	0.0010	0.0747	0.1017
t-Stat	3.306	2.217	1.496	4.754	-0.2714	0.0383	0.9492	3.462
P-Value	0.0010	0.0268	0.1348	0.0000	0.7861	0.9695	0.3428	0.0005
d-FIGARCH	0.4765	0.3621	0.4399	0.3885	0.4826	0.2839	0.4482	0.3124
t-Stat.	9.183	8.204	6.879	10.66	9.322	7.660	7.989	7.143
P-Value	0.0000	0.0000	0.0000	0.0000	0.0000	0.0000	0.0000	0.0000
ARCH	0.4949	0.8857	0.3027	0.0847	0.4591	0.8853	0.9249	0.9479
t-Stat.	5.432	22.50	2.283	0.6297	6.380	21.95	48.73	17.77
P-Value	0.0000	0.0000	0.0226	0.5290	0.0000	0.0000	0.0000	0.0000
GARCH	0.6668	0.9408	0.5405	0.3488	0.6247	0.8715	0.9713	0.9598
t-Stat.	9.947	45.83	4.340	2.678	7.553	21.51	126.0	28.52
P-Value	0.0000	0.0000	0.0000	0.0075	0.0000	0.0000	0.0000	0.0000
df	8.5597	5.8316	5.0600	6.7618	4.2141	3.3625	4.0938	6.5017
t-Stat.	8.003	9.122	10.23	7.386	12.32	27.04	12.11	13.69
P-Value	0.0000	0.0000	0.0000	0.0000	0.0000	0.0000	0.0000	0.0000

to increase while it diminishes in periods of low volatility. This shows that the VaR measure varies with the market volatility. The ARFIMA(1,d,0)–FIGARCH(1,d,1)-POT model captured successfully the large losses occurred during the periods of high volatility, especially over the post-crisis hand. All of these findings show the efficacy of our procedure and again display the deterioration of financial conditions following the occurrence of dramatic country-specific events.

The degree of pertinence of the model used can be investigated by the implementation of the backtesting procedure. The empirical results for the backtesting test are reported in Table 6.

From Table 6, the backtesting procedure is summarized by the unconditional coverage (LRuc) of Kupiec (1995), the conditional coverage (DQcc) of Engle and Manganelli (2004). From Table 6, we show that model of VaR measure well performs during the two sub-periods. More explicitly, the p-value of backtesting is generally significant for all indices over the two sub-periods for different significance levels.

CONCLUSION

This chapter attempts to investigate the impact of the long memory property on the market risk across two sub-periods based on four country-specific events (Tunisian revolution; Egyptian revolution; assassination of Prime Minister Rafik El Hariri and series of suicide bomber in Morocco).

Figure 3. Dynamic VaR for MENA markets

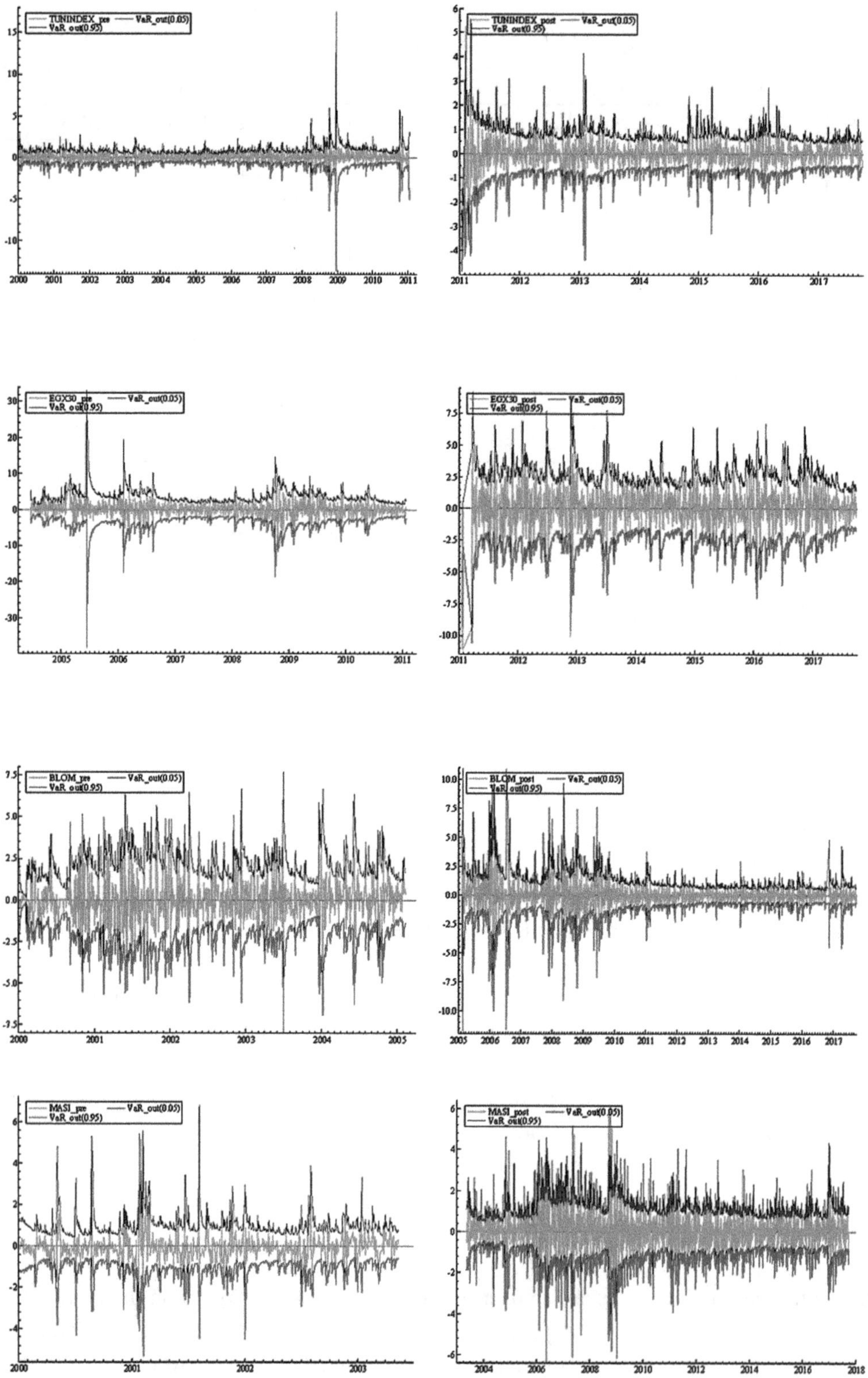

For a more accurate representation see the electronic version.

Table 6. Backtesting test for pre- and post-crisis periods

	Pre-Crisis Period						Post-Crisis Period					
	pvalue-LRuc			*pvalue-DQcc*			*pvalue-LRuc*			*pvalue-DQcc*		
	5%	2.5%	1%	5%	2.5%	1%	5%	2.5%	1%	5%	2.5%	1%
Tunindex	0.4613	0.0555	0.5109	0.3799	0.3222	0.0043	0.8211	0.8125	0.5628	0.7306	0.9999	0.9650
EGX30	0.8417	0.5417	0.4720	0.1182	0.2549	0.9468	0.0165	0.0003	0.0919	0.0002	0.0000	0.6784
BLOM	0.0018	0.0027	0.0425	0.1238	0.0644	0.4492	0.6852	0.2785	0.4529	0.4752	0.5496	0.6296
MASI	0.0828	0.1300	0.3755	0.3765	0.3763	0.9627	0.0318	0.0020	0.0018	0.1722	0.0935	0.0356

Using the modified R/S test and the GPH test, we show long-range dependence behavior which changes over time. In particular, the long memory property in returns and volatility tends to be significant in the pre-crisis period for all emerging markets. As well, the results clearly reveal that the long memory property in the post-crisis period seems to be as significant as in the pre-crisis period. Market thinness and nonsynchronous trading, among others, as well as the country-specific circumstances can be considered as factors responsible of such behavior of long memory property before and after political events. The empirical results also show that long memory models maybe useful for depicting cyclical fluctuations in index returns. Dávid and Baros (2007) and Bardos (2008), among others, reveal that the long range dependence indicates that the information from "today" is not instantaneously absorbed by the price in the stock market and also investors react with delay to information.

The aforementioned empirical findings then lead to apply ARFIMA(1,d,0)–FIGARCH(1,d,1) model on time series over the two sub-periods (the pre-and post-crisis periods) to investigate the dual long memory property in the emerging stock markets. Unlike the ARFIMA model or the FIGARCH model, the ARFIMA–FIGARCH model obviously provides the robustness of long memory test results. This leads to better explanation for long memory dynamics in both the conditional mean and variance. The empirical results reveal that the dual long memory property is well-documented in returns and volatility of the emerging stock markets during the pre-crisis period. However, during the post-crisis period, the dual long memory process is uniquely detected in the Moroccan market. Using the conditional EVT, we also document the dynamic behavior of VaR during the two sub-periods. As well, using the backtesting test clearly shows that VaR well performs during the pre-crisis period as well as the post-crisis period.

Our findings can have several financial implications on market efficiency, portfolio allocations and risk management. Lastly, there are several clear directions for future research. We propose to re-examine the concept of informational efficiency in MENA markets during specific-country events.

REFERENCES

Al Janabi, M. A. M. (2011a). A generalized theoretical modeling approach for the assessment of economic capital under asset market liquidity risk constraints. *Service Industries Journal*, *31*(13), 2193–2221. doi:10.1080/02642069.2010.503878

Allen, D. E., Singh, A. K., & Pownell, R. (2011). Extreme market risk-An extreme value theory approach. *Econometrics Society Australasian Meeting*, 1-26.

Andersen, T. G., & Bollerslev, T. (1997). Heterogeneous information arrivals and return volatility dynamics: Uncovering the long-run in high frequency returns. *The Journal of Finance*, *52*(3), 975–1005. doi:10.1111/j.1540-6261.1997.tb02722.x

Andersen, T. G., & Bollerslev, T. (1998). Answering the skeptics: Yes, standard volatility models do provide accurate forecasts. *International Economic Review*, *39*(4), 885–905. doi:10.2307/2527343

Baillie, R. T., Bollerslev, T., & Mikkelsen, H. O. (1996). Fractionally Integrated Generalized Autoregressive Conditional Heteroskedasticity. *Journal of Econometrics*, *73*(1), 3–20. doi:10.1016/S0304-4076(95)01749-6

Bali, T. G., & Neftçi, S. N. (2003). Disturbing extremal behaviour of spot price dynamics. *Journal of Empirical Finance*, *10*(4), 455–477. doi:10.1016/S0927-5398(02)00070-1

Bardos, S. (2008). Academy of economic studies. Bucharest Doctoral School of Finance and Banking, DOFIN, Bucharest.

Barkoulas, J. T., Baum, C. F., & Travlos, N. (2000). Long memory in the Greek stock market. *Applied Financial Economics*, *10*(2), 177–184. doi:10.1080/096031000331815

Bekaert, G., & Harvey, C. (1995). Time-varing world market integration. *The Journal of Finance*, *50*(2), 403–444. doi:10.1111/j.1540-6261.1995.tb04790.x

Bekaert, G., & Harvey, C. (1997). Emerging equity market volatility. *Journal of Financial Economics*, *48*(1), 29–78. doi:10.1016/S0304-405X(96)00889-6

Bekaert, G., & Harvey, C. (2003). Emerging markets finance. *Journal of Empirical Finance*, *10*(1-2), 3–55. doi:10.1016/S0927-5398(02)00054-3

Bekaert, G., Harvey, C., & Lundblad, C. (2007). Liquidity and expected returns: Lessons from emerging markets. *Journal of Financial Studies*, *20*(6), 1783–1831. doi:10.1093/rfs/hhm030

Bekaert, G., & Hodrick, R. (1992). Characterizing predictable components in excess returns on equity and foreign exchange markets. *The Journal of Finance*, *47*(2), 467–509. doi:10.1111/j.1540-6261.1992.tb04399.x

Bhattacharryya, M., & Ritolia, G. (2008). Conditional VaR using EVT-towards: A planned margin scheme. *International Review of Financial Analysis*, *17*(2), 382–395. doi:10.1016/j.irfa.2006.08.004

Bhattacharya, S. N., & Bhattacharya, M. (2012). Long Memory in Stock Returns: A study of emerging markets. *Iranian Journal of Management Studies*, *5*, 67–88.

Billio, M., & Pelizzon, L. (2000). Value-at-risk: A multivariate switching regime Approach. *Journal of Empirical Finance*, *7*(5), 531–554. doi:10.1016/S0927-5398(00)00022-0

Blavy, R. (2002, Autumn). Changing volatility in emerging markets: A case study of two Middle Eastern stock exchanges. *Revue Entente Cordiale*, 1-35.

Bollerslev, T. (1986). Generalized autoregressive conditional heteroskedasticity. *Journal of Econometrics*, *31*(3), 307–327. doi:10.1016/0304-4076(86)90063-1

Breidt, F. J., Crato, N., & de Lima, P. (1998). On the detection and estimation of long memory in stochastic volatility. *Journal of Econometrics*, *83*(1-2), 325–348. doi:10.1016/S0304-4076(97)00072-9

Byström, H. (2004). Managing extreme risks in tranquil and volatile markets using conditional extreme value theory. *International Review of Financial Analysis*, *13*(2), 133–152. doi:10.1016/j.irfa.2004.02.003

Cavalcante, J., & Assaf, A. (2005). Long-range dependence in the returns and volatility of the Brazilian stock market. *European Review of Economics and Finance*, *5*, 5–20.

Cheung, Y., & Lai, K. (1995). A search for long memory in international stock market returns. *Journal of International Money and Finance*, *14*(4), 597–615. doi:10.1016/0261-5606(95)93616-U

Crato, N. (1994). Some international evidence regarding the stochastic behavior of stock returns. *Applied Financial Economics*, *4*(1), 33–39. doi:10.1080/758522123

Crato, N., & de Lima, P. (1994). Long range dependence in the conditional variance of stock returns. *Economics Letters*, *45*(3), 281–285. doi:10.1016/0165-1765(94)90024-8

Dávid, L., & Baros, Z. (2007). A possible use of indicators for sustainable development in tourism Anatolia. *An International Journal of Tourism and Hospitality Research*, *18*, 349–355.

Davies, R. B., & Harte, D. S. (1987). Tests for Hurst effect. *Biometrika*, *74*(1), 95–101. doi:10.1093/biomet/74.1.95

De Santis, G., & Imrohoroglu, S. (1997). Stock returns and volatility in emerging financial markets. *Journal of International Money and Finance*, *16*(4), 561–579. doi:10.1016/S0261-5606(97)00020-X

Ding, Z., Granger, C. W. J., & Engle, R. F. (1993). A long memory property of stock market returns and a new model. *Journal of Empirical Finance*, *1*(1), 83–106. doi:10.1016/0927-5398(93)90006-D

Disario, R., Saraoglu, H., McCarthy, J., & Li, H. (2008). Long memory in the volatility of an emerging equity market: The case of Turkey. *Journal of International Financial Markets, Institutions and Money*, *18*(4), 305–312. doi:10.1016/j.intfin.2007.01.001

Eldera, J., & Serletis, A. (2008). Long memory in energy futures prices. *Review of Financial Economics*, *17*(2), 146–155. doi:10.1016/j.rfe.2006.10.002

Engle, R. F. (1982). Autoregressive conditional heteroskedasticity with estimates of the variance of U.K. inflation. *Econometrica*, *50*(4), 987–1008. doi:10.2307/1912773

Engle, R. F., & Bollerslev, T. (1986). Modeling the persistence of conditional variances. *Econometric Reviews*, *5*(1), 1–50. doi:10.1080/07474938608800095

Engle, R. F., & Manganelli, S. (2004). CAViaR: Conditional autoregressive Value-at-Risk by regression quantiles. *Journal of Business & Economic Statistics*, *22*(4), 367–381. doi:10.1198/073500104000000370

Granger, C., & Newbold, N. (1986). *Forecasting economic time series* (2nd ed.). San Diego, CA: Academic Press.

Granger, C. W., Ding, J. Z., & Spear, S. (1997). *Stylized facts of the temporal and distributional properties of daily speculative markets* (Unpublished manuscript). San Diego, CA.

Granger, C. W. J., & Joyeux, R. (1980). An introduction to long memory time series models and fractional differencing. *Journal of Time Series Analysis, 1*(1), 15–39. doi:10.1111/j.1467-9892.1980.tb00297.x

Greene, M. T., & Fielitz, B. D. (1977). Long-term dependence in common stock returns. *Journal of Financial Economics, 4*(3), 339–349. doi:10.1016/0304-405X(77)90006-X

Gurgul, H., & Wojtowicz, T. (2006). Long memory on the German stock exchange. *Czeck Journal of Economics and Finance, 56*, 447–468.

Harvey, A. C. (1993). Long memory in stochastic volatility. In J. Knight & S. Satchell (Eds.), *Forecasting Volatility in Financial Markets* (pp. 307–320). Oxford, UK: Butterworth-Haineman.

Harvey, C. (1995). Preditable risk and returns in emerging markets. *Journal of Financial Studies, 8*(3), 773–816. doi:10.1093/rfs/8.3.773

Harvey, C., & Viskanta, T. (1996). Expected returns and volatility in 135 countries. *Journal of Portfolio Management, 22*(3), 46–58. doi:10.3905/jpm.1996.409554

Hiremath, G. S., & Bandi, K. (2010). Long memory in stock Market volatility: Evidence from India. *Artha Vijnana, 52*(4), 332–345. doi:10.21648/arthavij/2010/v52/i4/115316

Hosking, J. R. M. (1981). Fractional differencing. *Biometrika, 68*(1), 165–176. doi:10.1093/biomet/68.1.165

Jefferis, K., & Thupayagale, P. (2008). Long memory in Southern African stock markets. *The South African Journal of Economics, 75*(3), 384–398. doi:10.1111/j.1813-6982.2008.00200.x

Kang, S. H., & Yoon, S. M. (2008). Long memory features in the high frequency data of the Korean stock market. *Physica A, 5*, 189–196.

Kawata, R., & Kijima, M. (2007). Value-at-risk in a market subject to regime switching. *Quantitative Finance, 7*(6), 609–619. doi:10.1080/14697680601161795

Koong, C. S., Tsui, A. K., & Chan, W. S. (1997). On tests for long memory in Pacific Basin stock returns. *Mathematics and Computers in Simulation, 43*(3-6), 445–449. doi:10.1016/S0378-4754(97)00030-X

Korkmaz, T., Cevik, E. I., & Ozatae, N. (2009). Testing for Long Memory in ISE using ARFIMA-FIGARCH Model and Structural Break Test. *International Research Journal of Finance and Economics, 26*, 187–191.

Kuester, K., Mittnik, S., & Paolella, M. (2006). Value-at-risk prediction: A comparison of alternative strategies. *Journal of Financial Econometrics, 4*(1), 53–89. doi:10.1093/jjfinec/nbj002

Kupiec, P. (1995). Techniques for verifying the accuracy of risk management models. *Journal of Derivatives, 3*(2), 73–84. doi:10.3905/jod.1995.407942

Lanouar, C., & Ajmi, A. N. (2013). The Tunisian stock market index volatility: Long memory vs. switching regime. *Emerging Markets Review, 16*, 170–182. doi:10.1016/j.ememar.2013.05.003

Li, M.-Y. L., & Lin, H.-W. W. (2004). Estimating value-at-risk via Markov-switching ARCH models-an empirical study on stock index returns. *Applied Economics Letters, 11*(11), 679–691. doi:10.1080/1350485042000236539

Lin, X., & Fei, F. (2013). Long memory revisit in Chinese stock markets: Based on GARCH-class models and multiscale analysis. *Economic Modelling, 31*, 265–275. doi:10.1016/j.econmod.2012.11.037

Lo, A. W. (1991). Long-term memory in stock market prices. *Econometrica, 59*(5), 1279–1313. doi:10.2307/2938368

Mandelbrot, B. B. (1971). A fast fractional Gaussian noise generator. *Water Resources Research, 7*(3), 543–553. doi:10.1029/WR007i003p00543

Mandelbrot, B. B. (1972). Possible refinement of the lognormal hypothesis concerning the distribution of energy dissipation in intermittent turbulence. In M. Rosenblatt & C. Van Atta (Eds.), *Statistical Models and Turbulence*. New York: Springer. doi:10.1007/3-540-05716-1_20

Mandelbrot, B. B. (1997). *Fractals and scaling in finance: Discontinuity, concentration, risk*. New York: Springer-Verlag. doi:10.1007/978-1-4757-2763-0

Marimoutou, V., Raggad, B., & Trabelsi, A. (2009). Extreme value theory and value-at-risk: Application to oil market. *Energy Economics, 31*(4), 519–530. doi:10.1016/j.eneco.2009.02.005

Mc Neil, A., & Frey, R. (2000). Estimation of tail related risk measures for heteroscedastic financial time series: An extreme value approach. *Journal of Empirical Finance, 7*(3-4), 271–300. doi:10.1016/S0927-5398(00)00012-8

McMillan, D. G., & Thupayagale, P. (2009). Efficiency of the South African equity market. *Applied Financial Economics Letters, 4*(5), 327–330. doi:10.1080/17446540701720717

Mendes, B. V. de M., & Kolev, N. (2006). *How Long Memory in Volatility Affects True Dependence Structure* (Working Paper). Department of Statistics, Federal University, Brazil.

Mukherjee, I., Sen, C., & Sarkar, A. (2011). Long memory in stock returns: Insights from the Indian market. *The International Journal of Applied Economics and Finance, 5*(1), 62–74. doi:10.3923/ijaef.2011.62.74

Nath, G. C. (2001). Long memory and Indian stock market: An empirical evidence. *Proceedings of UTI Institute of Capital Markets Conference*.

Petersen, C., & Plenborg, T. (2006). Voluntary disclosure and information asymmetry in Denmark. *Journal of International Accounting, Auditing & Taxation, 15*(2), 127–149. doi:10.1016/j.intaccaudtax.2006.08.004

Pitterle, I., Haufler, F., & Hong, P. (2015). Assessing emerging markets' vulnerability to financial crisis. *Journal of Policy Modeling*, 1–32.

Pownall, R. A. J., & Koedijk, K. G. (1999). Capturing downside risk in financial markets: The case of the Asian crisis. *Journal of International Money and Finance, 18*(6), 853–870. doi:10.1016/S0261-5606(99)00040-6

Robinson, P. M. (2003). Long memory time series. In M. R. Peter (Ed.), *Time series with long memory* (pp. 4–32). Oxford University Press.

Sadique, S., & Silvapulle, P. (2001). Long-term memory in stock market returns: International evidence. *International Journal of Finance & Economics, 6*(1), 59–67. doi:10.1002/ijfe.143

Samuel, Z.-T. (2008). Value-at-risk and conditional extreme value theory via Markov regime switching models. *Journal of Futures Markets*, *28*(2), 155–181. doi:10.1002/fut.20293

Venkatraman, S. (1997). Value-at-risk for a mixture of normal distributions: The use of quasi-bayesian estimation techniques. *Economic Perspectives*, 2–13.

Zangari, P. (1996). An improved methodology for measuring VaR. In *Risk Metrics Monitor*. Reuters/ JP Morgan.

ENDNOTE

[1] Empirical researches disclosed an important feature for the financial markets' volatilities; they are meant to be persistent and to model them by applying the classical GARCH models. These latter perform moderately (e.g. Hamilton and Susmel, 1994; Gray, 1996). As a result, many enhanced models have been proposed, mainly the Markov-switching ARCH, SWARCH, (Hamilton and Susmel, 1994) and Markov-switching GARCH, MS-GARCH, (Gray, 1996) models.

Chapter 4
Predicting Financial Crises in a Globalized World:
The Case of the Turkish Banking Sector

Asli Yuksel Mermod
Marmara University, Turkey

Ülkü Yüksel
The University of Sydney, Australia

Catherine Sutton-Brady
The University of Sydney, Australia

ABSTRACT

This chapter highlights the facts about financial crises and their fundamental causes on specific incidents, including the 1929 Great Depression that lasted until the early-1940s, 1997 Asian Financial Crises, 1998 Russian Financial Crises, and the Liquidity Crises of 2008, and makes a comparison among them and their various outcomes. In doing so, the study specifies the cues that emerge in the financial system that may help governments predict upcoming financial crises through those early warning signals. This case study specifically analyses the Turkish Banking System that was restructured after the enormous financial crises in Turkey in 2001, which caused many Turkish banks to collapse. However, the precautions taken in the aftermath of the financial turmoil allowed them to survive the liquidity crises in 2008. The indicators of an upcoming crisis are examined, the lessons learned from this case are analyzed, and important recommendations to overcome banking crises are provided.

INTRODUCTION

Globalization and deregulation processes worldwide accelerated the integration procedure of national and international financial markets, particularly after the 1990s. As a result of globalization, investment decisions and locations of investments have changed, and various countries' regulations influenced other countries. In some countries, authority over the financial markets has been lessened by eliminating regu-

DOI: 10.4018/978-1-5225-3767-0.ch004

lations to ease transactions and thus improve the ease of doing business. On one hand, the liberalization of financial markets significantly increased the flow of the capital between countries. Yet, on the other hand, this tendency caused long term financial crises whose violence and scope are rapidly expanding. Financial crises take place in many different countries contagiously on a regular basis, and are triggered by continuously varying factors.

The term financial crisis is applied broadly to a variety of situations in which some financial institutions or financial assets suddenly lose a large part of their value. Many financial crises in the 19th and early 20th centuries were caused by panics pertaining to banks and banking systems (i.e., banking panics), which then ended with recession in the global economy (Brunnermeier, 2016). There are also other types of financial crises stemming from stock market crashes, bursting financial bubbles, currency crises and sovereign defaults. Many economists have offered theories about how financial crises develop and how they could be prevented. Yet there is little consensus, as to which theory holds true with reference to the development and prevention of financial crises. Financial crises are still a regular occurrence around the world, and continue to have negative impacts on the growth rates of the countries that encounter and have to deal with crises. The Great Depression between the late 1929s and the early 1940s, The Asian Financial Crises in 1997, the Russian Financial Crisis in 1998 and the Liquidity Crisis of 2008 are examples of crises that have affected the growth of the major economies (Castiglionesi & Lorenzoni, 2017).

GLOBALIZATION AND FINANCIAL CRISES

Globalization is accepted as a process that removes geographical boundaries and enables the economic integration and interdependence of national economies into the international economy through trade, foreign direct investments, capital flows and migration. Globalization initiates the widening of international trade, transformation of financial sources, increasing foreign investment and joint enterprises, generating interdependency. Today, countries become dependent on each other economically due to technological advancement, digitalization, and easy and fast communication. Since the1990s, financial globalization removed boundaries, opened up financial markets to international competition, and increased international capital flows. Yet, the increase in the short-term capital movements as an inevitable result of globalization increased the volumes of financial crises.

Economic crises are unexpected, powerful, and sudden events that can have serious implications, which may arise with consequences, seriously affecting s country's' macro economy and the firms in the micro economy (Aktan & Huseyin, 2001). The United Kingdom's Department for Business, Enterprise and Regulatory Reform (2008) describes a crisis as *"an abnormal situation, or even perception, which is beyond the scope of everyday business, and which threatens the operation, safety, and reputation of an organization."* A crisis is a major, unpredictable event that threatens an organization and its stakeholders. Although crises are unpredictable, they are not unexpected.

Crises can affect businesses, educational institutions, families, non-profits, the governments, religious institutions, and are caused by a wide range of reasons. Crises can be separated into two according to their impact: real sector crises and financial sector crises. While the former impacts on production or employment, the latter, financial crises, are major distractions in financial markets that are characterized by sharp declines in asset prices and the failures of many financial and non-financial firms (Mishkin, 2015).

Financial crises can cause gigantic shocks to a country's economic, social and political mechanisms. The causes which trigger financial crises can be classified into four groups: (i) Increases in interest rates due to short term liabilities; (ii) Increases in uncertainty and currency problems; (iii) Macroeconomic factors; and (iv) Banking panics (i.e., panics pertaining to banks and banking systems; Berger & Bouwman, 2013). The vicious circle of financial crises is shown in Figure 1.

WORLD FINANCIAL CRISES

The 1929 Great Depression can still be considered as the biggest financial crisis in the developing process of capitalism. The world did not face another real global financial crisis after until the end of 1960's. Beginning with the collapse of Bretton Woods System, triggering the oil crisis in the early 1970s, the world started to experience many financial crises. The international dynamism of capital in/out flows came along with the financial crises at an accelerating frequency. These crises had badly influenced non-financial sectors, international investors, banks and governments. One of the most comprehensive studies by Kaminski and Reinhart (1996) sets forth the connection of financial liberalization and crises. Financial liberalization enabled the international capital to travel easily between countries as portfolio investments. This foreign capital called also hot money' can be a temporary relief for developing countries' economies (Anderson, Bordo, & Duca, 2017; Temin, 2016).

Monetary history has been full of many financial crises. As investors' optimism increases depending on the expansion of the economies, the growth rate of credit increases, the economic growth accelerates, and an increasing number of individuals begin to invest for short-term capital gains rather than for the returns associated with the productivity of the assets they were requiring (Kindleberger & Aliber, 2005).

Occurring in different geographical areas and in different macroeconomic circumstances, it is surprising that whatever are the causes, the consequences of financial crises are always the same. Because of the decrease in national income the countries' economy become poor and distressed, and because of the volatility in the financial markets the investors lose money and unemployment increases (Ataman and Tulay, 1999).

1997 Asian Financial Crisis

For many years the Asian countries including Korea, Thailand, Indonesia and Malaysia had proved miracles with average annual growth rates ranging from 6.9 percent to 8.4 percent. The transformation

Figure 1. The vicious circle of financial crises
Krugman 2008.

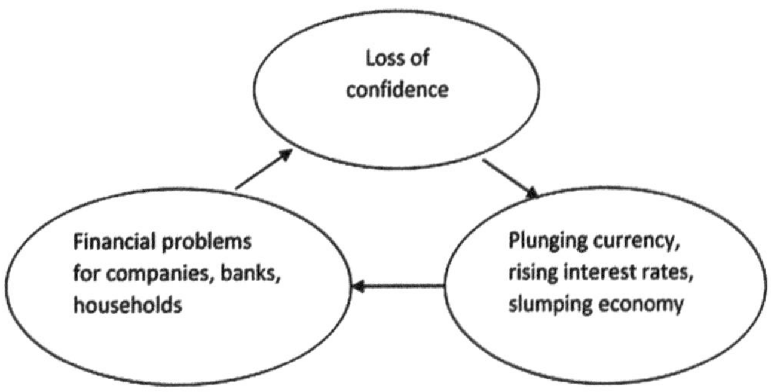

of these economies from poor, largely rural less-developed countries to middle income emerging markets has been one of the most remarkable success stories in history (Berg, 1999). This was mainly because Asia had attracted most of the capital from developing countries with high interest rates until 1997. Accordingly, Thailand, Malaysia, Indonesia, Singapore, and South Korea economies had high growth rates, known as the "Asian Economic Miracle". This growth, which was the result of capital investments, unfortunately, did not increase the total factor productivity. South Korea, Thailand and Indonesia had current account deficits and became dependent on foreign creditors because of low exchange rates. After the US recession in the mid-1990s, the US raised interest rates to attract money; as a result, East Asian countries were affected negatively (Mera & Renaud, 2016).

Following the Thai Baht's devaluation in mid-1997, the region entered severe economic crisis. Growth was negative in most of the countries in the region, and the recession was the deepest since World War II. The real GDP declined in 1998 by 13.7 percent in Indonesia, 9.4 percent in Thailand, 6.7 percent in Malaysia and 5.8 percent in Korea.

The East Asian countries had records of growth before the crisis. GDP growth has turned to negative sharply in 1998. The deep decline in output was driven by a tough decrease in private capital investment and reduction in private consumption. The largest decreases in investment happened in Indonesia and Malaysia. In addition, declining inventories played an important role in Korea's GDP reduction. Exports rate decreased in 1998 for most of the crisis countries. Although export volumes increased only in 1999, prices were not competitive. Most countries had achieved macroeconomic stability in 1998 after exchange rates became constant and interest rates started to decline. By the end of 1998, interest rates were below pre-crisis levels in all countries except Indonesia.

Thus, the East Asian miracle had been dependent on export growth. But, while exports began to rise with trade liberalization in the short term, imports also tended to increase strongly, especially because the domestic currency appreciated in real terms. The ratio data in Table 1 show that these East Asian countries had good economic conditions but after the crisis their financial positions were reversed.

After the crisis, the budget balance of the countries turned to negative in 1998. Thailand and Malaysia's situations have not changed as seen in Figure 2. Macroeconomic and financial systems' weaknesses can cause crisis that have negative effects on growth. As a result of globalization, investment decisions and locations have changed and many countries regulations affected other ones. Earning money through money does not provide long term growth; the point is to grow with productivity.

1998 Russian Financial Crisis

It is accepted that Russia's 1998 crisis occurred as a result of unsustainable public debt dynamics. In 1997, Russia compromised for the repayment of Soviet debt. Russia's credit ratios were good enough and the economic outlook appeared relatively optimistic. But the Asian financial crisis and declines in the demand and price of oil and metals impacted Russian foreign exchange reserves. The crisis wwas costly to Russia. GDP shrank by 4.9 percent in 1998. By December that year, 12 month inflation reached 84 percent. $30 billion in total foreign exchange-equivalent to one sixth of 1999 GDP- was used to defend the fixed exchange in 1998. Most of the big Moscow-based banks failed, while retail depositors took a big hit.

Table 1. Macroeconomic indicators

EAST ASIA FOUR: MACROECONOMIC INDICATORS, 1990–1999

	Unemployment rate					*Savings/GDP*				
	1990	*1996*	*1997*	*1998*	*1999*	*1990–95*	*1996*	*1997*	*1998*	*1999*
Indonesia	n.a.	4.1	4.6	5.5	6.3	31.0	26.2	26.4	26.1	23.7
Malaysia	6.0	2.5	2.4	3.2	3.0	36.6	37.1	37.3	39.6	38.0
Rep. of Korea	2.4	3.0	2.6	6.8	6.3	35.6	33.7	33.3	33.8	33.5
Thailand	4.9	1.1	0.9	3.5	4.1	34.4	33.0	32.5	34.9	31.0

	Investment/GDP					*(Savings-investment)/GDP*				
	1990–95	*1996*	*1997*	*1998*	*1999*	*1990–95*	*1996*	*1997*	*1998*	*1999*
Indonesia	31.3	29.6	28.7	22.1	19.3	-0.3	-3.4	-2.3	4.0	4.4
Malaysia	37.5	42.5	43.1	26.8	22.3	-0.9	-5.4	-5.8	12.8	15.7
Rep. of Korea	36.8	36.8	35.1	29.8	28.0	-1.2	-3.1	-1.8	4.1	5.5
Thailand	41.0	41.1	33.3	22.2	21.0	-5.6	-8.1	-0.9	12.8	10.0

	Incremental capital-output ratios					
	1987–89	*1990–92*	*1993–95*	*1997*	*1998*	*1999*
Indonesia	4.0	3.9	4.4	1.7	0.4	1.8
Malaysia	3.6	4.4	5.0	3.9	28.2	4.3
Rep. of Korea	3.5	5.1	5.1	4.2	-15.1	3.2
Thailand	2.9	4.6	5.2	12.9	-11.5	14.5

	Fiscal balance/GDP				
	1990–95	*1996*	*1997*	*1998*	*1999*
Indonesia	0.2	1.4	1.3	-2.6	-3.4
Malaysia	-0.4	0.7	2.4	-1.8	-3.2
Rep. of Korea	0.2	0.5	-1.4	-4.2	-2.9
Thailand	3.2	2.4	-0.9	-3.4	-3.0

Source: Radelet and Sachs (1998: table 11); ADB (1999); Bank of Thailand, Bank Indonesia, Bank of Korea, Bank Negara Malaysia. http://ksghome.harvard.edu/~drodrik/g24-jomo.pdf pg13

2008 Liquidity Crisis

Beginning in 2007 the world faced one of the deepest financial crises of history due to the collapse of sub-prime mortgage markets. This crisis first existed in credit markets but as it continued, it became a serious liquidity crisis that affected the world's largest banks, investment institutions and the hedge funds so destructively that many of them went bankrupt or seized by governments (Wagner, 2007). The crisis started with creating risky sub-prime loans from borrowers without any collateral. Then, these loans were put together and packaged by some financial institutions. After that, these institutions started to sell these loans to investors of various hedge funds. Although these packages also included some high quality credits, the investors holding these loans could not understand the risk that they were holding. When the repayment problems with these loans started to expand, the institutions which held these loans started to sell their assets to provide cash. At this time, the liquidity crisis boomed and international banks stopped financing each other.

Figure 2. After the Asian crisis
Source: Growth after the Asian Crisis: What Remains of the East Asian Model? Jomo K.S. http://ksghome.harvard.edu/~drodrik/g24-jomo.pdf pg31.

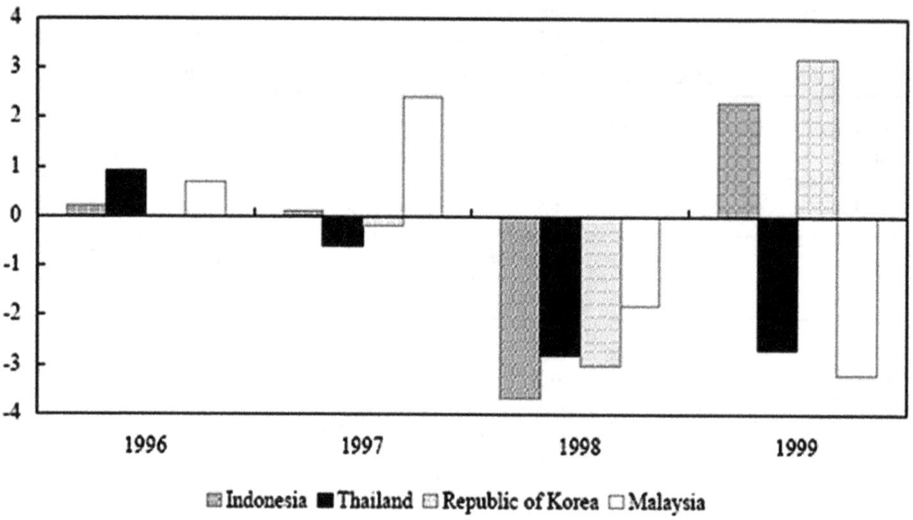

The one very obvious fact of this legendary financial crisis is that history will record it as unbelievable. The big question is how financial markets which were so unregulated had people believing they were so secure to invest in.

There has been a financial trend of 'more free markets with less regulations' since 1987 stock market meltdown. The US Treasury was advised at that time not to interfere in financial markets. So it had started passing its role of controlling the market to the market players themselves. This trend was followed during the 1990s. There was a stream of company mergers and acquisitions in that decade, however, many of them were breaking the law of the Glass-Steagall Act, which was put in place in 1933 after the 1929 crisis. The content of this act was corruption, manipulation and insider trading. Those illegal mergers and acquisition created big financial giants which were holding the control over the market. The speculation or manipulation was easy for a few large corporations compared to many small ones. During the 1990s, those financial conglomerates made very good gains with both regulated and unregulated financial instruments only by attracting funds and feeding the market. This system is called a 'Ponzi Scheme'.

A Ponzi scheme is a form of pyramid scheme in which earlier investors are paid with the money of later investors rather than from real profits. Charles Ponzi was an engaging Boston ex-convict who defrauded investors out of $6 million in the 1920s by promising them a 400 percent return on redeemed postal reply coupons. When he finally could not pay, he was put in jail for ten years (Brown, 2009).

The Ponzi scheme was just what the market players were running. They were simply finding more funds out of the money ocean to keep the system working. And just a few years before 2000, they found the biggest fish, which is the US housing market. Banks had started opening easy consumer credit lines and giving mortgages with very low rates to the customers. The market started expanding very fast, this time it was a really big catch.

Coming back to housing bubble, which we became so familiar with since it burst, with the 'credit derivatives'; this was the biggest game ever played in the Wall Street. Every dollar received from a customer was converted into several dollars with the help of the derivatives. It is important here to stress that derivatives are the instruments creating money out of nothing real but the paper assets. It works as long as the mortgage borrowers keep paying their mortgages, and it really worked that way. Because the early borrowers were creditworthy customers; there was no problem at the start. But when there were no other creditworthy borrowers left in the market, the credit institutions had to absorb the 'subprime borrowers' to keep the Ponzi scheme running. It was then when the sub primers began to fail to meet their repayment obligations.

Great Depressions in 1929 and 2008

After the World War I, the US had important economic growth until 1929 with the increasing demand caused by the war. The employment, production, wages and gains increased systematically. The Wall Street crash of 1929, the "Black Thursday", was 'the event' that propelled the global economy into panic. After a huge speculative rise in the late 1920s, based partly on the rise of new industries such as radio broadcasting and car making, shares fell by 13 percent on one day, and despite efforts of market authorities, fell by another 11 percent on the next day. By the time the market had reached the bottom in 1932, 90 percent of the value of shares had been wiped off. The effect on the real economy was severe, the US economy had declined by half, and one-third of the workforce was unemployed. The whole US financial system also went into meltdown, with a shutdown of the entire banking system in March 1933. The US central bank actually raised interest rates to protect the value of dollar and preserve gold standard, while the US government raised tariffs and ran a budget surplus, but the US economy did not fully recover until World War II, when massive military spending eliminated unemployment and boosted growth (Bordo and Haubrich, 2017).

Today, we are not anywhere near to the level of unemployment in the 1930s. Nevertheless, there are many similarities between today's environment and the Great Depression, with things happening at present that we have not seen since then. First of all, there is the magnitude of the stock market's move up and down. The real (inflation-corrected) value of the Standard & Poor's 500 (S&P 500) nearly tripled from 1995 to 2000, and by November 2008 was down nearly to 60% from its 2000 peak. The only other comparable event was the one in the 1920s. First, the real stock prices more than tripled from 1924 to 1929 and then fell 80% from 1929 to 1932. Second, we have experienced the biggest housing bust since the Great Depression. Third, we have seen 0% interest rate; actually, negative short-term interest rate briefly; this has not happened since 1941. There was a period from 1938 to 1941 when interest rates were bouncing around at zero and sometimes negative (Shiller, 2007).

Turkish Banking Crises of 1994 and 2001

Banks are basically fragile institutions and banking is a frail system. Banks borrow from depositors with a promise to repay in full and on demand and mostly invest those deposits in longer term loans. If there is a sudden withdrawal demand from the depositors, say, due to a panic or uncertainty, the banking system or that specific bank will face some major problems. If one bank is perceived to be in danger, other banks are likely to come under suspicion too. When we talk about banking crises, the main reason, usually, depends on mass withdrawals which cause liquidity problems in the financial system. When a bank gives the sign of a threat of being in a difficult financial situation, the other banks may also be affected by the signal as this causes the mistrust of public who would like to withdraw their deposits and 'run' from the system.

Other causes of financial (i.e., banking) crises can be microeconomic or macroeconomic, or a combination of both, including: Lax lending practices, weak risk control systems, mismatches in loans and deposits, in other words, poor asset control and poor banking practices in combination with poor regulatory/supervisory practices can be considered as microeconomic causes. Macroeconomic imbalances, such as large current account deficit, high degree of dollarization, unhedged borrowers and high and volatile interest rates can be considered as macroeconomic causes which make the banks vulnerable to any shock.

The scope of this study's case is the fragile banking system in Turkey before 2001 and banking crises the Turkish banks experienced up to that point. To understand the financial history and the mistakes made in the banking system in Turkey, the past two decades will be examined.

The starting point of the banking crises of Turkey in 2001 was due to liberalization in the 1980s. In 1980, a financial liberalization program was launched. This financial liberalization program included: (i) Elimination of interest rate controls; (ii) High permission to new entries for banking activities; and (iii) Allowance to banks for issuing certificates of deposits (CDs).

The economic stabilization and structural adjustment program announced in 1980, largely known as the "January the 24th Decisions", adopted policies giving the priority to economic growth based on export promotion and to structural reforms, including, deregulation and liberalization of the financial markets. The implementation of this program eliminated quantitative and price controls and put emphasis on free market approach, relying on the price mechanism.

The following outcomes of this "financial deregulation and liberalization program" were observed in 1982:

1. Banks changed their business strategies and moved away from traditional banking activities, the need for maintaining large branch networks in order to attract more deposits decreased, they started to borrow foreign currency dominated funds from abroad and lend them to the Turkish government at high rates. The number of foreign banks in Turkey increased from 4 to 20. Similarly, the number of private banks went up from 24 to 32.
2. Real exchange rate depreciation and export promotion strategies led to strong export growth. Imports were liberalized gradually after 1983, and in 1989 tariffs were lowered and in 1990 nearly all quantity and price restrictions were removed. Starting in 1988 and by the end of 1989, the process of capital account liberalization was completed; capital flows were fully liberalized in the external accounts.
3. The liberalization of capital flows increased interest rates as well, as is the case in many financial liberalization episodes. The public sector borrowing requirement (PSBR) of Turkey rose steadily between 1988 and 1993, the gap between public sector revenue and expenditure was widening between 1989 and 1993, and the PSBR was increasing along with the budget deficit.
4. The financing of public sector deficits was shifted to domestic borrowing, and the share of external borrowing was to be reduced. External borrowing was delegated to the private financial institutions, mainly to commercial banks, which were the main source of demand for domestic debt instruments.
5. The share of domestic debt (as opposed to external financing) increased until 1992. In the crisis year of 1994, however, domestic borrowing rebounded not only to finance the government deficit but also the repayments of accumulated external debt.

McKinnon-Shaw (Shaw, Edward and McKinnon 1973) states that financial liberalization provides financial deepness and this provides economic growth. However, in Turkey, financial liberalization had notprovided economic development due to inadequate legal infrastructure, absence of institutional regulators and uncontrolled short term profit aimed at investors who had come for interest arbitrage.

The 1994 Currency Crisis in Turkey

Turkey experienced large and growing fiscal and external imbalances following the capital account liberalization in 1989. Until the first quarter of 1994, the real exchange rate appreciation was no less than 20%. Against this background of rising and very high Public Sector Borrowing Requirement (PSBR; about 10 and 12% in 1992 and 1993 respectively), there were remarkable policy mistakes committed on the monetary front (Celasun, 1994).

The commercial banks made offshore borrowing in 1992-93 and held mainly TL denominated assets. This accelerated the process of acquiring foreign currency to close their open foreign currency positions. However, capital flight began and as a result of the Central Bank's aim to defend the currency, the foreign currency reserves melted, which started to heavily intervene in the interbank market and raised the overnight rate to record levels.

In the first quarter of 1994, the Turkish Lira (TL) was devalued more than 50% against the US$. The Central Bank was defending the parity by selling foreign currency at rates below the market rate. As a result, the Central Bank lost half of its reserves while the commercial banks closed their short positions (Figure 3), interest rates sharply increased (Figure 4), and the inflation rate reached three digit levels (Figure 5). A stabilization program, later supported by an IMF Stand-By was launched on April 5th, 1994, but no success had been achieved.

There were large withdrawals of TL and foreign currency denominated deposits from the system. Following that, 100% of bank deposits were insured by the Central Bank in April, 1994. Although the amounts of reserves were at their all-time lowest level, this action of the Central Bank seems to have helped to stop the withdrawals. This insurance on bank deposits has not been removed to date, signifying the fragility of the system. That is, 100% of bank deposits are still insured by the Central Bank in Turkey.

Figure 3. Foreign exchange reserves of the central bank of Turkey
Source: http://evds.tcmb.gov.tr/yeni/cbt-uk.html

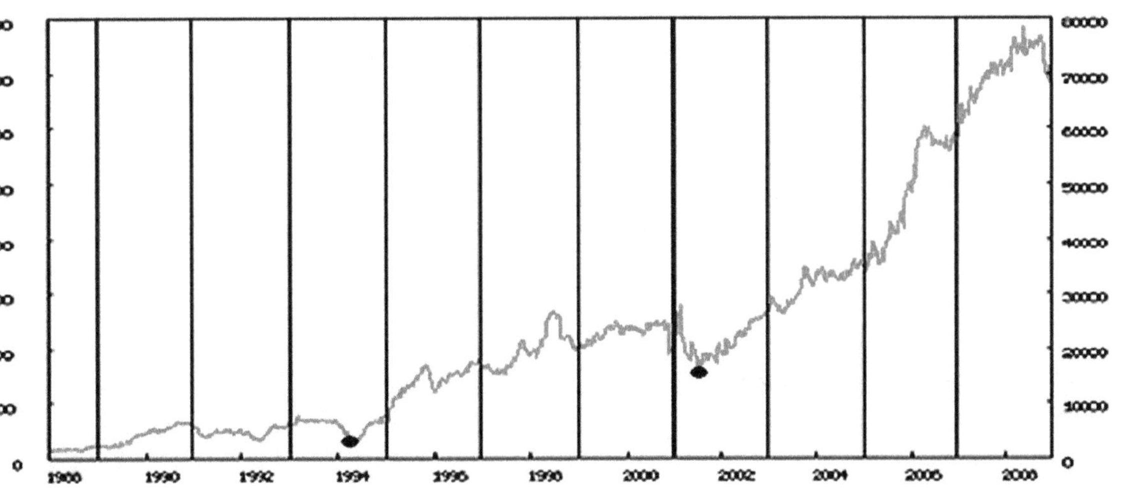

Figure 4. Short term interest rates in Turkey
Source: http://evds.tcmb.gov.tr/yeni/cbt-uk.html

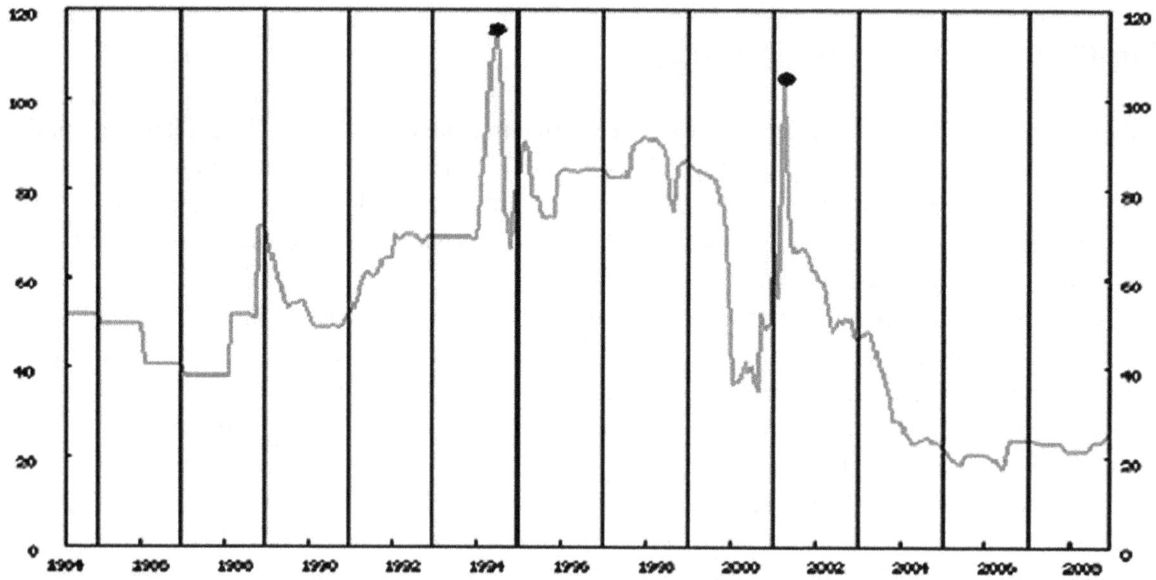

Figure 5. Turkish inflation rates
Source: the World Bank (formal web site of The World Bank) at http://www.worldbank.org.tr/WBSITE/EXTERNAL/COUNTRIES/ECAEXT/TURKEYEXTN/0,,menuPK:361756~pagePK:141132~piPK:64025829~theSitePK:361712,00.html

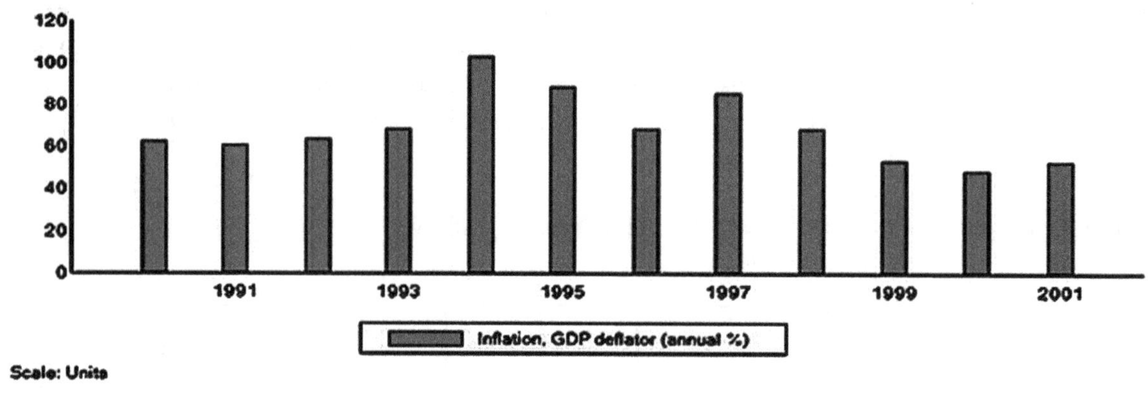

The main underlying reason behind the crisis of 1994 was the uncontrollable growth in domestic debt. The risks of fast capital account liberalization without any fiscal adjustment had become evident in the former stages of Turkey's financial liberalization and many economists gave early warnings of this problem.

The 2001 Currency Crisis in Turkey

The decisions made after the 1994 crisis became the main reason for the 2001 crisis. 100% bank deposits' guarantees provided by the government and the excessive increase in interest rates on government debt instruments encouraged banks to take risky positions and have high profits. Consequently, Banks could and did offer the public the highest interest rates for their deposits and the public choose the bank that offered the highest return because all the deposits were under the guarantee of the government anyway. Banking business was not conducted properly anymore, especially by private banks. The Central Bank's full insurance meant that there would be no more risk and return measures for customers. Accordingly, many corporations and holdings started to establish their own banks, exploiting the gaps in the system, to raise funds from the public for funding their own entities. There was no longer a need to make cost and benefit calculations for the credits extended for those private banks. In fact, many fraud cases were observed during this period caused by those who abused the system.

The risk depended on the exchange rate fluctuations as banks borrowed cheap funds from abroad to invest in local government securities in order to gain high interest yields. Specifically, these banks borrowed syndicated loans from abroad at relatively low interest rates and invested these funds in government securities. This increased the banks' open foreign currency positions and caused a mismatch in the maturity of loan and deposits because they borrowed these funds in foreign currencies and converted them into the local currency to buy government securities with high interests. The currency risk remained unhedged by either the banks or the government and caused severe imbalance in Turkey's national accounts which transferred significant currency risk to the government.

However, the sustainability of this financing mechanism depended upon the continuation of demand for government securities. If there would be a decline in demand for government securities that would force authorities to monetize and hence cause a jump, both, in the exchange rate and the inflation rate. The upward trend in government debt instruments portfolios of the banks increased the vulnerability of the banking system. In the end the banks could not sustain this burden and collapsed in November 2000 and February 2001. This was a result of the floating exchange regime and many of the banks were confiscated by the Savings Deposit Insurance Fund (SDIF) (TMSF - Tasarruf Mevduatı Sigorta Fonu (Turkish).

The trigger of the 2001 crises was politics. On February 19th, 2001, when the Prime Minister announced that there was a severe political crisis, this announcement caused a sudden outflow of capital from the country. Over-night the interest rate rose abruptly up to 2058% and to 4019% the next day. Rising interest rates, with overnight rates reaching 5000%, could not stop the rapid flight from the TL. Not only non-residents but also residents and especially the banking sector rushed to foreign currency (dollarization). The Central Bank lost a significant portion of its foreign reserves in defending the currency peg. The initial optimism had worn out. The rapid flight of capital from the country started because of: (i) Disappointing inflation results for October; (ii) Unexpectedly high monthly trade deficits, (iii) Political difficulties faced in privatization, (iv) Worsening relations with the EU, (v) The economic crisis in Argentina, and (vi) Rising the US official interest rate.

The outflow was followed by the rise in the interest rates, and the market risk in Turkey increased. Banks who were in need of short term funding began to sell their securities, which caused a decline in the value of government securities. The rising sovereign risk led foreign investors to exit the Turkish Market. The dream was over. Within a single day, the local currency lost about *one-third* of its value against the dollar. Finally, the exchange rate system collapsed and Turkey declared that it was going to implement a floating exchange rate system.

PRE-CRISIS EARLY INDICATORS IN NUMBERS

After the liberalization process, Turkey started to face almost all the risks which cause financial crises, including;

Macroeconomic and Political Risks

1. High chronic inflation could be seen nearly 70% in 1998 and 1999, 39% in 2000, 69% in 2001.
2. High and very volatile nominal (Nearly 100%) and Real Interest Rates nearly 30% or more were being used in practice.
3. The exchange rate was falling and unpredictable.
4. High and growing overall fiscal deficit, 15% of GNP in 1998, 24% in 1999, 20% in 2000, 17% in 2001.
5. Large and growing public debt --57% of GDP in 2000, 91% in 2001.
6. Uncertain economic growth, -6.4% in 1999, 6.3% in 2000, --8.5% in 2001.
7. Lack of clear economic strategy.
8. Lack of political will and consensus to undertake serious reforms.
9. Fragmented political coalitions leading to a loss of public confidence in the government.

Regulatory and Institutional Framework

1. Fragmented banking regulation and supervision authority—Lack of focus, coordination and responsibility.
2. Lack of independence of the financial institutions and political interference.
3. Outdated and overly lenient prudential regulations.
4. Enforcement very weak, widespread regulatory forbearance.
5. Deficient accounting and auditing practices, poor public disclosure.

Private Banks: Pre-Reform Situation

1. High forex exposure.
2. High interest rate exposure.
3. High level of non-performing loans (NPLs).
4. High connected lending.
5. Poor capital adequacy.
6. A number of insolvent banks in the system.

State Banks: Pre-Reform Situation

1. Politically directed subsidized lending- Ziraat and Halkbank
2. Huge losses (duty losses nearly $21Billion) hidden in the balance sheet as government dues.
3. Insolvent housing bank, the EmlakBank
4. Excessive number of branches in Agriculture bank, the Ziraat Bank, and Public Bank, the Halkbank.
5. Excessive staff in both these State Banks.

Causes and Outcomes of These Crises

1. As a result of anxiety in the financial sector, Turkey's credit rating was downgraded.
2. The slow decrease of inflation led to reduction in confidence to the disinflation program being followed.
3. The uncertain economic environment and interest rates made it harder for banks to borrow from abroad.
4. Banks ended up with criminal investigations about them.
5. Problems in privatization of some industries created obstacles in IMF's scheduled fund transfers.

POST CRISIS BANKING ACTIVITIES

1980 - 1994 Crisis

The Crisis of 1994 showed that interest rates needed to be controlled and the private banks in the system should be strengthened. Operational and financial restructuring of state banks were made. New legal and institutional framework, which will increase supervision and audit in the sector is developed.

1999 - 2001 Crisis

Regulatory and Institutional Framework

1. A newly structured banking law was adopted creating an Independent Banking Regulation and Supervision Entity (BRSA), and strengthening bank failure resolution provisions.
2. The "Banking Sector Restructuring and Rehabilitation Program" was implemented by BRSA. The main goal of the Program was to eliminate the distortions in the financial sector and to adopt regulations to promote an efficient, competitive and sound banking sector. The strategy under this program rested on four main pillars: the financial and operational restructuring of state banks, the resolution of the banks under management of the Savings Deposit Insurance Fund (SDIF) banks, the strengthening of private banking; and, the strengthening of the legal and regulatory environment.
3. Prudential regulations were upgraded to EU/BIS standards.
4. Enforcement of new regulatory standards was strengthened.
5. Banking Discipline was restored through broad SDIF interventions in insolvent banks.

Private Banks: Present Situation

Net forex exposure was reduced to prudent levels and they were forced to reduce their short-term liabilities. Detailed portfolio audits carried out to reflect NPLs more accurately. Connected lending exposure reduced and is being tightly monitored. 20 insolvent private banks have been intervened and taken over by Savings Deposit Insurance Fund (SDIF) .

State Banks: Present Situation

The duty losses of the state banks were eliminated; their balance sheets were restructured, short-term liabilities reduced. Some of them were closed down. Excess branches (nearly 900) of state banks were closed down. Excess staff was reduced, 16,000 either retired or transferred to other public entities.

SOLUTIONS AND RECOMMENDATIONS

After experiences of two destructive banking crises, Turkey learnt some lessons from its mistakes, which protected the country's financial position against the attacks from the global financial crisis of 2008. Turkey found out that the costs of banking crises are really high, including the fiscal cost of restructuring, impacts on lending and deposits, impacts on economic growth. Therefore they implemented very strict restructuring in its banking regularities and practices.

The root cause of the crisis in Turkey was mostly the fragile banking sector. After analyzing the data for 2000, one would immediately observe the poor macroeconomic performance. Public sector borrowing requirement, public debt to GNP ratio, current account deficit, inflation level and liabilities of financial sector to official reserves ratio were all high. Moreover the lira was appreciated in real terms. But still we argue that without a fragile banking system and triggering factors, high current account deficit and real appreciation of the lira would not have been enough to precipitate the 2000-2001 crises. It should be noted that the root cause of the fragility of the banking system was the high public sector borrowing requirement and the way it was financed. This is a general reason for crises in emerging markets; we observe the same result when we analyze the Asian and Russian crises. There are many similarities, such as, high interest rates causing capital inflows in the countries which after a sudden panic, generating capital outflows, leaves the country in a difficult situation. In the Turkish case, that budget deficits were mainly financed through government securities was a major issue. However, the sustainability of this financing mechanism was conditional on the continuation of demand for government securities.

The following are the crises prevention measures that any emerging country can utilize. Turkey realized most of those precautions correctly and on time in 2001 and strengthened its financial system, especially the structure of its banking system. Accordingly, Turkey had less negative effects from the 2008 global liquidity crises when compared to other emerging countries:

1. The most typical measure that the governments can take against the abundance of liquidity problems is tightening monetary policy. On the other hand, when there is a shortage of liquidity, central banks should provide high amounts of liquidity to the markets or demand liquidity support from other countries and from IMF.
2. Create strong and politically independent financial management institutions like Central Bank, Banking Regulation and Supervision authority.
3. Develop early warning systems based on crisis indicators.
4. Contingency planning, including what-if scenarios, should be in place before crisis actually happens.
5. Under macroeconomic stress situations the level of capital required should be increased. Risk management regulations should be enforced to assess individual bank exposures and capital requirements.
6. Consolidated supervision should be used to assess financial risks on a group basis

7. Deposit Insurance premiums to be increased in times of high macro risks.
8. Fit and proper ownership/ management assessment should be made to weed out incompetent/ dishonest bank owners.
9. Accounting/Auditing Standards and Practices should be thoroughly enforced to ensure disclosure of reliable financial information.
10. Insolvency and creditor rights legislation should be strengthened for efficient restructuring of corporate debtors.
11. In times of crisis, an economic crisis management team should be urgently formed between the relevant institutions to ensure coordinated rapid response.
12. In case of a liquidity distress the Central Bank liquidity should be provided early and quickly to avoid a systemic contagion effect.
13. Urgent intervention should be undertaken to stop further bleeding in loss making banks.
14. Institutional capacity including financial, human and technical resources should be mobilized to deal with the crisis.
15. Prompt restructuring of the banking sector should be initiated to restore confidence in the banking system.
16. Bad banks should be removed urgently from the system to reduce subsequent overall resolution costs.

FUTURE RESEARCH DIRECTIONS

As a result of global financial crises and the lessons learned central banks are now taking serious measures like injecting high amounts of liquidity into markets and decreasing interest rates in many countries in order to escape the liquidity crisis. But, according to many finance authorities, these measures provide temporary solutions to the liquidity risks. Because, the amount of risky assets in the international markets and the risky capital inflows cannot be controlled, as the transmission mechanism of money has been deteriorated with the existence of many financial institutions and funds which have no relations with central banks. It would be interesting to investigate how these stricter regulations are impacting banking systems worldwide.

Subsequently, at this point, the world has agreed to make stricter regulations to control the risky capital movements in the international markets, just as Turkey did after 2001, but there is still no specific solutions created by any authority. Research could be undertaken into why this has not taken place to date. Given that the World financial system has to be also regulated strictly by authorities who have accurate knowledge about the markets and the products and the financial system has to be restructured so that contagious financial crises do not occur anymore. An analysis of how this is progressing is also a worthy future research direction. Additionally it would be interesting to investigate how these stricter regulations are impacting banking systems worldwide.

CONCLUSION

Since the 2001 crisis, the banking sector has been purified from unsound and relatively fragile members and financial sector reform was made to implement a well-suited and high-controlling surveillance system (Yeldan, 2009). Therefore, the Turkish banking system was not so badly affected from the 2008 crisis. The Turkish banks were ready to fight against crises with their strong capital structure. They were being very strictly audited by regulators, and were very careful in credit extensions. Credit derivatives and derivative transactions in general, have a small portion in the banks' balance sheets. Additionally, a developed mortgage market did not exist in Turkey, this also protected the Turkish Banking System from the global financial boom. Nevertheless, we cannot estimate the same results for the real sector. Because of the tightening of international credit markets and capital outflows due to global crises, the stock markets were badly affected and the national currencies have depreciated in developing countries. Since Turkey is an emerging market, its economy depends heavily on international markets; the most powerful effects are on international trade, liquidity, external borrowing and production.

According to McKinnon-Shaw (Shaw, Edward & McKinnon 1973), finance sector empowered by liberalization would be the engine of the growth. We witnessed a relative progress in depth of financial markets but the financial liberalization did not provide economic development. On the contrary, the result of the process caused foreign dependency and instable reel production structure. This would be the beginning of the end especially for the developing or underdeveloped countries.

REFERENCES

Aktan, C., & Huseyin, C. (2001). Ekonomik Kriz; Nedenler ve Çözüm Önerileri. *Yeni Türkiye Dergisi, Ekonomik Kriz Özel Sayısı. Y, 7*, 42.

Anderson, R. G., Bordo, M., & Duca, J. V. (2017). Money and velocity during financial crises: From the Great Depression to the Great Recession. *Journal of Economic Dynamics & Control, 81*, 32–49. doi:10.1016/j.jedc.2017.03.014

Berg, A. (1999). *The Asia Crisis, Causes, Policy Responses and Outcomes.* NBER Working Paper.

Berger, A. N., & Bouwman, C. (2013). How does capital affect bank performance during financial crises? *Journal of Financial Economics, 109*(1), 146–176. doi:10.1016/j.jfineco.2013.02.008

Bordo, M. D., & Haubrich, J. G. (2017). Deep recessions, fast recoveries, and financial crises: Evidence from the American record. *Economic Inquiry, 55*(1), 527–541. doi:10.1111/ecin.12374

Brown, E. (2009). *The Wall Street Ponzi Scheme called Fractional Reserve Banking, Borrowing from Peter to Pay Paul.* Retrieved from www.globalresearch.ca/index.php?context=va&aid=11600

Brunnermeier, M. K. (2016). *"Bubbles." Banking Crises.* Palgrave Macmillan.

Burnside, C., Eichenbaum, M., & Rebello, S. (2001). Prospective Deficits and the Asian Currency Crisis. *Journal of Political Economy, 109*(6), 1155–1197. doi:10.1086/323271

Bustelo, P., Garcia, C., & Olive, I. (1999). *Global and Domestic Factors of Financial Crises in Emerging Economies: Lessons from the East Asian Episodes (1997-1999).* Retrieved from http://papers.ssrn.com/sol3/papers.cfm?abstract_id=201150

Castiglionesi, F., Feriozzi, F., & Lorenzoni, G. (2017). *Financial integration and liquidity crises. No. w23359.* National Bureau of Economic Research. doi:10.3386/w23359

Celasun, O. (1994). Macroeconomics and Growth Group Development Research Department. *The World Bank Review*.

Cheung, Y. L., Cheung, Y. W., & Ng, K. C. (2003). *East Asian Equity Markets Financial Crises, and The Japanese Currency*. Retrieved from http://papers.ssrn.com/sol3/papers.cfm?abstract_id=1009369

Crafts, N. (1999). East Asian Growth Before and After the Crisis. *IMF Staff Papers*, *46*(2), 139–166.

Edison, H. J., Luangaram, P., & Miller, M. M. (1998). *Asset Bubbles, Domino Effects and 'Lifeboats': Elements of the East Asian Crisis*. Retrieved from http://papers.ssrn.com/sol3/papers.cfm?abstract_id=93688

Eichengreen, B. (2001). *Crisis Prevention and Management: Any New Lessons from Argentina and Turkey?* Background Paper for the World Bank's Global Development Finance.

Erdonmez, A, P. & Tulay, B. (1999). Küresel Krizlere Yeni Yaklaşımlar. *Bankacılar Dergisi, S 31*, 64-76.

International Monetary Fund IMF. (1999). *Financial Sector Crises and Reconstruction, Lessons From Asia*. Washington, DC: World Economic Outlook.

Jomo, K. S. (2001). *Growth after the Asian Crisis What Remains of the East Asian Model?* Retrieved from http://ksghome.harvard.edu/~drodrik/g24-jomo.pdf

Kaminsky, G., & Reinhart, C. M. (1999). *The Twin Crises: The Causes of Banking and Balance-of-Payments Problems*. Retrieved from http://home.gwu.edu/~graciela/HOME-PAGE/RESEARCH-WORK/WORKING-PAPERS/twin-crises.pdf

Kindleberger, C., & Aliber, R. Z. (2005). *Manias, Panics and Crashes - A History of Financial Crises* (5th ed.). Palgrave Macmillan. doi:10.1057/9780230628045

Krugman, P. (1979). A Model of Balance-of-Payments Crises. *Journal of Money, Credit and Banking*, *11*(3), 311–325. doi:10.2307/1991793

Krugman, P. (2008). *The Return of Depression Economics* (1st ed.). New York: W.W. Norton & Company.

Mera, K. M., & Renaud, B. (2016). Asia's financial crisis and the role of real estate. New York: Routledge.

Mishkin, F. (1996). Understanding Financial Crises: A Developing Country Perspective. In *Annual World Bank Conference on Development Economics*. The World Bank. 10.3386/w5600

Mishkin, F. (2001). *Financial Policies and the Prevention of Financial Crises in Emerging Market Countries*. NBER Working Paper, 8087.

Mishkin, F. (2015). *The Economics of Money, Banking and Financial Markets* (11th ed.). London: Pearson Education.

Mundell, R. (2008). The international financial crisis and international monetary system. Paper presentation at Bahcesehir University, İstanbul, Turkey.

Rodrik, D. (1999). Where Did All the Growth Go? External Shocks, Social Conflict and Growth Collapses. *Journal of Economic Growth*, *4*(4), 385–412. doi:10.1023/A:1009863208706

Roubini, N. (2008). 2008 US and Global Economic Outlook and Implications for Financial Markets. In *RGE Monitor*. Stern School of Business, NYU.

Sakulrat, M., & Ramkishen, S. R. (2001). *Financial Crisis, Bank Restructuring and Foreign Bank Entry: An Analytic Case Study of Thailand.* CIES Working Paper No. 131. 10.2139/ssrn.294579

Shaw, E., & Mckinnon, R. (1973). *Financial Deepening in Economic Development*. New York: Oxford University Press.

Shedlock, M. (2007). *Russia liquidity problems and other warning signs*. Retrieved from http://globaleconomicanalysis.blogspot.com/2007/10/russia-liquidity-problems-and-other.html

Temin, P. (2016). *"Great Depression." Banking Crises*. Palgrave Macmillan.

Wagner, H. (2007). *Liquidity problems and the stock market.* Retrieved from http://www.marketoracle.co.uk/Article1811.html

Yeldan, E. (2009). *Küresel Kriz ve Türkiye Ekonomisine Olası Etkileri*. Retrieved from http://www.bianet.org/bianet/kategori/biamag/110856/kuresel-kriz-ve-turkiye-ekonomisine-olasi-etkileri

KEY TERMS AND DEFINITIONS

Deregulation: The removal of rules and regulations governing an industry or sector.

Financial Crisis: A situation where assets and money lose their value rapidly. Financial crises are often associated with banking and the panic that ensues can lead to recession.

Globalization: A process that removes geographical boundaries and enables the economic integration and interdependence of national economies into the international economy through trade, foreign direct investments, capital flows, and migration.

Great Depression: In 1929 a major financial crisis led to the largest economic downturn the world had ever seen. It effects were felt for years afterwards.

Inflation: Occurs when prices for goods and services are rising rapidly over a sustained period of time and as a result the purchasing power is falling.

Interest Rates: The proportion of a loan that a lender charges as interest to a borrower.

Liberalization: The relaxation of rules and regulation, particularly pertains to government policy.

Chapter 5
Global Crisis and Financial Distress Likelihood of SMEs:
Some Evidence From Panel Data Regression

Andrea Quintiliani
Pegaso Telematic University, Italy

ABSTRACT

In the aftermath of the global financial crisis, this chapter sheds light on the determinants of the financial distress costs between Italian and German small and medium enterprises (SMEs). The authors propose an innovative formulation of the expected costs originated by financial distress expressed as the product of the expected financial distress likelihood times the total amount of the financial distress costs if insolvency does occur. The model is estimated using panel data methodology on samples from two European countries (Italy and Germany). The results indicate that the amount of ex-post costs depends on derivative financial instruments, intangible assets, and relation with local banks (small local banks rather than large banking groups).

INTRODUCTION

This study is the continuation of a search (Quintiliani, 2017b) finalized to estimate the expected costs of financial distress, quantified in terms of product among financial distress likelihood and costs sustained by the insolvent society.

Compared with the previous one, the survey model analyzes SME samples from two European countries, Italy and Germany, whose financial and entrepreneurial systems show significant differences.

Data were extracted from Amadeus (a high quality European database), BvD databases (Aida and Mint Italy) and panel data methodology was used to control for potential endogeneity and unobservable heterogeneity.

The results obtained for each country taken individually, empirically validate our model, revealing that most parameters are significant and with the expected signs.

DOI: 10.4018/978-1-5225-3767-0.ch005

More specifically, the financial costs decrease in relation to the company's increased ability to use intangible assets, in relation to use of derivatives and in relation to the local roots of the banks (local banks rather than international banking groups).

Our paper provides valuable insights and contributions to the literature about the expected behaviour and value of a SME in financial distress.

We develop a definition of financial distress based on financial criteria (Keasey *et al.*, 2015; Quintiliani, 2017b; Quintiliani, 2017a): «Namely, a firm is defined in difficulty every time that the cash flows coming from the operational activities (EBITDA) are inferior to financial expenses for at least two consecutive exercises and it has a deteriorated financial structure (decreasing trend, among two periods, of the solvency ratio "NetWorth ÷ Total Debt" accompanied by a fall in its NetWorth)».

Considering that the expected (ex-ante) cost of financial distress is a function of both the probability of financial distress and of the consequences of running a firm in financial distress, we propose an innovative specification of the expected (ex-ante) costs of financial distress that shows the interplay of the financial distress likelihood and the variables which are the main determinants of the consequences of financial distress (total value lost by the enterprise at the end of a period of financial distress)».

The entity of the financial costs, ex-post crisis, is tightly connected with the immaterial endowment of the enterprise in financial difficulties, with the taking root level on the territory of the bank and with the use of derivatives; while the first two variables are useful to reduce informational asymmetries and thus to facilitate access to (re)financing, the third variable reduces the likelihood of financial crisis.

The remainder of the paper is organized as follows. Section 2 explains literature reviews. Section 3 explains the theoretical arguments behind our hypotheses and present the model (borrowed, partly, by Keasey *et al.*, 2015; Quintiliani, 2017b). The results and conclusions are discussed in Section 4.

BACKGROUND

The literature on the SMEs underlines that one of the principal characteristics of the SMEs are the elevated financial distress likelihood (Keasey & Watson, 1993; Andrade & Kaplan, 1998; Frank & Goyal, 2009; Mac & Bhaird, 2010) and it needs to investigate their financial structure to understand fully the reasons for the failure.

Capital structure theory begins with the Modigliani and Miller (1958) paradox of "capital structure irrelevance", where firm value is not affected by its financing mix. Since then, corporate finance literature has grown enormously and basically distinguishes between two main theoretical approaches (Balios *et al.*, 2015): i) the trade-off theory, ii) the pecking order theory.

The core of the trade-off theory refers to the balancing process of benefits of debt (tax shield, reduction of agency costs of equity, lower issuance costs) and costs of debt (direct and indirect financial distress costs, rising agency costs of debt) which leads to the concept of an optimal capital structure.

The second was developed mainly by Myers (1984) and Myers and Majluf (1984), based mainly on informational asymmetries, and states that firms do not typically aim at a target debt ratio, but their financing decisions follow a hierarchy, with a preference for internal over external finance and for debt over equity. Asymmetric information has generated various other approaches such as the signaling theory by Ross (1977), and the market timing approach developed by Lucas and Mcdonald (1990).

SMEs present distinct specificities that must be considered. For example, there are actually no (or very few) agency costs of equity, because managers are, most likely, also the owners of the SMEs.

On the other hand, agency costs of debt may be severe, because the level of asymmetric information between shareholders and lenders is considerably high (van der Wijst, 1989; Ang, 1992) which leads lenders to require guarantees materialized in collateral (Scott, 1977; Harris & Raviv, 1990; Frank and Goyal, 2009). Thus, firms with tangible assets are expected to have better access to debt financing and thus exhibit higher debt ratios.

However, Hall *et al.* (2004) and Sogorb-Mira (2005) find a negative relationship between short term debt and the level of tangible assets and a positive relationship between long-term debt and leverage, because SMEs that employ large holdings of tangible assets may be related with a stable source of internally generated returns. Bearing in mind that short-term financial obligations are usually more often used as an indirect means of financing by SMEs than long-term debt, a negative relationship could be observed.

The size of the firm is always an important determinant because larger enterprises are more diversified and are expected to go bankrupt less often than smaller ones (Ang *et al.*, 1982; Pettit & Singer, 1985; Frank & Goyal, 2009; Balios *et al.*, 2015). It is also worth noting that this positive relationship between size and leverage seems to go beyond institutional characteristics. For example, Hanousek and Shamshur (2011) study capital structure determination from seven Eastern European countries and find a positive and highly significant effect between size and leverage, while their findings regarding other determinants are not so robust. Therefore, larger firms are expected to show higher levels of financial leverage.

Myers (1977) argues that growth opportunities can produce moral hazard effects and can push firms to take more risk, thus firms with growth potential will tend to have lower leverage. This may explain why firms with important growth opportunities will be considered as risky and face difficulties in raising debt capital on favorable terms:

On the other hand, Michaelas *et al.* (1999) argue that growth will push firms into seeking external financing, as firms with high growth opportunities are more likely to exhaust internal funds and require additional capital.

Risk is also considered as another important determinant of financial leverage under the financial distress approach. Specifically, riskier firms, in terms of earnings volatility, will have to confront relatively higher levels of difficulty in accessing debt financing (DeAngelo & Masulis, 1980; Titman & Wessels, 1988; Opler & Titman, 1994; Balios *et al.*, 2015). This negative expected relationship is also explained in the pecking order perspective (Balios *et al.*, 2015): firms with high volatility on earnings will try to accumulate cash to avoid under-investment issues in the future.

Asymmetric information provides another theoretical approach of capital structure determination, mainly via the pecking order theory, which is a consequence of information asymmetries between insiders of the firm and outsiders. Specifically, internal funds have no informational costs, which are very high when issuing new capital, whereas debt lies in an intermediate position. Thus, a negative relationship between leverage and profitability should be expected (Rajan & Zingales, 1995; Boot, 2000; Booth *et al.*, 2001; Frank & Goyal, 2009; Balios *et al.*, 2015). Focusing on SMEs, this theory is especially appropriate (Ang, 1991; Holmes & Kent, 1991; Watson and Wilson, 2002), for a number of reasons. First, these firms are often opaque and have important adverse selection problems that are explained by credit rationing and therefore bear high information costs (Psillaki, 1995), considerably higher than large enterprises, where information asymmetry is lower.

Furthermore, SMEs are usually managed by owners whose main objective is to minimize the intrusion in their business and avoid the discipline inherent in financing options other than internal funds. That is why internal funds will lie in first place of their preference of financing. If internal funds are not sufficient, SMEs will prefer debt to new equity mainly because debt means lower levels of intrusion and, most importantly, lower risk of losing control and decision-making power than new equity. There is strong evidence that their financing decisions follow a hierarchy, with a preference for internal over external financing and for debt over equity (Michaelas *et al.*, 1999; Boot and Thakor, 2000; Daskalakis & Psillaki, 2008; Psillaki and Daskalakis, 2009; Balios *et al.*, 2015).

A further area of studies has dwelt on the findings that characterize the relationship Bank-Sme. Particularly, Berger and Udell (1998) show as information opacity, typical of the small and medium-sized enterprises (SMEs), is a crucial factor in the choice of the financial structure, directed to the long-term debt, with negative consequences about the risk profile and about the probability of default.

This informational opacity, in the form of costly verification, adverse selection, and moral hazard, typically affects the financial policy of small firms, specifically in terms of debt and external equity sources. Costly verification and adverse selection problems tend to favor debt contracts, whereas moral hazard problem stend to favor external equity contracts (Berger and Udell, 1998). Specifically, the effects of financial institutions on the various types of external finance differ so that the relative amount of debt may depend strictly on the abilities of these institutions to solve information problems, e.g., by engaging in screening, contracting, and monitoring activities (Beck *et al.* 2002).

With reference to the amplest theme of the distress likelihood of the SMEs, part of the literature has dwelt on the phenomenon of the credit rationing and its purposes; particularly, some studies have brought back the phenomenon to a series of variables, such as (Quintiliani, 2017b):

1. Banking localism (Diamond, 1984; Fama, 1985; Diamond, 1991; Berger & Mester, 1997; Cesarini & Ferri, 1997; Goddard et al., 2001; De Laurentis, 2001; De Bruyn & Ferri, 2005; De Laurentis, 2011).
2. Bank size (Jayaratne & Wolken, 1999; Berger & Udell, 2002; Stein, 2002; Takats, 2004; Berger et al., 2005).
3. The nature of the bank-enterprise relationship (Berger & Udell, 2006).
4. Banks'Internal Ratings Systems (Modina, 2015).

FINANCIAL DISTRESS COSTS: RESEARCH HYPOTHESIS, ANALYSIS MODEL, AND DATA SET

Present your perspective on the issues, controversies, problems, etc., as they relate to theme and arguments supporting your position. Compare and contrast with what has been, or is currently being done as it relates to the chapter's specific topic and the main theme of the book.

The (ex-ante) financial distress costs are esteemed on the basis of the financial distress likelihood and owing to the following consequences to the financial distress status, these, ex-post, constitute the total value lost by the enterprise at the end of a period of financial difficulties (Quintiliani, 2017b).

We propose a definition based on the failure of the society when it is not able to settle its financial obligations. This definition qualifies a society as financially in difficulties in presence of two conditions (Asquith *et al.*, 1994; Whitaker, 1999; Pindado *et al.*, 2008; Keasey *et al.*, 2015; Gupta *et al.*, 2015):

- Firstly, it is estimated if a society were able to cover (for at least two consecutive years) the financial expenses through operating income cash flows (EBITDA).

- Secondly, we classifyas deteriorated the financial structure that introduces a decreasing solvency ratio (Net Worth ÷ Total Debt) for at least two consecutive years.

From this classification, we can build a variable that captures the probability, ranging from 0 to 1, of a company becoming financially distressed. We expect this financial distress likelihood to have a positive impact on the cost of financial distress.

Therefore, our first hypothesis is (Keasey *et al.*, 2015): «There is a positive relationship between financial distress likelihood and (ex-ante) financial distress costs».

Continuing with the explanation of the research hypothesis, it is possible to list the elements that show a negative relationship with the expected financial distress costs (ExpFDC):

- Intangible assets (hypothesis 2).
- Banking localism (hypothesis 3).
- Derivative financial instruments (hypothesis 4).

With reference to the hypothesis 2 it is believed that the endowment of intangible assets (invisible as regards accounts) on the total assets of a firm [INTAN] is a measure of its ability to do and so to get (re)financing under economic conditions to restructure the business.

For the enterprises subject to restructuring, there are some focal aspects (Lenz, 1981; McNair *et al.*,1990; Neely, 1999; Bontisa *et al.*, 1999; Buckmaster, 2000; Unger, 2000; Joia, 2000; Ching & Yang, 2000; Harris & Ogbonna, 2001; Dubrovski, 2001; Hoque *et al.*, 2001; Schwarz *et al.*, 2002; Catasis & Gröjer, 2003; Katcher, 2003): organizational structure, age of the firm, clientele's quality, contractual power with the suppliers, reliability of the plans and sustainability of the investments, leadership style, R&D investment, Process Control Systems, management skills, credibility of management, brand development, customer loyalty, investment in training.

Such data have been picked after an access to the databases Amadeus, Aida and Mint Italy.

The data have opportunely been quantified through a scoring and normalization process and in order to get homogeneous values among 0 and 1 (Min. 0 - Max 1).

Accordingly, the second research hypothesis affirms as follows (Quintiliani, 2017b): «There is a negative relationship between SMEs' invisible intangible assets and expected financial distress costs (ExpFDC). The greatest endowment of intangible assets determines the reduction of the expected financial distress costs».

Based on this hypothesis there is a literature showing that high immaterial companies are less exposed to significant bankruptcy costs; such evidence depends on (Quintiliani, 2017b):

- Their propensity to have a low leverage (Harris & Raviv, 1991; Degryse & Ongena, 2002; Frank & Goyal, 2003; Parsons & Titman 2009; Degryse *et al.*, 2010; Campello & Giambona, 2011; Koksal *et al.*, 2013).
- The negative relationship between level of indirect financial distress costs and intangible capital intensity (Bulot *et al.*, 2015).
- The high capacity of intangible assets to generate cash flows and to increase the competitive capacity of SMEs (Berry, 2004; Cohen, 2005; Heiens *et al.*, 2007; Hulten & Hao, 2008; Fukao *et al.*, 2009; Roulstone, 2011; Marrocu *et al.*, 2012; Lim *et al.*, 2014; Cucculelli & Bettinelli, 2015).

With reference to the third hypothesis, it is believed that the dimension and the proximity of the bank to the enterprise are critical elements (particularly in the phases of restructuring for financial distress) in attenuating the opacity and the connected asymmetric information problems.

For further information on the "size of banks" we suggest the following studies: Jayaratne and Wolken (1999); Berger and Udell (2002); Stein (2002); Takats (2004); Berger *et al.* (2005); Modina (2015); Modina et *al.* (2013); Formisano (2016).

Among the major studies on "bank proximity" we suggest the following: Uzzi and Gillespie (1999) and DeYoung *et al.* (2004).

Starting from the foregoing considerations, the following search hypothesis is examined (Quintiliani, 2017b): «The negative relationship between SMEs' invisible intangible assets and expected financial distress costs is more evident for the SMEs that relate to local banks».

In order to verify the third search hypothesis, we use a binary variable, LOCALB, that assumes values equal to 1 if the enterprise is a client of a local bank and equal to 0 in case of presence of a national and/or international banking group. The typology of partner bank of the enterprise has been individualized by accessing to Amadeus and Aida databases in the section "financial partners".

This variable influences our model in the part that considers the weight of the invisible intangible assets on total assets [INTAN]; the dummy LOCALB variable assumes a critical role in estimating the effects of the intangibles on the expected financial distress costs.

With reference to the fourth and last hypothesis, Smith and Stulz (1985) show that bankruptcy costs is a further source of incentive for the use of derivatives. In fact, by reducing the fluctuations in corporate flows, hedging through derivatives makes it possible to reduce the likelihood of a financial crisis, which can lead to extremely critical situations such as bankruptcy, liquidation or, at the very least, corporate restructuring and the need to bear direct costs (legal costs, administrative costs and the lesser value attributed to assets at the time of liquidation) and indirect (loss of image and consequent reduction in contractual power) of a very high amount.

Based on this model, the likelihood of using derivatives is greater for companies with high financial loss costs. If these costs, as Nance, Smith and Smithson (1993) point out, are a fixed component, then smaller companies will be the ones that will most effectively use hedging derivatives because they have a stronger impact on the fixed component of costs.

In addition, given the ability of derivatives to reduce the company's default probability, they also make it possible to increase the level of indebtedness and thus the value of the enterprise.

In particular, this theory was developed by Stulz (1984), which suggests that by reducing the volatility of company profits and hence the likelihood of financial stress, the company is able to increase its potential debt capability.

If companies increase leverage in response to this higher debt capability, the associated increase in financial burdens will lead to a reduction in taxation and, consequently, an increase in the value of the company.

Starting from the foregoing considerations, the following search hypothesis is examined: «There is a negative relationship between SMEs' derivative financial instruments and expected financial distress costs (ExpFDC). The greatest endowment of derivatives determines the reduction of the expected financial distress costs».

With reference to the hypothesis 4 it is believed that the endowment of derivative financial instruments on the total assets of a firm [DER shows the amount of derivative financial instruments as a percentage of total assets] is a measure of its ability to reduce the likelihood of financial distress.

Our model was designed to provide a representation of the value loss of SMEs in financial crisis.

This section clarifies how "probability of financial suffering" and "ex-post financial aid costs" can be useful drivers in building a model for estimating ex-ante financial distress costs.

In literature we see a rich literary production engaged in the formulation of ex-ante models to estimate the financial distress likelihood (FDL).

Particularly, Grice and Ingram (2001), Pindado *et al.* (2008), Keasey *et al.* (2015) and Gupta *et al.* (2015) underline that the seminal models by Beaver (1966), Altman (1968, 1984) and Ohlson (1980) are revealed suitable in the classification of the sector (better for the manufacturing enterprises) but poorly sensitive to the typology of financial distress of the enterprise. To analogous conclusions the studies of Zmijewski (1984) are reached.

Given our definition of financial distress, we can estimate the FDL using the methodology developed by Pindado *et al.* (2008) and Keasey *et al.* (2015). These Authors obtained a model of the ex-ante FDL, with the following functional form (1):

$$\log\left(\frac{Prob\left(Insolvency\right)}{Prob\left(NoInsolvency\right)}\right) = \beta_0 + \beta_1\left(\frac{EBIT_{it}}{TA_{i-1}}\right) + \beta_2\left(\frac{FE_{it}}{TA_{i-1}}\right) + \beta_3\left(\frac{RE_{it-1}}{TA_{it-1}}\right) + \varepsilon_{it} \quad (1)$$

This model was chosen because the work of these Authors provides evidence that they could obtain a more stable model of FDL in terms of magnitude, sign and significance of the coefficients, using the proposed panel data methodology.

The model considers three main variables to estimate FDL (Keasey *et al.*, 2015). The first ratio can be interpreted as the capacity of the firm to generate operational funds from its assets, independently of any tax or leverage factors. It is also the main driver of liquidity, conditioning the capacity of the firm to extend credit or renegotiate payments. The second ratio portrays the impact of the weight of financial expenses on the financial distress likelihood and like the other variables in the model, it is scaled by total assets. When this ratio rises, the probability of not being able to comply with its financial obligations is expected to increase. Finally, the ratio of retained earnings represents the cumulative profitability over time, and highlights the usefulness of past profitability in predicting future results and the capacity for self-financing.

The application of the method developed by Pindado *et al.* (2008) and Keasey *et al.* (2015) allows to reach to estimate financial distress likelihood (FDL).

The FDL variable is born from the model of logistic regression above proposed; coherently with the expected values of a likelihood function, the variable records values among 0 and 1. It appears evident as the FDL variable has a positive impact on the (ex-ante) financial distress costs (ExaFDC).

The total value loss of a firm's assets during financial distress, measured at the end of the period of financial crisis is the ex-post financial distress cost (ExpFDC). We can formulate these ex-post costs as a discount factor of the Future Value of the firm, consistent with our formulation of ExaFDC.

Because the majority of these costs have an unobservable nature the discount rate is a function of variables defined earlier as the main determinants of the impact of financial crisis.

Furthermore, because of our expectation of the cost financial distress depends of the financial distress likelihood, we can write the expected costs of financial distress as the weighting of ExpFDC by the likelihood of financial distress (2):

$$ExaFDC \cong ExpFDC \times FDL \tag{2}$$

Since the value of the of Financial Distress Likelihood (FDL) always ranges from 0 to 1, the ex- ante estimated insolvency costs variable take the highest values when FDL and ExpFDC are high. The lowest values of ExaFDC are obtained when FDL is near zero and ExpFDC are low.

If we take into consideration our definitions of ExaFDC, we obtain a factor that stands for the expected proportion of value loss due to financial distress can be defined as the product between the FDL and a proportion value loss when it occurs (3):

$$ExaFDC = FDL \times e^{\beta_1 X_1 + \beta_2 X_2 + \beta_3 X_3 \ldots + \beta_n X_n} \tag{3}$$

The coefficients of the variables $\beta_1 X_1$, $\beta_2 X_2$, $\beta_3 X_{3,}$ $\beta_n X_n$ presented earlier, are the main determinants of the bargaining position of different claimants during financial distress.

The model experimented in this article, borrowed by the work by Keasey *et al.* (2015), considers that the main factors driving of the esteem of ExpFDC are the enterprise ability to equip itself by intangible assets (invisible as regards accounts), the privileged relationship with the local banks (reduce asymmetric information) and the use of derivative financial instruments.

Given that estimating a linear model has a lot advantages from an econometric point of view, we take natural logarithms of both sides of the last equation of the previous section and we obtain the following model (4):

$$\varphi = \frac{1}{\tau} LnFDL + \frac{\beta_1}{\tau} X_1 + \frac{\beta_2}{\tau} X_2 + \frac{\beta_3}{\tau} X_3 \ldots + \frac{\beta_n}{\tau} X_n \tag{4}$$

LnFDL is the natural logarithm of financial distress likelihood and τ is the horizon of time over which firms formulate expectations for the resolution of financial distress. Consequently, the estimated coefficients a_1, a_2 and a_3 represent the adjustment rate of the value of the firm due financial distress, devided by the number of periods till the resolution of financial distress (5):

$$\alpha_1 = \frac{1}{\tau}; \alpha_2 = \frac{\beta_1}{\tau}; \alpha_3 = \frac{\beta_2}{\tau}; \alpha_4 = \frac{\beta_3}{\tau} \tag{5}$$

The model, therefore, is so developed (6):

$$\varphi_{it} = \alpha_1 LnFDL_{it} + \alpha_2 X_1 + \alpha_3 X_2 + \alpha_4 X_3 + \varepsilon_{it} \tag{6}$$

The equation 6 is completed with the attainment of the influential variables about our model (7):

$$\varphi_{it} = \alpha_1 LnFDL_{it} + \alpha_2 INTAN_{it} + \left(\alpha_3 + \lambda_1 LOCALB_i\right) INTAN_{it} + \alpha_4 DER_{it} + \varepsilon_{it} \tag{7}$$

All the variables represented in formula 7 are expressed in terms of weight on total assets. The variable ε_{it} is composed by the following investigation elements: $\varepsilon_{it} = \eta_i + d_t + v_{it}$. In this, v_{it} is the random disturbance, d_t expresses the annual impact of macroeconomic factors while η_i captures the not observable effects on the single enterprise. As previously affirmed, the model is completed with the introduction of the LOCALB variable, that assumes values equal to one if the enterprise is client of a local bank and equal to zero in case of presence of a national and/or international banking group.

Data were extracted for companies (volume of sales < 20 mlns; number of employees < 250; organized in the form of Ltd) for which the information is available for at least six consecutive years between 1999 and 2006, the number of observations is calculated after removing the first-year data which is only used to construct several variables.

The empirical investigation has been effected through the use of Amadeus and Aida databases.

Table 1 shows the number of firms from each country.

The samples presents a good statistical significance since it covers the 46% of the universe of SMEs active (in Italy and in Germany) at the end of 2006; as a whole, the investigation has required 385.671 statistic observations.

The temporal trend, object of empirical analysis, finishes deliberately by the 2006 exercise; in effects, it is by the 2007 exercise that the burst of the world financial crisis is recorded.

The descriptive statistics of the model's variables are reported in Table 2.

Since there is no fixed rule for dealing with outliers we use the procedure of trimming the data to the 99th percentile, as a general rule of thumb.

FINDINGS AND CONCLUSION

As can be in Table 3, all the coefficients are statistically significant and of the expected sign. Additionally, the m_2 test guarantees that there is no second-order serial correlation.

Table 1. Structure of the samples by country

Country	Firms
Italy	25000
Germany	12072
Total	37072

Table 2. Descriptive statistics of the total sample (basic model)

Variables	Mean	SD	Minimum	Maximum
ExaFDC	-0.077	0.687	-2.798	0.579
LnFDL	-1.896	1.723	-6.988	-0.098
INTAN	0.457	0.158	0	1
DER	0.098	0.355	-2.987	0.689

Source: Our elaboration on Amadeus and Aida databases.

Table 3. Results of the models

Model	Basic (Total Sample)	LOCALB Adj. Model	Italy	Germany
N. of Firms	37072	37072	25000	12072
$LnFDL_{it}$	0.0999 (0.0000)	0.0883 (0.0000)	0.0856 (0.0000)	0.09001 (0.0000)
$INTAN_{it}$	-0.2258 (0.0000)	-1.7867 (0.0000)	-0.1890 (0.0000)	-0.1911 (0.0000)
$LOCALBINTAN_{it}$	-	-1.5678 (0.0001)	-	-
DER_{it}	-1.0006	-1.1445	-1.4879	-1.8874
z_1	758.2323	326.0089	510.2878	482.3666
z_2	206.2199	212.3456	201.4355	202.5589
t_1	-	9.0000	-	-
t_2	-	11.1025	-	-
m_1	0	0	0	0
m_2	0.0456	0.1022	0.0556	0.0489
m_3	0.5598	0.6589	0.5467	0.4589
Hansen	1658 (33)	1879 (105)	1145 (47)	1178 (78)

Source: Our elaboration on Amadeus and Aida databases.

We also performed a Wald test of the joint significance of the dummies which is always positive, providing evidence that we need to control for the macroeconomics events that tend to influence all firms.

The regressions are performed using the panel previously described (Normal: 68,2%; Financial distressed: 31,8%).

It should be noted that logistic regression is a powerful tool, especially in financial studies, as it allows multiple explanatory variables being analyzed simultaneously, meanwhile reducing the effect of confounding factors.

In future studies, we shall include further means such as Conic Multivariate Adaptive Regression Splines (CMARS), Robust Conic Multivariate Adaptive Regression Splines (RCMARS) and Conic Generalized Partial Linear Models (CGPLM). These methods can help to give new perspectives and developments in financial mathematics to make more accurate predictions about financial distress likelihood.

Several studies have been carried out to apply the above methods to various fields of study including finance, industry, business and environment. These include: Özmen *et al.* (2010), Özmen *et al.* (2012), Weber *et al.* (2012).

With reference to our study, the variables investigated are: LnFDL, INTAN, LOCALBINTAN and DER.

The empirical evidence obtained for our basic model (total sample) corroborates our initial conclusions.

As shown in column 1 of Table 3, the positive coefficient of the FDL variables confirms Hypothesis 1 and supports the evidences emerged by Keasey' studies (Keasey *et al.*, 2015) that the FDL is one of the main explanatory variables of financial distress costs.

Additionaly, to confirm the second hypothesis, we find empirical evidence supporting the negative relationship between SMEs' invisible intangible assets and expected financial distress costs, since the coefficient ($ß_1 \div \tau = -0.2258$) obtained for $INTAN_{it}$ is negative and significant.

The third hypothesis is tested using the dummy $LOCALB_{it}$.

The negative relationship between SMEs' invisible intangible assets and expected financial distress costs (ExpFDC) is more evident for the SMEs that relate to local banks: $ß_2 \div \tau + \lambda_1 \div \tau = -1.5678$. This finding confirms the results that emerge from the doctrine: the ability of local banks to attenuate information asymmetries that, especially in the phases of restructuring for financial distress, are critical in the bank-firm relationship.

As shown in column 1 of Table 3, the negative term derivative coefficient for firms ($ß_3 \div \tau = -1.0006$) confirms our hypothesis 4. The greatest endowment of derivatives determines the reduction of the expected financial distress costs.

Moreover, the results obtained from our cross-country comparison, provide additional evidence. Without prejudice to previous assumptions, Germany, compared to Italy, recorded more significant coefficients ($ß_1$ and $ß_3$).

In summary, German SMEs have their own peculiarities to Italian SMEs (Abel-Koch *et al.*, 2015).

The largest $ß_1$ coefficient is justified by Germany's highest R&D investments: 33% of German companies invest 1% in R&D (24% invest 2%, 18% invest 4%, and 25% invest more than 4%). A key to the success of the "Made in Germany" strategy is, moreover, the After Sales, organized efficiently, precisely and in a very short time. Another important factor contributing to the success of the German Industry is the Common Labor Market - synergies through industry training - which shows that there is less competition in Germany than Italian reality. Certainly, greater financial solidity has been a key driver of the creation of world-leading industry leaders. While the suffering in Italy is 3 times higher than in Germany.

Another interesting fact is that 61% of Mittlestand has an Advisory Board or Supervisory Board and 67% of the Advisory Board is made up of owners of other companies. Business continuity, entrepreneurship and management experience are also important drivers in generational steps.

With reference to the coefficient $ß_3$, the German financial system shows peculiarities respect to the Italian system. These peculiarities concern the use of derivatives. Compared with Italian companies, German companies are more likely to use derivatives: 78% of German SMEs than 24% of Italian SMEs.

Italian companies use less derivatives than German companies. The use of derivatives is more significant among large companies for each category of risk suggesting the presence of economies of scale in the use of such tools.

The most heavily hedged type of risk is the exchange rate risk followed by interest rate risk and ultimately credit risk.

As emerges from a study by Casey and O'toole (2014): «The category of financial risk hedging services presents the most modest use grade. It also emerges that the SME has a low ex-ante perception of exchange rate risk and, above all, the risk of interest. In addition, entrepreneurs are experiencing strong commercial pressure from banks on this product category, which is not balanced by proper coordination with financing operations».

The estimate of the variables LnFDL, INTAN, LOCALBINTAN and DER is useful in building a "matrix" capable of suggesting, at country level, policies for SMEs for tackling the crisis as well as to mitigate the costs of exit from the market (Figure 1).

In conclusion, the results of our work are in line with those emerging from the literature that quantifies the value creation of SMEs in relation to the following: i) intangible (Berry, 2004; Cohen, 2005; Heiens *et al.*, 2007; Hulten and Hao, 2008; Fukao *et al.*, 2009; Roulstone, 2011; Marrocu *et al.*, 2012; Lim *et al.*, 2014; Cucculelli and Bettinelli, 2015), ii) local banks (DeYoung *et al.*, 2004; Berger *et al.*,

Figure 1. Policy matrix
Source: Our elaboration.

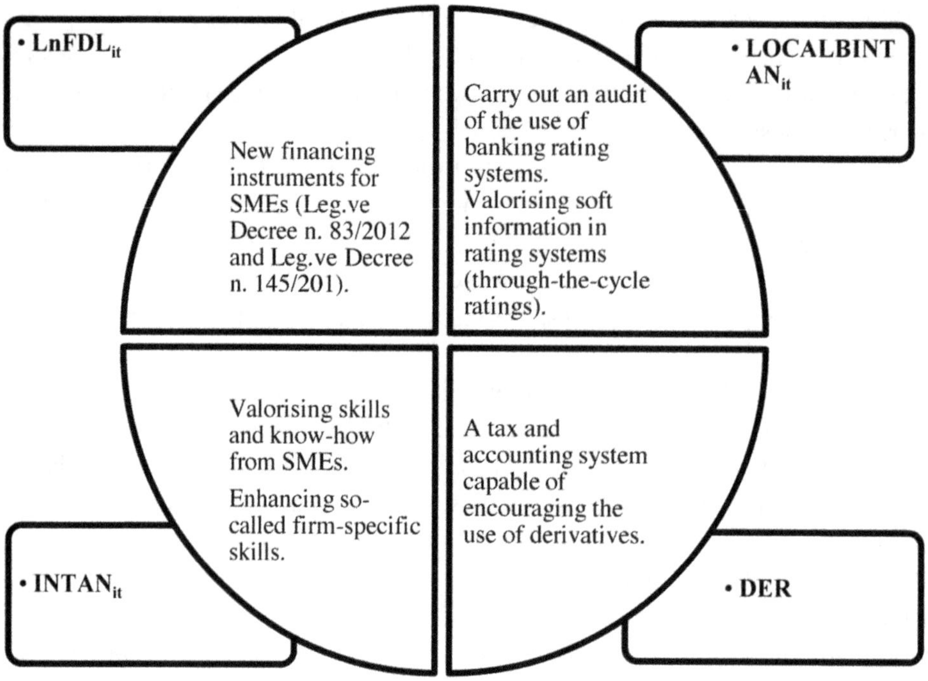

2005; Carter and McNulty, 2005; Udell, 2008; DeYoung *et al.*, 2008; Modina, 2015; Formisano, 2016), iii) derivative financial instruments (Stulz, 1984; Smith and Stulz, 1985; Nance et *al.*, 1993).

It should be emphasized that the present work limits its field of investigation to a few variables without fully addressing other elements of uncertainty which may adversely affect the financial distress likelihood and the value creation of the SME.

An interesting starting point for future research is indeed represented by application of additional methods (RCMARS and CGPLM) useful to mitigate the uncertainty of the forecast and capable to analyzing further variables nonlinearly (environmental, political, social and labor).

REFERENCES

Abel-Koch, J., Del Bufalo, G., Fernandez, M., Gerstenberger, J., Lo, V., Navarro, B., & Thornary, B. (2015). SME Investment and Innovation France, Germany, Italy and Spain. Maisons-Alfort Cedex, France: Bpifrance (BPI).

Altman, E. (1968). Financial ratios, discriminant analysis and the prediction of corporate bankruptcy. *The Journal of Finance*, *23*(4), 589–609. doi:10.1111/j.1540-6261.1968.tb00843.x

Altman, E. (1984). A further empirical investigation of the bankruptcy cost question. *The Journal of Finance*, *39*(4), 1067–1089. doi:10.1111/j.1540-6261.1984.tb03893.x

Andrade, G., & Kaplan, S. (1998). How costly is financial (not economic) distress? Evidence from highly leveraged transactions that become distressed. *The Journal of Finance*, *53*(5), 1443–1493. doi:10.1111/0022-1082.00062

Ang, J. S. (1991). Small business uniqueness and the theory of financial management. *Journal of Small Business Finance*, *1*(1), 11–13.

Ang, J. S. (1992). On the theory of finance for privately held firms. *Journal of Small Business Finance*, *1*(3), 185–203.

Ang, J. S., Chua, J. H., & McConnell, J. J. (1982). The administrative costs of corporate bankruptcy: A note. *The Journal of Finance*, *37*(1), 219–226. doi:10.1111/j.1540-6261.1982.tb01104.x

Asquith, P., Gertner, R., & Scharfstein, D. (1994). Anatomy of financial distress: An examination of junk-bond issuers. *The Quarterly Journal of Economics*, *109*(3), 625–658. doi:10.2307/2118416

Balios, D., Eriotis, N., Fragoudaki, A., & Giokas, D. (2015). Economic efficiency of Greek retail SMEs in a period of high fluctuations in economic activity: A DEA approach. *Applied Economics*, *47*(33), 3577–3593. doi:10.1080/00036846.2015.1019033

Beaver, W. H. (1966). Financial ratios as predictors of failures. *Journal of Accounting Research*, *4*, 71–111. doi:10.2307/2490171

Beck, T., Demirguc-Kunt, A., & Maksimovic, V. (2002). Financing patterns around the world: The role of institutions. *Proceedings of World Bank*, *1*, 1–40.

Berger, A. N., Frame, W. S., & Miller, N. H. (2005). Credit scoring and the availability, price, and risk of small business credit. *Journal of Money, Credit and Banking*, *37*(2), 191–222. doi:10.1353/mcb.2005.0019

Berger, A. N., & Mester, L. J. (1997). Inside the Black Box: What explains differences in the efficiencies of financial institutions? *Journal of Banking & Finance*, *21*(7), 895–947. doi:10.1016/S0378-4266(97)00010-1

Berger, A. N., & Udell, G. F. (1998). The economics of small business finance: The roles of private equity and debt markets in the financial growth cycle. *Journal of Banking & Finance*, *22*(6-8), 613–673. doi:10.1016/S0378-4266(98)00038-7

Berger, A. N., & Udell, G. F. (2002). Small business credit availability and relationship lending: The importance of bank organisational structure. *Economic Journal (London)*, *112*(477), 32–53. doi:10.1111/1468-0297.00682

Berger, A. N., & Udell, G. F. (2006). A more complete conceptual framework for SME finance. *Journal of Banking & Finance*, *30*(11), 2945–2966. doi:10.1016/j.jbankfin.2006.05.008

Berry, J. (2004). *Tangible Strategies for Intangible Assets*. New York, NY: McGraw-Hill Professional.

Bontisa, N., Dragonetti, N. C., Jacobsen, K., & Roosc, G. (1999). The knowledge toolbox: A review of the tools available to measure and manage intangible resources. *European Management Journal*, *17*(4), 391–402.

Boot, A. W. A. (2000). Relationship banking: What do we know? *Journal of Financial Intermediation*, *9*(1), 7–25. doi:10.1006/jfin.2000.0282

Boot, A. W. A., & Thakor, A. V. (2000). Can relationship banking survive competition. *The Journal of Finance*, *55*(2), 679–713. doi:10.1111/0022-1082.00223

Booth, L., Aivazian, V., Demirguc-Kunt, A., & Maksimovic, V. (2001). Capital structures in developing countries. *The Journal of Finance*, *56*(1), 87–130. doi:10.1111/0022-1082.00320

Buckmaster, N. (2000). The performance measurement panacea. *Accounting Forum*, *24*(3), 264–277. doi:10.1111/1467-6303.00041

Bulot, N., Salamudin, N., & Abd Aziz, R. (2015). Size of indirect financial distress costs: Which factors are reliably important? *Jurnal Intelek*, *12*(1), 12–20.

Campello, M., & Giambona, E. (2011). *Capital structure and the redeployability of tangible assets*. Retrieved November 30, 2011, from https://papers.tinbergen.nl/11091.pdf

Carter, D. A., & McNulty, J. E. (2005). Deregulation, technological change, and the business-lending performance of large and small banks. *Journal of Banking & Finance*, *29*(5), 1113–1130. doi:10.1016/j.jbankfin.2004.05.033

Casey, E., & O'Toole, C. (2014). Bank lending constraints, trade credit and alternative financing during the financial crisis: Evidence from European SMEs. *Journal of Corporate Finance*, *27*, 173–193. doi:10.1016/j.jcorpfin.2014.05.001

Catasis, B., & Gröjer, J. E. (2003). Intangibles and credit decisions: Results from an experiment. *European Accounting Review*, *12*(2), 327–356. doi:10.1080/0963818032000089418

Cesarini, F., & Ferri, G. (1997). *Credit and development: Local cooperative banks and small businesses*. Bologna, Italy: Il Mulino.

Ching, C. L., & Yang, J. (2000). Knowledge value chain. *Journal of Management Development*, *19*(9), 783–793. doi:10.1108/02621710010378228

Cohen, J. A. (2005). *Intangible assets: valuation and economic benefit*. Hoboken, NJ: John Wiley & Sons.

Cucculelli, M., & Bettinelli, C. (2015). Business models, intangibles and firm performance: Evidence on corporate entrepreneurship from Italian manufacturing SMEs. *Small Business Economics*, *45*(2), 329–350. doi:10.100711187-015-9631-7

Daskalakis, N., & Psillaki, M. (2008). Do country or firm factors explain capital structure? Evidence from SMEs in France and Greece. *Applied Financial Economics*, *18*(2), 87–97. doi:10.1080/09603100601018864

De Bruyn, R. (2005). Le ragioni delle Banche Popolari: motivi teorici ed evidenze empiriche. Genova, Italy: Disefin.

De Laurentis, G. (2001). *Internal ratings and credit risk management*. Rome, Italy: Bancaria Editrice.

De Laurentis, G. (2011). *Business credit after the crisis*. Rome, Italy: Bancaria Editrice.

DeAngelo, H., & Masulis, R. W. (1980). Optimal capital structure under corporate and personal taxation. *Journal of Financial Economics*, *8*(3), 3–29. doi:10.1016/0304-405X(80)90019-7

Degryse, H., Goeij, P., & Kappert, P. (2010). The impact of firm and industry characteristics on small firms capital structure. *Small Business Economics*, *38*(4), 431–447. doi:10.100711187-010-9281-8

Degryse, H., & Ongena, S. (2002). Bank-firm relationships and international banking markets. *International Journal of the Economics of Business*, *9*(3), 401–417. doi:10.1080/13571510210000010000A

DeYoung, R., Glennon, D., & Nigro, P. (2008). Borrower-lender distance, credit scoring, and loan performance: Evidence from informational-opaque small business borrowers. *Journal of Financial Intermediation*, *17*(1), 113–143. doi:10.1016/j.jfi.2007.07.002

DeYoung, R., Hunter, W. C., & Udell, G. F. (2004). The past, present, and probable future for community banks. *Journal of Financial Services Research*, *25*(2-3), 85–133. doi:10.1023/B:FINA.0000020656.65653.79

Diamond, D. (1984). Financial Intermediation and Delegated Monitoring. *The Review of Economic Studies*, *51*(3), 393–414. doi:10.2307/2297430

Diamond, D. (1991). Monitoring and reputation: The choice between bank loans and directly placed debt. *Journal of Political Economy*, *99*(4), 689–721. doi:10.1086/261775

Dubrovski, D. (2001). The role of customer satisfaction in achieving business excellence. *Total Quality Management*, *12*(7-8), 920–925. doi:10.1080/09544120100000016

Fama, E. (1985). What's different about banks? *Journal of Monetary Economics*, *15*(1), 29–39. doi:10.1016/0304-3932(85)90051-0

Formisano, V. (2016). *Non-Knowledge Risk and Bank-Company Management. The Role of Intangibles in Rating Models*. Basingstoke, UK: Palgrave Macmillan.

Frank, M., & Goyal, V. (2003). Testing the pecking order theory of capital structure. *Journal of Financial Economics*, *67*(2), 217–248. doi:10.1016/S0304-405X(02)00252-0

Frank, M., & Goyal, V. (2009). Capital structure decisions: Which factors are reliably important? *Financial Management*, *1*(38), 1–37. doi:10.1111/j.1755-053X.2009.01026.x

Fukao, K., Miyagawa, T., Mukai, K., Shinoda, Y., & Tonogi, K. (2009). Intangible investment in Japan: Measurement and contribution to economic growth. *Review of Income and Wealth*, *55*(3), 717–736. doi:10.1111/j.1475-4991.2009.00345.x

Goddard, J., Molyneux, P., & Wilson, J. (2001). *European banking efficiency, technology and growth*. Chichester, UK: Wiley Finance.

Grice, J., & Ingram, R. (2001). Tests of the generalizability of Altman s bankruptcy prediction model. *Journal of Business Research*, *54*(1), 53–61. doi:10.1016/S0148-2963(00)00126-0

Gupta, J., Gregoriou, A., & Ebrahimi, T. (2015). *Using hazard models correctly: a comparison employing different definitions of SMEs financial distress*. Retrieved September 10, 2015, from http://eprints.brighton.ac.uk/14274/

Hall, G. C., Hutchinson, P. J., & Michaelas, N. (2004). Determinants of the capital structures of European SMEs. *Journal of Business Finance & Accounting*, *31*(5-6), 711–728. doi:10.1111/j.0306-686X.2004.00554.x

Hanousek, J., & Shamshur, A. (2011). A stubborn persistence: Is the stability of leverage ratios determined by the stability of the economy? *Journal of Corporate Finance*, *17*(5), 1360–1376. doi:10.1016/j.jcorpfin.2011.07.004

Harris, L. C., & Ogbonna, E. (2001). Leadership style and market orientation: An empirical study. *European Journal of Marketing*, *35*(5-6), 744–764. doi:10.1108/03090560110388196

Harris, M., & Raviv, A. (1990). Capital structure and the informational role of debt. *The Journal of Finance*, *45*(2), 321–349. doi:10.1111/j.1540-6261.1990.tb03693.x

Harris, M., & Raviv, A. (1991). The theory of capital structure. *The Journal of Finance*, *46*(1), 297–355. doi:10.1111/j.1540-6261.1991.tb03753.x

Heiens, R. A., Leach, R. T., & McGrath, L. C. (2007). The contribution of intangible assets and expenditures to shareholder value. *Journal of Strategic Marketing*, *15*(3), 149–159. doi:10.1080/09652540701319011

Holmes, S., & Kent, P. (1991). An empirical analysis of the financial structure of small and large Australian manufacturing enterprises. *Journal of Entrepreneurial Finance*, *1*(2), 141–154.

Hoque, Z., Mia, L., & Alam, M. (2001). Market competition, computer-aided manufacturing and use of multiple performance measures: An empirical study. *The British Accounting Review*, *33*(1), 23–45. doi:10.1006/bare.2000.0149

Hulten, C. R., & Hao, X. (2008). *What is a company really worth? Intangible capital and the market to book value puzzle*. Retrieved January 15, 2009, from https://papers.ssrn.com/sol3/papers.cfm?abstract_id=1327221

Jayaratne, J., & Wolken, J. (1999). How important are small banks to small business lending? New evidence from a survey of small Firms. *Journal of Banking & Finance*, *23*(2-4), 427–458. doi:10.1016/S0378-4266(98)00085-5

Joia, L. A. (2000). Measuring intangible corporate assets: Linking business strategy with intellectual capital. *Journal of Intellectual Capital*, *1*(1), 68–84. doi:10.1108/14691930010371636

Katcher, B. L. (2003). Make more strategies use of customer satisfaction surveys. *The Journal of Business Strategy*, *24*(1), 34–37. doi:10.1108/02756660310734893

Keasey, K., Pindado, J., & Rodrigues, L. (2015). The determinants of the costs of financial distress in SMEs. *International Small Business Journal*, *33*(8), 862–881. doi:10.1177/0266242614529317

Keasey, K., & Watson, R. (1993). *Small Firm Management: Ownership, Finance and Performance*. Blackwell Business.

Koksal, B., Orman, C., & Oduncu, A. (2013). *Determinants of capital structure: evidence from a major emerging market economy*. Retrieved July 19, 2013, from https://mpra.ub.uni-muenchen.de/48415/

Lenz, R. T. (1981). Determinants' of organizational performance: An interdisciplinary review. *Strategic Management Journal*, *2*(2), 131–154. doi:10.1002mj.4250020204

Lim, S. C., Macias, A. J., & Moeller, T. (2014). *Intangible assets and capital structure*. Retrieved October 25, 2014, from https://papers.ssrn.com/sol3/Papers.cfm?abstract_id=2514551

Lucas, D. J., & McDonald, R. L. (1990). Equity issues and stock price dynamics. *The Journal of Finance*, *45*(4), 1019–1043. doi:10.1111/j.1540-6261.1990.tb02425.x

Mac An Bhaird, C. (2010). *Resourcing Small and Medium Sized Enterprises: A Financial Growth Life Cycle Approach*. Berlin, Germany: Springer Science & Business Media. doi:10.1007/978-3-7908-2399-8

Marrocu, E., Paci, R., & Pontis, M. (2012). Intangible capital and firms' productivity. *Industrial and Corporate Change*, *21*(2), 377–402. doi:10.1093/icc/dtr042

McNair, C. J., Lynch, R. L., & Cross, K. F. (1990). Do financial and non financial performance measures have to agree? *Management Accounting Research*, *75*(5), 28.

Michaelas, N., Chittenden, F., & Poutziouris, P. (1999). Financial policy and capital structure choice in UK SMEs: Empirical evidence from company panel data. *Small Business Economics*, *12*(2), 113–130. doi:10.1023/A:1008010724051

Modigliani, F., & Miller, M. H. (1958). The Cost of Capital, Corporation Finance and the Theory of Investment. *The American Economic Review*, *48*(3), 261–297.

Modina, M. (2015). *Credit Rating and Bank-Firm Relationship: New Models to Better Evaluate SMES*. Basingstoke, UK: Palgrave Macmillan. doi:10.1057/9781137496225

Modina, M., Quintiliani, A., & Arnone, M. (2013). Il ruolo delle banche locali e del relationship lending in scenari economici in profondo cambiamento. *Arethuse Rivista di studi economico-gestionali, 2*, 91-106.

Myers, S. C. (1977). Determinants of corporate borrowing. *Journal of Financial Economics*, *5*(2), 147–175. doi:10.1016/0304-405X(77)90015-0

Myers, S. C. (1984). The capital structure puzzle. *The Journal of Finance*, *39*(3), 575–592. doi:10.2307/2327916

Myers, S. C., & Majluf, N. S. (1984). Corporate financing and investment decisions when firms have information that investors do not have. *Journal of Financial Economics*, *13*(2), 187–221. doi:10.1016/0304-405X(84)90023-0

Nance, D. R., Smith, C. W. Jr, & Smithson, C. W. (1993). On the Determination of Corporate Hedging. *The Journal of Finance*, *48*(1), 267–284. doi:10.1111/j.1540-6261.1993.tb04709.x

Neely, A. (1999). The performance measurement revolution: Why now and what next? *International Journal of Operations & Production Management*, *19*(2), 205–228. doi:10.1108/01443579910247437

Ohlson, J. (1980). Financial ratios and the probabilistic prediction of bankruptcy. *Journal of Accounting Research*, *18*(1), 109–131. doi:10.2307/2490395

Opler, T., & Titman, S. (1994). Financial Distress and Corporate Performance. *The Journal of Finance, 49*(3), 1015–1040. doi:10.1111/j.1540-6261.1994.tb00086.x

Özmen, A., Weber, G. W., & Batmaz, I. (2010). The new Robust CMARS (RCMARS) method. In *Proceedings of 24th Mini EURO Conference* (vol. 12, pp. 362-368). Izmir, Turkey: Izmir University of Economics Press.

Özmen, A., Weber, G. W., Çavuşoğlu, Z., & Defterli, O. (2012). The new robust conic GPLM method with an application to finance: Prediction of credit default. *Journal of Global Optimization, 56*(2), 233–249. doi:10.100710898-012-9902-7

Parsons, C., & Titman, S. (2009). Empirical capital structure: A review. *Foundations and Trends in Finance, 3*(1), 1–93. doi:10.1561/0500000018

Pettit, R. R., & Singer, R. F. (1985). Small business finance: A research agenda. *Financial Management, 14*(3), 47–60. doi:10.2307/3665059

Pindado, J., Rodrigues, L., & Torre, C. (2008). Estimating financial distress likelihood. *Journal of Business Research, 61*(9), 995–1003. doi:10.1016/j.jbusres.2007.10.006

Psillaki, M. (1995). Credit rationing and small and medium sized firms: A tentative of linkage. *Revue Internationale PME, 8*(3-4), 67–90. doi:10.7202/1008359ar

Psillaki, M., & Daskalakis, N. (2009). Are the determinants of capital structure country or firm specific? *Small Business Economics, 33*(3), 319–333. doi:10.100711187-008-9103-4

Quintiliani, A. (2017a). Financial Crisis and Local Banks: a Review of Bank-Firm Relationships in Italy. *Corporate Ownership & Control, 14*(3-1), 249-258.

Quintiliani, A. (2017b). Financial distress cost of Italian Small and Medium Enterprises: A predictive and interpretative model. *The Journal of Risk Finance.*

Rajan, R. G., & Zingales, L. (1995). What do we know about capital structure? Some evidence from international data. *The Journal of Finance, 50*(5), 1421–1460. doi:10.1111/j.1540-6261.1995.tb05184.x

Ross, S. A. (1977). The determinants of financial structure: The incentive-signaling approach. *The Bell Journal of Economics, 8*(1), 23–40. doi:10.2307/3003485

Roulstone, D. T. (2011). Discussion of intangible investment and the importance of firm-specific factors in the determination of earnings. *Review of Accounting Studies, 16*(3), 574–580. doi:10.100711142-011-9149-4

Schwarz, J., Beloff, B., & Beaver, E. (2002). Use of sustainability metrics to guide decision-making. *Chemical Engineering Progress, 98*(7), 58–63.

Scott, J. H. Jr. (1977). Bankruptcy, secured debt, and optimal capital structure. *The Journal of Finance, 32*(1), 1–19. doi:10.1111/j.1540-6261.1977.tb03237.x

Smith, C. W., & Stulz, R. M. (1985). The Determinants of Firms' Hedging Policies. *Journal of Financial and Quantitative Analysis, 20*(4), 391–404. doi:10.2307/2330757

Sogorb-Mira, F. (2005). How SME uniqueness affects capital structure: Evidence from A 1994–1998 Spanish data panel. *Small Business Economics*, *25*(5), 447–457. doi:10.100711187-004-6486-8

Stein, J. C. (2002). Information production and capital allocation: Decentralized versus hierarchical firms. *The Journal of Finance*, *57*(5), 1891–1921. doi:10.1111/0022-1082.00483

Stulz, R. M. (1984). Optimal hedging policies. *Journal of Financial and Quantitative Analysis*, *19*(2), 127–140. doi:10.2307/2330894

Takats, E. (2004). *Banking consolidation and small business lending*. Retrieved December 13, 2004, from https://papers.ssrn.com/sol3/papers.cfm?abstract_id=601027

Titman, S., & Wessels, R. (1988). The determinants of capital structure choice. *The Journal of Finance*, *43*(1), 1–19. doi:10.1111/j.1540-6261.1988.tb02585.x

Udell, G. F. (2008). What's in a relationship? The case of commercial lending. *Business Horizons*, *51*(2), 93–103. doi:10.1016/j.bushor.2007.10.005

Unger, B. (2000). Innovation Systems and Innovative Performance: Voice Systems. *Organization Studies*, *21*(5), 941–969. doi:10.1177/0170840600215005

Uzzi, B., & Gillespie, J. J. (1999). What small firms get capital and what cost: Notes on the role of social capital and banking networks. *Proceedings of Federal Reserve Bank of Chicago*, *778*, 413–444.

Van Der Wijst, D. (1989). *Financial structure in small business, theory, tests and applications*. Berlin, Germany: Springer-Verlag. doi:10.1007/978-3-642-45656-5

Watson, R., & Wilson, N. (2002). Small and medium size enterprise financing: A note on some of the empirical implications of a pecking order. *Journal of Business Finance & Accounting*, *29*(3-4), 557–578. doi:10.1111/1468-5957.00443

Weber, G. W., Batmaz, I., Köksal, G., Taylan, P., & Yerlikaya-Özkurt, F. (2012). CMARS: A new contribution to nonparametric regression with multivariate adaptive regression splines supported by continuous optimization. *Inverse Problems in Science and Engineering*, *20*(3), 371–400. doi:10.1080/17415977.2011.624770

Whitaker, R. (1999). The early stages of financial distress. *Journal of Economics and Finance*, *23*(2), 123–132. doi:10.1007/BF02745946

Zmijewski, M. (1984). Methodological issues related to the estimation of financial distress prediction models. *Journal of Accounting Research*, *22*, 59–82. doi:10.2307/2490859

Chapter 6
The Effects of Behavioral Factors on the Creditworthiness of Small-Scale Enterprises

Dmitry Shevchenko
Southern Federal University, Russia

Ellah Igoche Godwin
Nürtingen-Geislingen University, Germany

ABSTRACT

This chapter uses the relationship between behavioral factors and the creditworthiness of small-scale enterprises to increase access of SMEs to credit facilities. The inability of several small businesses to secure loans cannot be overemphasized. Heuristics affecting entrepreneurs are explained in this chapter, and a regression model showing the dependence of creditworthiness on behavioral factors is proposed. If banks consider using psychometric tools in testing for creditworthiness of small-scale entrepreneurs, access to credit facilities will be significantly increased and businesses will flourish. Regressesion models such as the one explained in this chapter may be imbedded in psychometric tools to enhance creditworthiness testing and improve the quality of loans that banks give.

INTRODUCTION

One of the major challenges faced by small and medium scale enterprises is the inability to secure loans from commercial banks because these banks fear small scale entrepreneurs will not meet the terms and conditions of the loans when due, and thus default. Banks have devised various methods to determine the creditworthiness of small businesses. However, one cannot state categorically that these methods are without flaws. Most decisions in small businesses are made by the proprietors and it will not be out of point to assume that the financial progress of the business depends on the owners and decisions made by them.

Entrepreneurs can make suboptimal, not totally rational decisions that may undermine the financial result of their businesses. Individual decision making is a process affected by behavioral factors. For

DOI: 10.4018/978-1-5225-3767-0.ch006

the purpose of this research, behavioral factors shall refer to all cognitive and psychological factors that reduce the rationality of an entrepreneur's decision (anchoring, risk aversion, framing etc.) They also include heuristics which are systematic methods based on psychological factors adopted in the process of decision making. If banks can properly measure the effects of an entrepreneur's behavioral factors on the financial results of his business, they will most likely predict the creditworthiness of his business more accurately. Hence reduce the risk of giving out bad loans. The transdisciplinary task of combining economics with psychology and sociology is a challenging but surmountable one.

This research seeks to find the relationship between behavioral finance and the creditworthiness of small scale enterprises

BACKGROUND

This topic was chosen because in the study of behavioral finance, few researchers have sought to find out the underlying relationship between the behavior of small scale entrepreneurs and the financial success of the enterprise they control. Behavioral factors in economic decision-making are well described and studied in the works of Fox and Tversky (1995), Genesove and Mayer (2001), Heath and Tversky (1991), Hirshleifer (2001), etc. Some scientists have studied the effect of individual behavioral and psychological characteristics of people in the decision-making processes in organizations (Fraedrich and Ferrell, 1992). Quantitative studies of the effect of behavioral factors on economic variables can be found in the works of Bharat Singh Thapa (2013), Evstigneev et al. (2013). Currently, the authors of this paper are not aware of any empirical study on the analysis of the creditworthiness of small businesses with the inclusion of behavioral aspects of economic decision-making.

MAIN FOCUS OF THE CHAPTER

The management of a small business is significantly different from that of a large corporation. Besides the existence of differences in the number of staff employed in the company, there are a number of peculiarities to small business finance that make them unique. One is that most financial and strategic decisions are made by a single individual who is most likely the owner of the business.

This research shall find out how credit worthy small businesses are, then compare creditworthiness to business owner's rationality. In determining the creditworthiness of a borrower, financial condition, financial stability, level of profitability, working capital turnover, performance, liquidity, production plans, the amount and terms of overdue loans amongst other indicators are considered.

It can be argued that small scale businesses contribute immensely to the socioeconomic and political development in any given country. In terms of living standards and social status, people affiliated to small scale businesses make up majority of the population in developing countries. They occupy a special niche in society by virtue of the fact that they act as both direct producers and consumers of goods and services. Small businesses operate mainly in national markets and are directly related to mass consumer of goods and services. Their small size, level of technological advancement, production methods and management flexibility allows them to respond to the changing market conditions in a timely manner.

An important feature of small business finance is the small amount of financial resources needed to develop the company. As a rule, the owner of a small enterprise must make financial decisions quickly and any wrong decision can destroy the company as they mostly have not insurance cover.

Additionally, financial reports provided by small scale enterprises to government agencies, is also different from the financial statements of large enterprises. Many small businesses have the right to a simplified tax regime, which allows them to pass fewer documents to the tax authorities so, they have very simplified financial reporting method.

Most small businesses do not have access to long-term credit facilities and they have a seasonally oriented business model. Most private entrepreneurs are more worried about how to "stretch" production to the next season, rather than about how to purchase modern equipment or even introduce more advanced technologies

One of the main differences between small scale enterprises and big companies lies in the means of financing. In small firms, it involves the use of, first and foremost, internal sources such as company revenue, personal funds, family, friends and fools. In well-developed market conditions, there exist another source- 'bank loans' provided through special programs supported by the government or by banks.

Access to financial and money markets is extremely difficult for small scale enterprises and in most cases, it is not possible for them to obtain a loan, not only because of its high cost, but also because of their risky financial position lack of credit history. In such circumstances, small businesses are left to rely only on the talent of the leader, his agility and ingenuity in using available resources.

The value of small businesses is not publicly evaluated by the financial markets (capitalization), and it is very difficult to determine their true and fair value and as a result, financial institutions find it more difficult to transact with them.

It can be argued that the success or failure of a small business depends significantly on the well-being and even the fate of the owner of a small business which is not good news for banks. What happens if the owner dies?

In general, the paradoxical combination of the greatest risk and the least profitability accompanies a small-scale enterprise at the first stages of its existence. Therefore, securing financing (bank loans, private equity capital) is very difficult for small scale companies and behavioral finance might just be able to solve this problem.

Most financial agencies today determine if a client is creditworthy or not by carrying out thorough financial analysis of the balance sheet, P&L, credit history of the clients. Also, availability of collateral plays a significant role in the determination of creditworthiness.

For this research, creditworthiness shall be defined as the willingness and ability to payback a loan without contravening any section of the loan agreement. For an enterprise to be considered creditworthy, it should be able to payback its loans completely plus interest within the time frame stipulated in the loan agreement without going bankrupt.

It is common knowledge that most small-scale enterprises do not have a well-defined credit history as they may never have received loans in the past, they also lack collateral, robust balance sheets and a P&L that shows liquidity. However, they may have the capacity to run a business properly and make profit if funds are made available to them but banks will not give loans to them for fear that they will default. This is caused primarily because the method used to test for creditworthiness currently does not take into consideration small 'garage' startups.

Traditional finance theory generally sees all financial agents including small scale entrepreneurs as rational egoistic beings 'Homo Economicus' who is not affected by any form of irrationality

However, in the field of finance, as well as in all other areas of human activities, people make decisions under the influence of stereotypes, illusions of perception, biased preconceived notions, wrong information and heuristics, which all together have a strong influence on the decision-making process of financial agents.

Behavioral finance is a relatively new area of research that seeks to combine behavioral and cognitive psychological theories with traditional economic and financial management theory to provide an explanation to why people make irrational financial decisions.

Numerous studies in recent years have confirmed that, when acting in uncertain and risky conditions, people are influenced by several illusions, emotions, misperception of information and other "irrational" factors. If people are affected by irrationality and the level of irrationality can be measured just like creditworthiness can be measured, then, this research shall explore means of establishing a model to show a relationship between creditworthiness and these psychological factors.

In financial theory, it is generally accepted that all market agents behave rationally but the theory of behavioral finance has a skeptical view of this assumption and tries to look at the behavior of economic agents somewhat differently.

The efficient market hypothesis (EMH) was the cornerstone of classical financial theory for four decades. According to the various hypothesis of the efficient market theory, market prices always fully reflected by available information (Fama, 1970, pp.387-414). The practical importance of EMH is that investors cannot consistently "beat the market". If investors consistently make above average returns, it is viewed by proponents of EMH as mere luck. EMH is based on the assumptions that all investors in the market are rational.

EMH was initially supported by a significant number of researchers. Indeed, in 1978, Michael Jensen declared triumphantly that there is no other assumption in economics that has stronger empirical evidence supporting it than EMH. But in recent years, more and more theoretical and empirical evidence are challenging the two main assumptions of EMH: human rationality and arbitrage effectiveness. The rise of behavioral finance has provided evidence against human rationality in finance.

In reality investors react to unnecessary information, or they react excessively or insufficiently to fundamental information. Many investors trade because of "noise". Traders who trade because of "noise" follow the advice of experts and price models. Thus, market prices are distorted. Asset prices, especially stock prices, are unlikely to reflect fundamental information about expected earnings. As a result, the importance of market prices in the adoption of corporate decisions is significantly reduced (Charest and Guy, 1978, pp. 297-330).

The most well-known financial theories that speak of the rationality of a financial agent are, in addition to the hypothesis of Efficient Markets Hypothesis (EMH), the portfolio theory of G. Markowitz, Capital Asset Pricing Model – CAPM (William, 1964, pp. 425-442) and the Option Pricing Model (OPM) by Black F and Scholes M. (1973, pp. 637-654).

Besides human rationality and efficient markets, the fractal approach still exists. fractal refers to a geometric shape which may be separated into its constituent parts, and every part can be reduced to a scale version of the whole. In finance, this concept is not a rootless abstraction but a theoretical reformulation of a down-to-earth bit of market activities. The fractal approach posits that, the movements of a stock or tradable paper all look alike when the size of a market chart is magnified or reduced so that it fits the same time and price scale. The fractal markets hypothesis (FMH), on contrary, is based on empirically observed characteristics of the financial markets and thus considers the market to consist of heterogenous agents who react to the inflowing information with respect to their investment horizon. So also, we may consider the market to be made up of heterogeneous agents who are affected by varying psychological factors and as such they should not be regarded as a bunch of logical beings.

Critics of the behavioral approach, one of which is Eugene Fama, typically support the efficient market hypothesis. They contend that behavioral finance theory is just a collection of anomalies and not an actual sphere of financial science. These critics posit that these anomalies either quickly disappear from the market without any significant consequences or are explained by the structure of the market. On the other hand, proponents of behavioral finance posit that the anomalies studied in behavioral finance are not mere isolated cases that the market forces can take care of as some investors have systematically used these anomalies to gain huge returns from markets. These anomalies have been found to distort market efficiency. More studies on behavioral finance are required to properly establish its place in science.

An important feature of small business finance is the small amount of money required to carry out operations. As a rule, the proprietor of small enterprises makes quick decisions about the expenditure of limited available financial resources under uncertain conditions and any wrong decision made is capable of destroying the company.

The financial reports provided by small scale enterprises to the government (eg tax office), also differs from the financial statements of large and listed enterprises. Many small businesses have the right to a simplified taxation regime, which allows them to submit less detailed documents to the tax authorities and significantly simplify the calculation of taxes. As a result of the less details in their financial statements, lending housese find it difficult tohave a true and fair view on the state of their finance.

Most small businesses cannot finance operations with long-term investment, which as their business model is seasonally oriented and they are more worried about how to "stretch" to the next season, rather than about how to purchase new equipment or introduce more advanced technologies

A major difference between small businesses and big companies is the procedure of financing which involves the use of first and foremost, internal sources such as the entrepreneur's personal funds, and funds from family or friends. It is noteworthy that these sources are highly unreliable as they have no structure and can hardly ever be regulated by the existing laws. Friends and Family may decide to withdraw their funding from the business at any time leading to an untimely death of business operations. In developed market conditions, there is another source of capital for small businesses "bank loan". This may be provided directly from commercial bank to the entrepreneur after proper creditworthiness testing or through special programs supported by the government in partnership with specialized financial institutions to strengthen the growth of small businesses.

Access to financial and money markets for small businesses is extremely difficult if not impossible and the option of issuing shares is totally ruled out in most cases. The possibility of obtaining a loan is limited, not only because of its high cost, but also because of the risky financial position small scale enterprises. Also, the absence of a sound credit history may be a limiting factor. Most small businesses solely depend on the talent of the entrepreneur, his agility and ingenuity in sourcing for funds as financial institutions are not willing to provide funding to them.

Small enterprises usually do not have goals to maximize stock price, increase number of shares in circulation or increase dividend payout ratio. The value of small enterprises is not publicly evaluated by the financial market and thus it is very difficult to determine. However, the owners of small businesses are extremely interested in maximizing the market value of their enterprises: some dream sooner even dream to grow big and go public others consider profitable ways of exit such are M&A. On the other hand, A significant number of small scale entrepreneurs are motivated by the phrase: "I want to be my own master." They may tend to pay less attention to company growth, profit maximization or increase in company value. With such motivation, entrepreneurs usually avoid bringing in potential new owners, even if it is for the good of the company. This may have negative consequences on the company.

Small businesses are faced with a different set of risks compared to large companies. The health, well-being and even the fate of the owner of a small business is paramount. The success or failure of the business depends on him. The opportunities for diversifying the personal investment portfolio for the owner of a small business are also limited: any extra money he gets is immediately invested into his business. Thus, the owner of a small enterprise is exposed to a much greater risk than the holders of shares in the joint-stock company, which can at any time throw off shares and diversify his portfolio. Does the increased risk in small businesses pay off with an increased level of profitability?

In general, the paradoxical combination of the high risk and low profitability that accompanies small enterprises at the first stages of their existence makes it difficult to secure private financing.

Why are small enterprises less liquid?

Liquidity issues are severe in small firms. The most important objective in financial management of small businesses is the competent control of working capital: inventory, accounts receivable, cash and short-term liabilities. If the entrepreneur can manage all these, liquidity issues may be evaded. The need for effective working capital management is dictated by the difficulty in accessing capital markets by small enterprises: once there are few funds, they should all be used in such a way that each cent brings maximum economic return.

The low liquidity levels experienced by small businesses might also be related to the fact that entrepreneurs tend to take high risks with their little funds because of the general notion that it is easier to risk millions than to risk billions

Research from Russia indicates that the profitability of small firms has a variation of about 22% around the mean. In large enterprises, this figure is three times "smaller" - 7%. This signifies higher risk for small enterprises. The increased variation of financial results limits the possibility for portfolio diversification. Thus, small enterprises are forced "to put all their eggs in one basket" thereby exposing themselves to enormous risk of not just loss, but loss that cannot be compensated for by other investments, because these other investments simply do not exist. As for financial risk, when small businesses mange to obtain loans, they obtain mostly short-term liabilities which put their business in strain as they have to pay back in a short period of time.

The main method of financial analysis in small enterprises is the coefficient method. It is necessary to analyze the solvency, liquidity and profitability ratios to be able to have a true and fair view of the company. However, the data presented by small business are always in doubt as they are not strongly regulated by accounting principles. The business and market operations of the enterprise can also be scrutinized to have a fair view of the company.

The problem of reliability of accounting data and reporting leads to thoughts about the culture and ethics of business relations and taxation. It can be argued that this is a moral issue and it shall not be discussed in this paper. Researchers have established that careless, improper accounting and reporting accurately point to poor financial management skills. Indeed, it is impossible to make rational financial decisions based on distorted data or the absence data. On the other hand, however, this rule does not work: excellent accounting and reliable reporting is not a guaranteed evidence of the high professionalism in financial management. The quantitative values of the financial ratios calculated for small businesses usually indicate an increased entrepreneurial and / or financial and other risks faced by these enterprises. Therefore, the coefficient (quantitative) analysis should be accompanied by a qualitative study of the nature of the risks. It is necessary to answer at least three vital questions for small businesses:

1. Is the clientele sufficiently diversified? A large enterprise with numerous customers is unlikely to be knocked out of the saddle by the loss of one or two large buyers. For a small enterprise, this same scenario will be a disaster;
2. Is the range of products (services) diversified enough? Orientation to monotonous production brings entrepreneurial risk to a dangerous rate, since any slight deterioration in the market situation can plunge a small company into the loss-making zone. Moreover, in the context of the concept of the effect of the operating leverage, the margin of profitability and the "financial strength" of an enterprise, the orientation on one "star" product is always not justified by the lowering of the profitability threshold and the strengthening of financial stability, because fixed costs are not spread into several commodities, but put a heavy burden on one product. The prospects of increasing revenue from one product are often very problematic because of limited demand.
3. Will the business continue its operations in the event of a sudden loss of the "soul and brain (current manager or owner)" of the enterprise? Or will the business die with him, causing loss profits for creditors, suppliers and other co-owners.

Financial analysis of small businesses should be about assessing the viability of the business - an assessment in which the results are of interest not only to the owners of the enterprise, but also to its creditors. In addition to the calculation of the coefficients, Western bankers usually conduct an analysis of the activities of a small business in five main parameters (such a rating system in the US is called the "five C system"):

1. **Personality of the Key Figure of the Enterprise (Character):** Does the trustee (director) deserve trust? What is his business reputation? Does he always fulfill his obligations?
2. **Capacity to Manage:** Can the entrepreneur ensure quality management and lead his enterprise to success?
3. **Cash Flows:** Is the company able to generate enough cash to repay the loan and interest?
4. **Assets That Can Serve as Collateral (Collateral):** What is the cost, structure, quality, liquidity of assets owned by the enterprise? Can they serve as collateral? Is it possible, if necessary, to compensate the creditor for his loss of this loan using the collateral provided?
5. **Conditions of Business:** Does the general economic, industry, market and tax situation favor the success of the enterprise? If no, it might be a case of swimming against the tides.

When applying for a loan, the head of the enterprise should not only clearly define the target use of the loan (how will the money be spent?), But also be prepared to respond to the bank's questions such as below:

- Why is additional money is required? Why is the company not generating enough funds for its activities currently? Indeed, it is one thing if the loan is objectively necessary for increasing turnover, for expanding the business when it is having profitability or liquidity challenges, and quite another - if it involves patching holes in the enterprise budget, generated by unprofessionalism, wastefulness or negligence. How will the loan and interest be?
- Assuming revenue growth is expected, then how realistic is the sales forecast? If sales are subject to seasonal fluctuations, then is the loan repayment period consistent with the time of cash inflow to the enterprise? How efficiently and effectively are accounts receivable, and inventory manage-

ment? If the revenue is not subject to serious seasonal fluctuations, is the gross margin sufficient to repay the loan and interest?
- What is the biggest problem of the enterprise? It is believed that business without problems does not exist per say, and if the manager does not want to point to the main difficulties faced by the enterprise, then he shows signs of closedness towards the banker and is not trustworthy.

Most of the questions asked by the bank are not easy but is the client's duty to choose a bank that is reliable, has sufficient capital and high reputation, a ban that shows interest in the prosperity of the client's enterprise, where there is sufficient experience of the operations of small business and where they not only provide credit but also offer additional banking services. In addition, it is advisable for entrepreneurs to try to get loans from different banks, rather than from one. Firstly, because focusing on a single lender puts the borrower in exclusive dependence on this lender. Secondly, by turning to different banks, the entrepreneur raises his chances to get a larger amount of money.

Creditworthiness and Effective Business Management

A small enterprise is created for the organization of entrepreneurial activity, and its social and economic goal is to ensure that the wants of the public are satisfied and to make profits (income). The activities of enterprises include the production and sale of products and the rendering of services. An enterprise can carry out any of these activities, or simultaneously all of them.

Enterprise finance may be defined as a set of economic relations that arise in the process of the sourcing for funds, distribution and use of available funds. Enterprise finance entails all forms of financing and lending to business aimed at achieving the common goal of effective development of the enterprises. The state of finance of an enterprise is the ability of the enterprise to finance its activities and it is characterized by:

- Availability of financial resources necessary to carry out the day to day operations of the enterprise;
- Effective and efficient distribution and use of capital;
- Sound financial relationships with other legal entities and individuals;
- Solvency and financial stability.

Creditworthiness is defined as a financial condition that permits an entrepreneur to obtain a loan and return it on time without bridging any of the terms of the loan contract. When testing for the creditworthiness of enterprises, banks take into account the borrower's reputation, size and composition of assets on its balance sheet, the state of economic and market conditions and the financial stability of the enterprise. In the first stage of the analysis of creditworthiness banks study diagnostic information about the client, which includes the accuracy in paying back accounts payable, the company's development trends, motives behind loan application, the composition and size of the debts of the enterprise.

Solvency refers to the ability of an enterprise to repay its loans in the long run and still retain its ability to perform its functions without going bankrupt. Solvency plays a significant role in the decision of the bank to either grant the loan or turn down the client. However, the decision whether to grant a loan or refusal of banks needs to be based on more information. The necessary information to assess the creditworthiness should not only be the state of assets and liabilities, but also sales, profits and the psychological factors affecting the entrepreneur.

The main aim of credit analysis to determine the ability and willingness of the borrower to return the loan requested in accordance with the terms of the loan agreement. The bank in every case determines the degree of risk that it is willing to take the size of loan which can be provided under the circumstances.

Due to the fact that the companies differ greatly in the nature of their production and financial activities, to create a single, universal and comprehensive guideline for the determination of creditworthiness and the calculation of its indicators is difficult.

In determining the estimated borrower's financial condition and its stability, the profitability of production, working capital turnover, performance of production plans, the amount and timing of overdue loans, etc. should not be below a specific range according to the principles used by banks today.

A common technique of determining the degree of creditworthiness of an organization used by many banks, which evaluate short term creditworthiness based on the calculation of five coefficients is explained below:

K1: Absolute liquidity ratio is calculated as the ratio of cash and short-term investments to short-term liabilities, net of deferred income and accrued expenses;
K2: Critical liquidity ratio shows what portion of short-term obligations of the organization can be immediately redeemed for cash, funds in short-term securities, as well as receipts for payments;
K3: Current ratio, calculated as the ratio of current assets, net of deferred expenses to short-term liabilities, net of deferred income and accrued expenses;
K4: Coefficient debt to equity ratio, calculated as the ratio of equity to debt;
K5: Profitability of sales, calculated as the ratio of profit to sales revenue (net) from sales of goods

After these ratios are calculated, companies that fall within certain thresholds are pronounced creditworthy and those who do not meet the requirements are not granted loan. A disadvantage to this system is the fact that small companies with barely any operations are left out and might not be able to secure loans to start up operations or expand already existing operations. It is proposed that psychometric tools which evaluate the entrepreneur's psychology based on mathematical analysis be used to test creditworthiness so as to increase the chances of obtaining lands by those who do not have big balance sheets.

In a nutshell, commonly used methods to manage SME credit risk include (OECD, 2013b):

1. Requests for high equity contributions by prospective borrowers
2. Requirements for collateral. i.e. an asset of the borrower, the possessive right of which is provided
3. to the lender in case of default
4. Credit guarantees, whereby should the borrower default the guarantor compensates a pre-defined
5. share of the outstanding loan
6. Loan covenants, i.e. a condition imposed by the lender with which the borrower must comply in
7. order to adhere to the terms in the loan agreement. Common loan covenants include:
    a. Hazard insurance/ content insurance, under which the entrepreneur is required to keep insurance coverage on the plant/equipment or inventory in order to safeguard against the catastrophic loss of collateral;
    b. Key-man life insurance, which insures the life of the indispensable entrepreneur without whom the company could not continue. The lender usually gets an assignment of the policy;
    c. Requirements for payment of taxes / fees / licenses, whereby the borrower agrees to keep those expenses up to date. In fact, failure to pay would result in the assets of the company

being encumbered by a lien (i.e. legal claim on property) from the government, which would take precedence to the one from the bank;

d. Provision of financial information on the borrower and guarantor, whereby the borrower agrees to submit financial statements for the continuing assessment by the bank;
e. Borrower prevented from taking specific actions without prior approval, such as: change in management or merger, demanding more loans, or distributing dividends.

In classical finance theory, it is a generally accepted assumption that all agents behave rationally in the market (Fama 1970). However, the theory of behavioral finance has a skeptical view of this assumption. It argues that economic agents are not always totally rational and are influenced by psychological and social factors. Behavioral finance attempts to incorporate psychology, sociology into the study of traditional finance so as to improve our understanding of the activities of economic agents.

The efficient market hypothesis (EMH) has been and is still the cornerstone of classical financial theory for more than four decades. In accordance with the various forms of this hypothesis, available information is always totally embedded in market prices (Fama, 1970)

In real life, people have limited time and cognitive resources, and they cannot optimally analyze all available information. In order to solve problems faster, people tend to formulate algorithms or mental systems in their minds. In psychology, these systems are called heuristics. They make it possible to quickly respond to certain stimuli. Heuristics refer to various methods of analysis of economic phenomena and making financial decisions based on intuition, analogies, experience, creativity, based on the features of the human mind and the human capacity to solve problems, for which formal logical solutions cannot be used intuitively. These heuristics are extremely effective in dealing with specific sets of problems. But when heuristics are used outside of their domain, they give rise to systematic errors (biases). Unfortunately, people tend to use heuristics in situations where they are not supposed to because they are unable to differentiate scenarios where heuristics will work from those where they will not work (Gilovich et al. 2002).

Since heuristics are formed naturally, they are shared by all humans and mistakes can systematically accumulate in the financial system. Numerous studies in psychology have proven the existence of a significant number of heuristics and described many kinds of systematic errors caused by them when making financial decisions. Heuristics are classified into two types: those that make decision making faster and those that make decision making seem easy (Gilovich et al., 2002).

Some heuristics are explained below:

- **Disposition Effect:** This refers to the tendency to go short on securities that have gained value ('winners') and go long of securities which have lost value ('losers'). Disposition effect may be explained by the notion that people evaluate gains and losses relative to a reference point (the initial purchase price of shares), and the tendency to assume risk when faced with possible losses, and be risk averse when a certain gain is possible (CAMERER, 1998).

Daniel et al. 2001 posit that overconfidence refers to over rating one's capability to effectively execute a given task. Individuals are said to be overconfident when they overate the quality of the skills they possess knowingly or unknowingly.

- **Regret Aversion Bias:** According to Bell 1982, individuals affected by regret aversion avoid making concrete decisions because of their fear that, in hindsight, any decision they make will be suboptimal. This bias seeks to forestall the negative feeling of regret associated with wrong decision-making.
- **Mental Accounting:** This refers to the propensity for individuals to allocate their wealth into separate accounts based on subjective reasons such as the source of the wealth or the intent for every account (Barberis and Huang, 2001)

In explaining Anchoring heuristics, Chapman and Johnson 2002 posit that human decisions should be based on relevant facts for them to be seen as rational. However, this is not always the case. The concept of anchoring refers to the propensity to attach or "anchor" our thoughts to a certain point of reference – even at times when the reference point has no logical connection to the decision being made

- **Ambiguity Aversion:** The main idea is that individuals overwhelmingly prefer to assume risk when they are aware of the odds and are more risk averse when the odds are uncertain. That is, given a choice of risks to take (eg.bets), people "prefer the devil they know" to assuming a risk where odds are not easy or not possible to calculate (Elsberg and Daniel, 1961)

Assuming data on heuristics are collected and statistically analyzed to determine the average number of heuristics that affected a single entrepreneur and the frequency with which the entrepreneurs in the sample make decisions using heuristics. Furthermore, simple statistical indicators such as modei s calculated to determine the most occurring heuristic.

The data on the financial results, leverage, credit history, liquidity, profitability of the sample is obtained and analyzed using the scoring and other methods to determine the performance and creditworthiness of individual firms (Emel et al., 2003). The relative values that show the performance of these individual firms is then compared with the frequency of the heuristics used by the company owners to create a regression model which shows the relationship between business performance and the behavior of the main decision maker in the small business. Thus, it is anticipated that the relationship between the psychological aspects of the process of decision making, the creditworthiness of small businesses and the prediction of business performance can be established. (a similar approach was applied by Bharat Singh Thapa in his survey titled "Investment Behaviour of Individual Investors in the Stock Market of Nepal").

$$P = \alpha + \beta 1H1 + \beta 2H2 + \beta 3H3 + et$$

where:

P = performance of the business,
H1 = first studied heuristic
H2 = second studied heuristic
H3 = third studied heuristic
et = error term

This model can be used by credit houses to predict the performance of a company and thus ascertain creditworthiness.

The Regression model obtained at the end of this study may be imbedded in a computer software to form the basis of a psychometric tool which may be used to test the creditworthiness of upcoming entrepreneurs who do not have either collateral or credit history. Entrepreneurs will stand a better chance to receive loans if a good psychometric tool can check the probability for them to default. The risk of giving out bad loans can be significantly reduced if the findings of this research is implemented by financial institutions. The precision of the model can be increased by constantly improving it with every new data obtained. Time is a critical factor here as it takes a long time to obtain data. If the model is adjusted with newly obtained data frequently and its does not remain static over time, creditworthiness will be predicted with more accuracy.

Some of the foreseeable advantages of inculcating the use of psychometric tools in creditworthiness testing are explained as follows.

It provides access to funding for small scale enterprises not only from banks but also from other alternative lenders because the psychometric tool will provide alternative lenders a means to properly test creditworthiness. Funds can be provided to SMEs with by alternative lenders where banks do not have the ability or are unwilling to do so. This can bring about a general increase in the volume of loans given to SMEs. According to a survey carried out by Nesta, 33% of SME borrowers from marketplace lending platforms were certain that they would not have been able to get funds elsewhere.

Alternative lenders can additionally extend the availability of credit to more risky classes by grouping them together with investors with a higher appetite for risk, especially institutional investors that ordinarily do not have access to SME otherwise. If marketplace lenders adopt the use of this tool, they will be able to provide credit to high-quality prime borrowers when traditional banks are not able to do so, because of their lack of data for assessing the risk appropriately.

Creditworthiness can typically be detected within few minutes and sound decisions can be made faster.

Also, the test process might not necessarily require a physical visit of the entrepreneur to the bank. This also reduces cost for both parties. The bank saves as it does not need to pay staff to carry out checks on the financial statements of the borrower, also, as no physical visit is required, rent costs are eroded. The entrepreneur on the other hand saves transport cost and time to travel to the bank (time is money).

The use of this tool can significantly close the gap that exist between business and consumer loans.

In lending for consumption purposes, China has been making astronomical strides. For instance, Tencent's WeChat now allows users to apply for uncollateralized loans of up to $30,000 and get a decision in under a minute. Such can be replicated for business loans.

Finally, loan applications through online platforms with psychometric tools Will increase convenience for both banks and entrepreneurs. Fewer branch visits are required, and loan applications can be completed from the entrepreneur's home or office. Typically, less paperwork, identity verification can be processed rather seamlessly with the level of advanced technology available.

LIMITATIONS AND DISCUSSION

Limitations refer to the weakness or deficiency in a certain research, some of the restrictions and downsides of this research are exposed below.

This research will require a very large smaple of respondents for its findings to be strongly suported. A solution to this limitation is to capture a large sample and repeat the research with cross sectional data severally. The area where data is obtained from should be large, the number of respondents should be increased and the experiment should be repeated severally.

Additionally, if the researchers are proponents of behavioral finance and it is possible to be bias when asking respondents interview questions and setting up the questions for the questionnaire. A solution to this limitation is to have neutral individuals who are neither proponents nor opponents of behavioral finance screen the questionnaire and possibly carry out the interview.

The Regression model obtained at the end of the comprehensive study may be imbedded in a computer software to form the basis of a psychometric tool which may be used to test the creditworthiness of upcoming entrepreneurs who do not have either collateral or credit history. Entrepreneurs will stand a better chance to receive loans if a good psychometric tool can check the probability for them to default. The risk of giving out bad loans can be significantly reduced if the findings of this research is implemented by financial institutions.

REFERENCES

Alkaraan, F., & Northcott, D. (2006). Strategic capital investment decision making: A role for emergent analysis tool? A study of practice in large UK manufacturing companies. *The British Accounting Review*, *38*, 140–173. doi:10.1016/j.bar.2005.10.003

Avery, C., & Zemsky, P. (1998). Multidimensional uncertainty and herd behavior in financial markets. *The American Economic Review*, *88*, 724–748.

Barber, B. M., & Odean, T. (2000). Trading is Hazardous to Your Wealth: The Common Stock Investment Performance of Individual Investors. *The Journal of Finance*, *55*(2), 773–806. doi:10.1111/0022-1082.00226

Barberis, N., & Huang, M. (2001). Mental accounting, loss aversion and individual stock returns. *The Journal of Finance*, *56*(4), 1247–1292. doi:10.1111/0022-1082.00367

Bell, D. (1982). Regret in decision making under uncertainty. *Operations Research*, *30*(5), 961–981. doi:10.1287/opre.30.5.961

Berger, A. N., & Udell, G. E. (2005). *A More Complete Conceptual Framework for Financing of Small and Medium Enterprises.* Policy Research Working Paper- 4481. The World Bank.

Bharat, S. T. (2013). *Investment Behaviour of Individual Investors in the Stock Market of Nepal*. Academic Press.

Black, F., & Scholes, M. (1973). The Pricing of Options and Corporate Liabilities. *Journal of Political Economy*, *81*, 637–654.

Chapman, G. B., & Johnson, E. J. (2002). Incorporating the irrelevant: Anchors in judgments of belief and value. In T. Gilovich, D. Griffin, & D. Kahneman (Eds.), Heuristics and biases: The psychology of intuitive judgment (pp. 120–138). Cambridge, UK: Cambridge University Press.

Charest, G. (1978). Dividend Information, Stock Returns and Market Efficiency-II. *Journal of Financial Economics*, *6*(2-3), 297–330. doi:10.1016/0304-405X(78)90033-8

Claessens, S., Honohan, P., & Rojas-Suarez, L. (2009). *Policy Principles for Expanding Financial Access*. CGD. Available at www.cgdev.org/files/1422882_file_Financial_Access_Task_Force_Report_FINAL.pdf

Daniel, K., Hirshleifer, D., & Subrahmanyam, A. (2001). Overconfidence, arbitrage and equilibrium asset pricing. *The Journal of Finance*, *56*(3), 921–965. doi:10.1111/0022-1082.00350

Ellsberg, D. (1961). Risk, Ambiguity, and the Savage Axioms. *The Quarterly Journal of Economics*, *75*(4), 643–669. doi:10.2307/1884324

Emel, A. B., Oral, M., Reisman, A., & Yolalan, R. (2003). A credit scoring approach for the commercial banking sector. *Socio-Economic Planning Sciences*, *37*(2), 103–123. doi:10.1016/S0038-0121(02)00044-7

Evstigneev, I. V., Amir, R., Hens, T., & Schenk-Hoppé, K. R. (2013). *Mathematical Behavioral Finance*. Academic Press.

Falkenstein, E., Boral, A., & Carty, L. (2000). *RiskCalc for Private Companies: Moody's Default Model*. Academic Press.

Fama, E. (1970). Efficient capital markets: A review of theory and empirical work. *The Journal of Finance*, *25*, 383–417.

Fama, E. (1998). Market efficiency, long-term returns, and behavioral finance. *Journal of Financial Economics*, *49*(3), 283–306. doi:10.1016/S0304-405X(98)00026-9

Fox, C., & Tversky, A. (1995). Ambiguity aversion and comparative ignorance. *The Quarterly Journal of Economics*, *110*(3), 585–603. doi:10.2307/2946693

Fraedrich, J. P., & Ferrell, O. C. (1992). The Impact of Perceived Risk and Moral Philosophy Type on Ethical Decision Making in Organizations. *Journal of Business Research*, *24*(4), 283–295. doi:10.1016/0148-2963(92)90035-A

Genesove, D., & Mayer, C. (2001). Loss aversion and seller behavior: Evidence from the housing market. *The Quarterly Journal of Economics*, *116*(4), 1233–1260. doi:10.1162/003355301753265561

Gilovich, T., Griffin, D., & Kahneman, D. (Eds.). (2002). *Heuristics and Biases: The Psychology of Intuitive Judgment*. Cambridge University Press. doi:10.1017/CBO9780511808098

Graham, J. R. (1999). Herding among Investment Newsletters: Theory and Evidence. *The Journal of Finance*, *54*(1), 237–268. doi:10.1111/0022-1082.00103

Heath, C., & Tversky, A. (1991). Preference and belief: Ambiguity and competence in choice under uncertainty. *Journal of Risk and Uncertainty*, *4*(1), 5–28. doi:10.1007/BF00057884

Hirshleifer, D. (2001). Investor psychology and asset pricing. *The Journal of Finance*, *56*(4), 1533–1597. doi:10.1111/0022-1082.00379

Hong, H., & Stein, J. (1999). A unified theory of underreaction, momentum trading and overreaction in asset markets. *The Journal of Finance*, *54*(6), 2143–2184. doi:10.1111/0022-1082.00184

Kahnemam, D., Slovic, P., & Tversky, A. (1982). *Judgment Under Uncertainty: Heuristics and Biases*. New York: Cambridge University Press. doi:10.1017/CBO9780511809477

Nesta. (2014). *Understanding Alternative Finance*. University of Cambridge. Available online at https://www.nesta.org.uk/sites/default/files/understanding-alternative-finance-2014.pdf

OECD. (2013a). SME financing: Roadmap for future work – Addendum. Selected issues and overview of past and current work on SME finance at the OECD. OECD, DAF/CMF(2013)3/ADD1.

Thaler, R., & Johnson, E. J. (1990). Gambling with the house money and trying to breakeven: The effects of prior outcomes or risky choice. *Management Science*, *36*(6), 643–660. doi:10.1287/mnsc.36.6.643

Tversky, A., & Kahneman, D. (1974). Judgment under uncertainty: Heuristics and biases. *Science*, *185*(4157), 1124–1130. doi:10.1126cience.185.4157.1124 PMID:17835457

Weber, M., & Camerer, C. F. (1998). The disposition effect in securities trading: An experimental analysis. Journal of Economic Behavior and Organization. *Journal of Economic Behavior & Organization*, *33*(2), 167–184. doi:10.1016/S0167-2681(97)00089-9

Chapter 7
Modeling the Impact of Crises on Evolution of Pension Systems

Larisa Yakimova
Yuriy Fedkovych Chernivtsi National University, Ukraine

ABSTRACT

The purpose of this chapter is twofold: to assess the relationship of the nonlinear dynamics of pension systems and economic cycles, and to develop a descriptive evolutionary model of the stages of pension systems. Hodrick-Prescott filter is used to identify cycles in the pension and economic dynamics. The study proved empirically that the evolutionary dynamics of pension systems depends on the cyclicity of national and world economies. In addition, the bifurcation points associate with big Kondratiev cycles, and the fluctuations of the indicators of pension systems correlate with the medium-term Juglar cycles. The crisis starts pension reforms. The results of this study indicate that public pension spending growth is countercyclical and coincident indicator relative to the global business cycle in 13 countries from 21 of the OECD countries studied. The amount and volatility of public pension spending depends on the basic pension model and has higher values in Bismarck-model countries.

INTRODUCTION

The degree of reduction of the financial sustainability of pension systems that invariably follows soon after major financial and economic crises depends on the depth and duration of the slump of national and global economies. Pension systems of many countries are still feeling the consequences of the global financial and economic crisis of 2008–2009. The imbalance between pension contributions and benefits, which is associated with profound declines in output and employment, is slowly recovering in the face of slow economic growth and, as a consequence, public pension spending remains high. These current challenges are compounded by both the phenomenon of "population aging" and "institutional aging" of the public redistributive pension systems known as PAYG-systems (financed on a pay-as-you-go basis).

The systemic pension crisis is long-term and compels governments to carry out not only parametric, but also structural pension reforms. Making decisions on pension reforming requires knowledge of the general patterns and the possible scenarios of the development of national pension systems, as well as

DOI: 10.4018/978-1-5225-3767-0.ch007

the degree of correlation between pension dynamics and fluctuation of the economy. The understanding which stage of the pension evolution and economic cycles the pension system is on will allow governments to make timely strategic decisions in the field of pension reforming.

The purpose of this study is twofold. First, this study will assess the relationship of the nonlinear dynamics of pension systems and economic cycles. Secondly, in this study will develop a descriptive evolutionary model of the stages of pension systems. The results of the modeling will show the degree of impact of crises on the evolution of pension systems.

BACKGROUND

Multiple research studies have examined the certain aspects of the evolution of pension systems. For instance, Orenstein (2003) treats pension reform as a political innovation and applies the diffusion of innovations theory. The study convincingly shows that international diffusion of pension systems follows the usual distribution pattern for adoption of innovations—a few countries are pioneers, followed by a steep increase in the rate of adoption, with a few laggards filling in at the end. The author examines the prospects of modern pension reforms by comparing the current spread of multi-pillar reform with the earlier diffusion of first pension system adoptions between 1889 and 1994. This approach helps to better understand the global context of pension reform, as well as the reasons and mechanisms of its spread. However, such global studies of the diffusion of pension systems and reforms are relatively few.

Lindeman et al. (2000) show how the pension systems of Eastern Europe and Central Asia responded to the new political and economic realities of the 1990s. Most of the countries in the region have replaced the state monopoly PAYG-system by a multi-pillar system that consists of a PAYG pillar and funded pillars. The authors evaluated the labor market effects and the long-term balance of pension systems and the size of the second pillar and financing the transition. However, the study only covers post-socialist countries and a short historical period.

Research in the United State (The Evolution of Public Pension Plans, 2008) found that demographic and economic pressures have led some public defined benefit (DB) plans to reexamine their plan designs, and consider blending DB and defined contribution (DC) plan features. Another research carried out by Yermo and Severinson (2010) examined the impact of the financial crisis on DB plans and proposed to develop countercyclical funding rules to enhance the sustainability and security of DB plans. However, it is important to distinguish between temporary impacts of the economic cycle on sponsor cash flows and long-term, structural changes to strengthen the plan funding status. However, the authors mainly focus only on the discussion of some regulatory initiatives that could be considered to promote procyclicality of funding rules for DB plans. It must be added that in the OECD Recommendation on Core Principles of Occupational Pension Regulation (OECD, 2009) it was noted that the "funding rules should aim to be countercyclical, providing incentives to build reserves against market downturns" (guideline 3.18).

Hinz et al. (2009, October) used World Bank's PROST1 (The Pension Reform Options Simulation Toolkit) projection model to undertake a series of "stress test" simulations that evaluate the impact of the crisis on both the short and long term financial status and benefit outcomes of several prototypical pension systems of the three regions that have been developed (Central Europe (CE), Latin America and the Caribbean (LAC), and the Middle East and North Africa (MENA)). The results of the simulation demonstrate the differing magnitude of the effect of the financial crisis on the pension systems (depth of fall and recovery time) from minimum in the LAC region to maximum in the CE region. The authors concluded that the overall effect of the crisis is likely to be less than the impact of an additional one year gain in life expectancy over the next 40 years.

Natali (2011) analyzed the impact of the economic and financial crisis on pension policy in Europe, and also identified general trends on reducing public pension spending and differences in approaches to private pension schemes (France, Sweden, and the United Kingdom vs. Central Eastern European (CEE) countries).

Many studies examine the impact of the last crisis and post-crisis pension reforms in the countries of CEE, since these are not fully formed multi-pillar pension systems. Since 2008, some CEE countries have implemented changes related to the second and third pillars. This matter is discussed by European scholars, e.g., Chybalski (2011) analyzes the consequences of these measures and argues for multidimensional diversification in mandatory pension systems and an assessment of this diversification in CEE countries. Chlon-Dominczak (2016) shows that net outcome of post-crisis pension system modifications depends both on the magnitude of fully-funded contribution reduction, but also on the design of PAYG component and the way individual pension rights are accrued. Mladen (2012) accomplished a comparative analysis of the pension systems performance based on calculating a composite indicator that includes three indicators (adequacy, financial sustainability, and modernity) to identify best practices for successful reform.

On the other hand, Sun and Hu (2014) studied the impact of pension systems on financial development. The authors constructed regression models on the data of countries of four types of pension systems (Anglo-American model, Continental model, Nordic European model, and Chilean model) from 2001 to 2008 and found that impact on financial development is less significant in the Anglo-American countries where the financial markets are well developed and pension assets are at a larger scale. To the contrary, the impact is stronger in the less development countries with smaller pension funds and less development financial markets.

Thus, without reducing the importance of researches, it should be noted that the scientists mainly assess only the impact of the recent crisis; there are no studies devoted to the study of the dependence of long-term dynamics of pension systems and cyclical economic dynamics. The formal, model side of the evolution and it is only it that can serve the basis for forecasting and selecting the best variants of the development, are avoided by the researchers.

MAIN FOCUS OF THE CHAPTER

The general theory of the evolution of complex nonlinear systems, which include socio-economic system and the pension system in particular, gives a general idea about the possible behavior of such systems. For nonlinear systems is essential bifurcational nature of their evolution, which suppose that the system developing randomly accumulates new properties quantitatively first and then moves into a new quality. The point of bifurcation has a relatively short duration and separates more lengthy sustainable modes of the system. In the context of the theory of nonequilibrium socio-economic systems pension crises should be seen as bifurcation points, i.e., the points of transition to a new state or points of evolutionary changes of the pension system and the forming of fundamentally new system with different structure and properties. At the bifurcation point the system has a choice in which there is an element of randomness, this disabled the accurate predict the further development of the system. The spectrum of possible perspectives should be formed on the basis of a study of the factors and causes of the nonlinear development of pension systems in a long-term retrospective.

In conducting this study will proceed from the following hypotheses. Hypothesis 1: The evolutionary dynamics of pension systems depends on the cyclicity of national and world economies. Hypothesis 2: The bifurcation points are associated with big Kondratiev cycles, and the fluctuations of the indicators of pension systems correlate with the medium-term Juglar cycles; the pension reforms are carried out or planned in the depression and recovery phases. Hypothesis 3: The demographic trend influences the amplitude of cycles and the power of the pension crisis.

For a correct understanding of this study of the evolution of pension systems from the perspective of the theory of economic cycles, it is necessary to determine the terminology that will be used. First, Nikolai Kondratiev (1922) dated three (incomplete) big cycles of conjunctural waves from 1789 to 1921. In each big cycle, the upswing and downswing waves were determined. Later, Joseph A. Schumpeter (1939) called these cycles "a Long Cycle or Kondratieff" and "Long Wave". Schumpeter also proposed a four-phase model of a big cycle: "We will refer to prosperity and revival as the positive phases of a cycle, to recession and depression as the negative phases" (1939, p. 156).

In addition to big cycles of about 50 years, Kondratiev (1922) revealed a small industrial-capitalist cycle 8–11 years in length. Cycles of similar length were identified by Clement Juglar in 1862 and therefore Schumpeter (1939) called these cycles as Juglar cycles. Today, the terms "cycle of medium-term fluctuations of business activity" or "business cycle" are widely used. In this study, the terms K-cycle and J-cycle (or business cycle) are used, respectively. As for the dating of post-Kondratiev and post-Juglar cycles, there is no single approach. This study is based on the scholarly papers of Grinin et al. (Grinin, Devezas, & Korotayev, 2014; Grinin, Korotayev, & Tausch, 2016).

SOLUTIONS AND RECOMMENDATIONS

The basis for testing the first hypothesis is analysis long time-series data of the annual growth rates of the world GDP from the 1870 to 2016 (Maddison, 2010; World Bank, 2017). Figure 1 displays the dynamics of world GDP growth.

However, figure 1 does not illustrate K-cycle inasmuch as the first half of the twentieth century are characterized by an enormous magnitude of fluctuations of world GDP growth rates. Therefore were calculated three-year averages (3y Avg) of the annual growth rates of world GDP. Then the Hodrick-Prescott filter (HP-filter) was applied to decompose of derived time series $Y = (y_1, y_2, \ldots, y_T)'$ into a trend $\tau = (\tau_1, \tau_2, \ldots, \tau_T)'$ and a cycle $C = (c_1, c_2, \ldots, c_T)'$. Note that formally the filter is defined as the solution to the following optimization problem for τ_t

$$S = \sum_{t=1}^{T}\left(y_t - \tau_t\right)^2 + \lambda\sum_{t=2}^{T-1}\left(\tau_{t+1} - 2\tau_t + \tau_{t-1}\right)^2 \rightarrow \min \qquad (1)$$

where the smoothing parameter λ is set fixed to a value.

Authors HP-filter used $\lambda = 1600$ for quarterly data and $\lambda = 100$ for yearly data, whereas in this study will be used $\lambda = 1600/4^4 = 6.25$. This approach proposed by Ravn and Uhlig (2002).

For a fixed λ, it can be shown that the trend component is calculated by

Figure 1. World GDP annual growth rate dynamics, 1870–2016
Source: Maddison, 2010; World Bank, 2017.

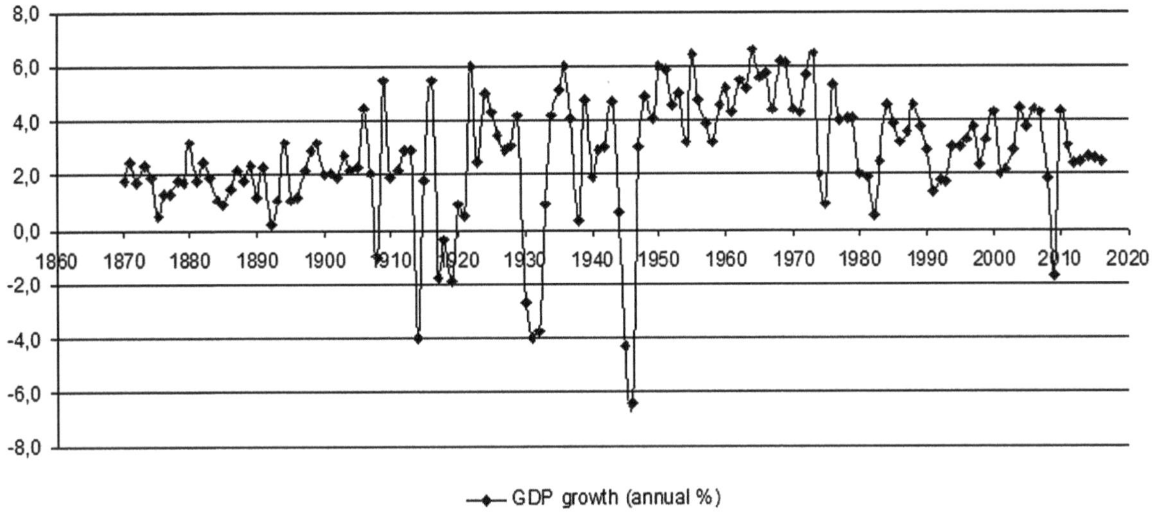

$$\hat{\tau} = \left(X'X\right)^{-1} X'Y \tag{2}$$

where X is the $T \times T$ matrix

$$X = \begin{pmatrix} 1+\lambda & -2\lambda & \lambda & 0 & \dots & 0 & 0 & 0 \\ -2\lambda & 1+5\lambda & -4\lambda & \lambda & \dots & 0 & 0 & 0 \\ \lambda & -4\lambda & 1+6\lambda & -4\lambda & \dots & 0 & 0 & 0 \\ 0 & \lambda & -4\lambda & 1+6\lambda & \dots & 0 & 0 & 0 \\ \dots & \dots & \dots & \dots & \dots & \dots & \dots & \dots \\ 0 & 0 & 0 & 0 & \dots & 1+6\lambda & -4\lambda & \lambda \\ 0 & 0 & 0 & 0 & \dots & -4\lambda & 1+5\lambda & -2\lambda \\ 0 & 0 & 0 & 0 & \dots & \lambda & -2\lambda & 1+\lambda \end{pmatrix}$$

and hence

$$\hat{N} = Y - \hat{\tau} = \left[\mathrm{I} - X^{-1}\right] \cdot Y \tag{3}$$

where I is the $T \times T$ identity matrix.

But derived time series did not show the K-cycle pattern, so the HP-filter was applied a second time to a trend component (2). Thus, fluctuations were removed from the time series and a long-term trend was found; after that, its cyclical component was highlighted. The empirical estimate of the cyclical component of the trend shows good enough a specific K-cycle pattern in the series (see Figure 2). Note

Figure 2. Correlation between K-cycles and events of pension's evolution, 1870–2016

Source: International Labour Organization [ILO], 2014; Maddison, 2010; OECD, 2013; OECD, 2015; Schwarz et al., 2014; World Bank, 2009; World Bank, 2017; own calculations.

Note 1: The introduction of pension systems (old-age pensions) in Germany (1889), Denmark (1891), New Zealand (1898), Belgium (1900), the Netherlands (1901), Austria, Slovakia (1906), Australia, the United Kingdom, Ireland (1908), Iceland (1909), France (1910), Luxembourg (1911), Romania (1912), Sweden (1913); 1[1]: Introduction of pension systems after World War I; 2: The introduction of pension systems in Canada (1927), Greece (1934), the United States, Portugal (1935), Norway (1936), Japan (1941), Switzerland (1946), Turkey (1949); reforming pension systems in France (1945) and the United Kingdom (1948); 2[1]: Parametric reforms (extension of pension rights and coverage of public pension schemes) are a typical trend; 3: Structural reforms (a move to a fully funded or multi-pillar system) in Chile (1981), the United Kingdom (1980-1986), Germany (1984), Italy (1992, 1995), Peru (1993), Colombia, Argentina, Sweden (1994), Uruguay, the Czech Republic (1995), Mexico, Bolivia (1997), El Salvador, Kazakhstan, Germany, Hungary (1998), Russia (1998, 2002), Belgium, Ukraine (2003); 3[1]: The slowing pace of reform is a typical trend; 4: Parametric and structural reforms (reduction of pension rights in public pension schemes, reduction of contribution flows into private pension savings accounts) are a typical trend.*

** Data of first law.*

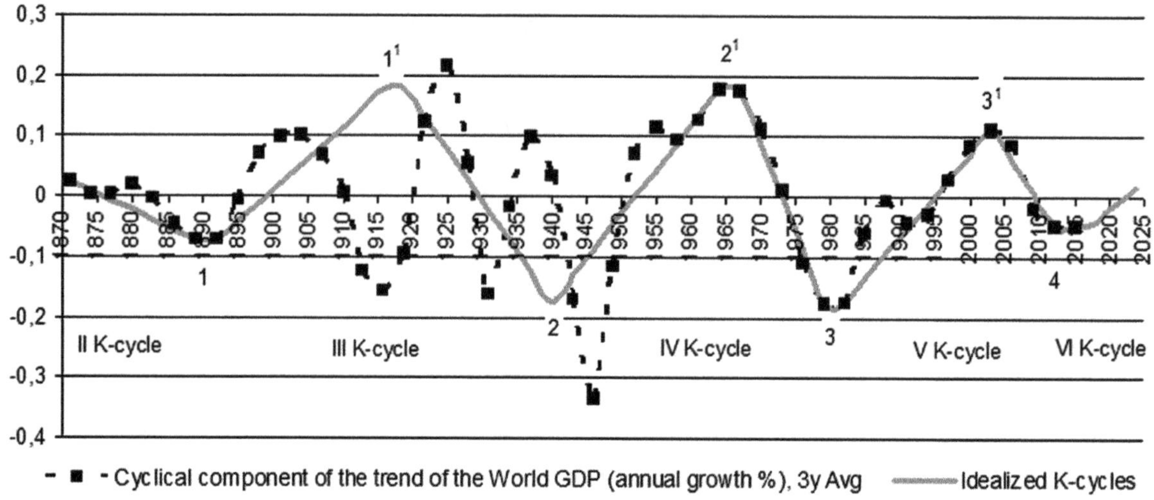

that the impact of the World War I and World War II on the global economy is not eliminated (in contrast to Grinin et al. (2014)), because the subject of the study is a comparison of actual economic cycles and the evolution of pension systems.

Based on the analysis of the evolution of pension systems, the major pension events and the K-cycles at the Figure 1 are synchronized. Historically, the first public pension system for all working population was created by the German chancellor Otto von Bismarck in 1889. Bismarck created the system of the obligatory pension provision for all working population as a redistributive PAYG-system. It means that pensions for current pensioners are financed by current workers by means of the insurance payments and their redistribution while the government guarantees financing pensions for future pensioners at the expense of future workers. In this way the principle of generation solidarity or the social contract of generations is realized.

The pension systems of the Anglo-Saxon model (now called "model of Beveridge" after the English economist William Henry Beveridge who developed the doctrine of social insurance in 1942) were introduced in Denmark in 1891 and in New Zealand in 1898. The Anglo-Saxon model is based on the principle that any person, irrespective of his economical activity, has the right for the minimum provision in his old age; it is based on the law of poor people: pensions are paid from the mutual tax income for people in need only.

Over the next 15 years countries such as Belgium (1900), the Netherlands (1901), Austria, Slovakia (1906), Australia, the UK, Ireland (1908), Iceland (1909), France (1910), Luxembourg (1911), Romania (1912), Sweden (1913) are introducing pension systems (old-age pensions, data of first law) (ILO, 2014). It is noteworthy that in subsequent years model of Bismarck was introduced in the continental law system countries (except the Netherlands) and in Finland, while model of Beveridge was introduced in the countries of the Anglo-Saxon law system, Scandinavian countries (except Finland) and the Netherlands.

On the other hand, Kondratiev (1922) defined the following scheme of conjunctural waves of big cycles for this period. Downswing wave of the II K-cycle began after the deep crisis of 1873 and continued until 1896. The new upswing wave (III K-cycle) began in 1896 and continued until 1920. The world economic crisis of 1920–1921 was the beginning of the downswing wave of the III K-cycle.

Thus, the first pension systems were introduced in the trough of II-III K-cycles or at the beginning of the upswing wave (or in the terminology of the four-phase K-cycle in the phase of depression and in the phase of recovery). This is due to the inertia of the legislative processes. It should be noted an exception in the revealed regularity. Pension systems in countries that were established (were successors or their parts) due to the collapse of the Austro-Hungarian and Russian empires after the World War I, were introduced in the 1920s, that is, at the beginning of the III K-cycle downswing wave. These are such modern countries: Italy (1919), Serbia, Slovenia, Croatia, Latvia, Lithuania, Russian Federation, Ukraine (1922), Bulgaria, Estonia (1924), Poland (1927), and Hungary (1928) (ILO, 2014). In a similar vein, it can also be interpreted as the response of governments to the world economic crisis of 1920–1921 against the background of military exhaustion.

In the next cavity of the big K-cycle, pension systems are introduced in Greece (1934),, the US, Portugal (1935), Norway (1936), Japan (1941), Switzerland (1946), Turkey (1949). Here and below, post-Kondratiev periodization of big cycles is used, which is presented in a study by Grinin et al. (2014, 2016). During the same period, the first reforms are carried out, which complicate the structure and diversify the financing of pension systems, in particular the UK (1948).

During the rapid economic growth of post-war economies and the favorable demographic structure of the 1950s and 1960s, the coverage of public pension systems and pension rights (reduction in retirement age and pension contributions, indexation and increase in pensions) expanded. This occurred on the upswing wave of the IV K-cycle and due to inertia at the beginning of the downswing wave.

But generous PAYG-systems are beginning to put pressure on public finances in the conditions of aging population, and the 1970s world energy crisis (the 1973 oil crisis and the 1979 energy crisis) exacerbated the problem. Therefore, in the depression phase of the IV K-cycle, reforms cover almost all countries. Structural reforms, characterized by a transition to fully funded or multi-pillar pension systems were carried from 1980 to the 2000s: Chile (1981), the UK (1980-1986), Germany (1984), Italy (1992, 1995), Peru (1993), Colombia, Argentina, Sweden (1994), Uruguay, the Czech Republic (1995), Mexico, Bolivia (1997), El Salvador, Kazakhstan, Germany, Hungary (1998), Russia (1998/2002), Belgium, Ukraine (2003).

The creation of a new pension model called the Chilean or Latin American model refers to this period. First in Chile, and then in other Latin America countries, a transition was made from defined benefit PAYG public systems to individual defined contribution accounts schemes. The introduction of the new pension model was implemented in different forms. The substitutive model closed the PAYG distribution scheme to new affiliates and replaced it with one based on privately-managed individual savings account plans. The parallel model preserved the PAYG-system which then competed with individual savings accounts (Tuesta, 2011). Subsequently, in Kazakhstan, the Chilean model was also adopted as the basis of the pension reform.

However, during the economic boom of the mid-2000s (V K-cycle), the pace of reforms slowed, and in rapidly aging societies, public pension spending began to increase again. Public spending on cash old-age pensions and survivors benefits in the OECD increased 27% faster than the growth in national income between 1990 and 2009, from an average of 6.1% of GDP to 7.8% (OECD, 2013).

The global financial and economic crisis of 2008-2009 have hit like a wrecking ball both on redistributive I pillar and funded II and III pillars. Therefore, in the phase of depression V K-cycle, reforms are being implemented to improve the financial sustainability of all pillars of national pension systems. According to the World Bank (World Bank, 2009; OECD, 2013) during this period, many countries have reduced pension rights in public pension programs, which manifested itself in the transition to less favorable indexation (Serbia, Hungary, Croatia, Macedonia, Austria, Greece, Portugal, Luxembourg, and Slovenia), in the raising taxes and contribution rates in DB systems (Romania, Russia, France, and Italy), in the raising the statutory retirement age (Hungary, Germany, Greece, Ireland, Italy, Spain, and Turkey), in the tightening early retirement provisions (Poland, Hungary, and Lithuania), etc. Governments are making changes to the 2nd pillar: the contributions paid to the 2nd pillar were fixed at a certain level (Bulgaria, Romania) or was reduced (Poland) or was lowered (Latvia, Lithuania,) or was moved to the 1st pillar (Estonia, Hungary). All these counter-crisis measures have made it possible to reduce public pension spending and lay the foundation for the financial sustainability of pension systems in the short term.

Based on Grinin et al. (2016), the VI K-cycle will probably begin approximately in the 2020s. There is reason to talk about slowing the process of pension reforming on its upswing wave. The question arises: how the next world crisis will affect the design of pension systems against the backdrop of a global population aging?

Thus, the analysis (see Figure 2) shows that the structural and parametric reforms that tightened pensioning rules were implemented in the troughs of global economic cycle and at the exit from them. On the other hand, during the phases of expansion and recession (due to inertia), the conditions of the pension provision improved. This shows that the evolutionary dynamics of pension systems, like economic conjuncture, is rhythmic. The period of expansion of pension rights is more or less sharply replaced by a period of "contraction" of pension rights. In a similar vein, the evolutionary process follows the path of complicating the structures of national pension systems and the convergence of basic pension models. However, as will be explained further, despite the similarity of the structures of modern multi-pillar models, differences remain, in particular in the amount of public pension spending and in the degree of participation in individual pension schemes.

Further it is expedient to investigate the relationship between the dynamics of the parameters of national pension systems and economic cycles. The study begins with the German pension system (as the oldest pension system). In this paper the contribution rate growth is used as a parameter of the pension system. The growth rates world GDP and GDP per capita in Germany are used as indicators of global and national economic cycles. The further analysis is carried out with normalized cyclical components and focuses on identification of attributes of indicators in relation to economic fluctuations and timing types of economic indicators. Figure 3 shows that take place both a coincidence of the turning points of the pension and world cycles (1916, 1925) and the leading behavior of the pension cycle (1940–1943).

Figure 4 displays that take place both a coincidence of the turning points of the pension and national economic cycles (1916, 1925) and the leading behavior of the pension cycle (1940–1943, 1964–1967).

Figure 3. K-cycles of the world economy and contribution rate in Germany (normalized cyclical components of trends of the annual growth rates), 1890–2016
Source: Federal Health Reporting Information System, 2016; Flegelskamp, 2008; Heide, 2016; Maddison, 2010; World Bank, 2017; own calculations.

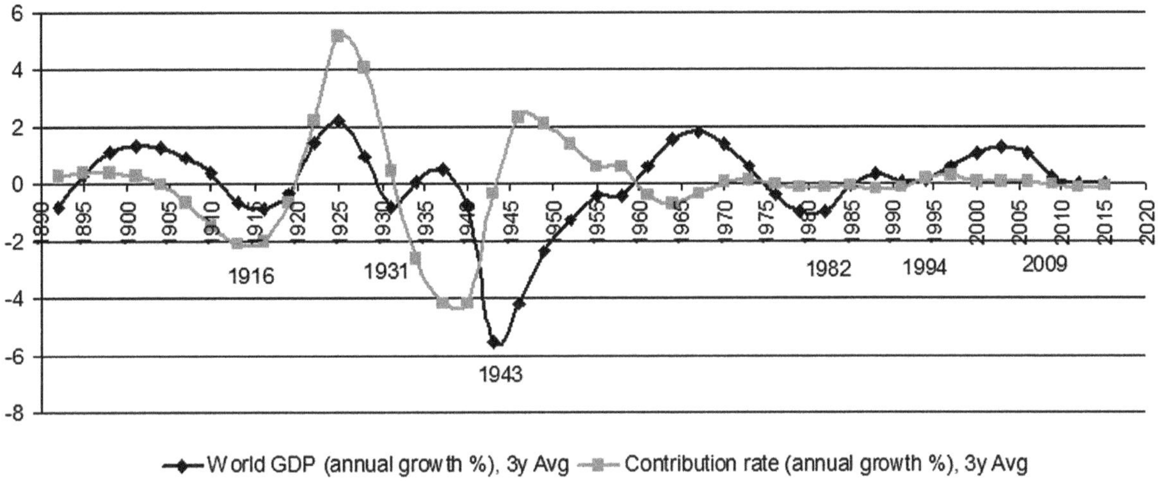

Figure 4. K-cycles and contribution rate in Germany (normalized cyclical components of trends of the annual growth rates), 1890–2016
Source: Federal Health Reporting Information System, 2016; Flegelskamp, 2008; Heide, 2016; Maddison, 2010; World Bank, 2017; own calculations.

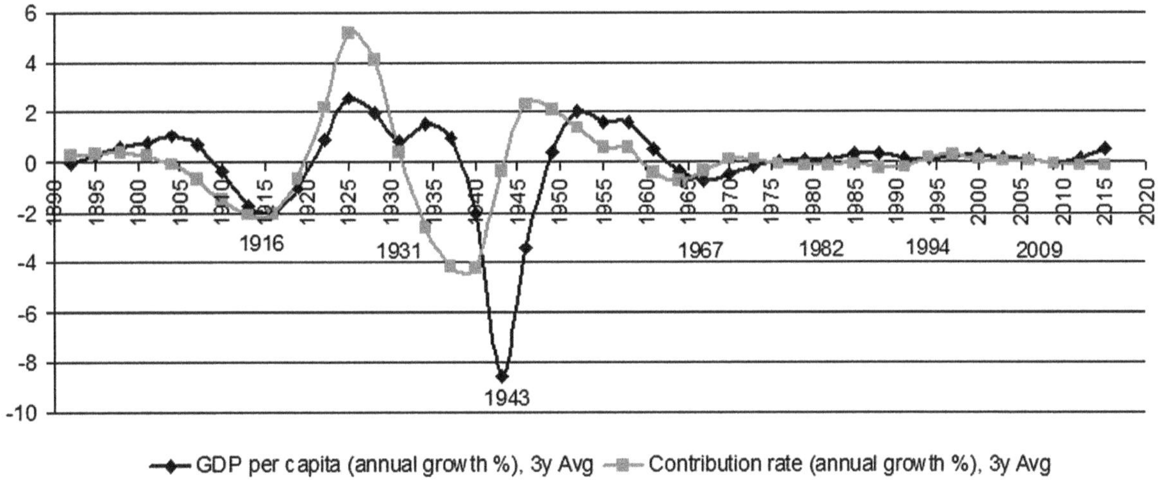

Algorithm for identifying indicators in relation to economic fluctuations is the standard testing of significance of the correlation coefficient $\rho(0) = r_{YX}$ using the Student t-test. Checks the null hypothesis H_0: $\rho(0) = 0$, i.e., the correlation coefficient is insignificant, and the indicator is acyclic. An alternative hypothesis (H_A: $\rho(0) \neq 0$) is divided into two hypotheses: H_{A1}: $\rho(0) > 0$, i.e., the correlation coefficient is significant, and the indicator is procyclical; H_{A2}: $\rho(0) < 0$, i.e., the correlation coefficient is significant, and the indicator is countercyclical. The probability of a type I error or significance level in this study is set at $\alpha \leq .05$.

The identification of timing types of economic indicators is based on the following rules: (i) if the cross-correlation coefficient reaches its maximum $\rho_{\max}\left(k\right) = \max_{k}\left\{\rho\left(k\right) = r_{Y_t X_{jt+k}}\right\}$ at $k < 0$, the X indicator is a leading indicator of Y-cycle turning points (X outstrips the Y-cycle for k periods) with a given level of significance α; (ii) if $\rho_{max}(k)$ at $k > 0$, the X indicator is a lagging indicator (X lags behind the Y-cycle for k periods); (iii) if $\rho_{max}(k)$ at $k = 0$, the X indicator is a coincident indicator.

Let's start by considering the results of evaluations relationship with K-cycles. As to indicators in relation to economic fluctuations, then the contribution rate growth is an acyclic indicator in relation to the K-cycles of the world and national economies. Table 1 shows the data in question.

One should note here that the contribution rate growth can be defined as pro-cyclical indicator in relation to the K-cycles of the national economy with 80% confidence level. However, in regard to timing types of economic indicators obtained multi-directional evaluations. The contribution rate growth is leading indicator ($k = 2$) K-cycles of national economy and lagging indicator ($k = 3$) K-cycles of world economy. The obtained estimates of timing indicators can be used to analyze the synchronicity of cycles only. When predicting turning points, it is incorrect to use $k > 1$, since time series of 3-year averages are analyzed.

The research interest is the answer to the question: are the indicators of national pension systems subject to medium-term business fluctuations? To study the statistical relationship pension cyclicity and medium-term J-cycles are analyzed the trend's cyclical components of the original time series for the period 1968–2016. HP-filter unequivocally identified medium-term Juglar cycles (J-cycles) in the dynamics of the world economy (1975–1982–1992–2002–2009) of 7–10 years (see Figure 5), and the German economy (1974–1982–1993–2003–2009) of 7 to 11 years (see Figure 6).

It should be noted that the dating of cycles of the world economy coincides with the dates of 18–21 J-cycles according to Grinin et al. (2016). As shown by Table 1, the correlation between the medium-term J-cycles of the world economy and contribution rate growth is not statistically significant. Nevertheless, contribution rate growth is countercyclical indicator in relation to the national business cycle. Figure 6 visualizes this conclusion. For example, the synchronicity of the downswing phase of the economy and

Table 1. The contribution rate growth in German pension system as an indicator of the direction / timing of world and national cycles

Economic Cycle	Year	Direction/Timing	Cross-Correlation Coefficient	Significance Level	Critical t-Value	Student t-Value
World K-cycle	1890–2016	acyclic	.0611	–	–	.3869
		lagging ($k = 3$)	.7079	.01	2.7154	6.0965
World J-cycle	1968–2016	acyclic	.2633	.10	1.6779	1.8710
		leading ($k = 2$)	.4192	.01	2.6896	3.0977
German K-cycle	1890–2016	acyclic	.2779	.20	1.3031	1.8294
		leading ($k = 2$)	.7045	.01	2.7116	6.1188
German J-cycle	1968–2016	countercyclical	–.3458	.05	2.0117	–2.5264
		leading ($k = 3$)	.5554	.01	2.6923	4.4297

Source: Federal Health Reporting Information System, 2016; Flegelskamp, 2008; Heide, 2016; Maddison, 2010; World Bank, 2017; own calculations.

Figure 5. J-cycles of the world economy and contribution rate in Germany (normalized cyclical components of trends of the annual growth rates), 1968–2016
Source: Federal Health Reporting Information System, 2016; Flegelskamp, 2008; Maddison, 2010; World Bank, 2017; own calculations.

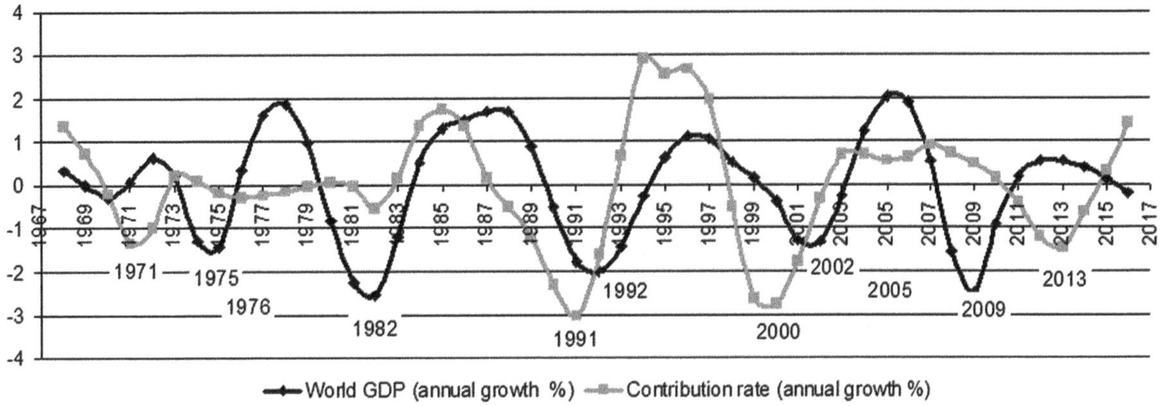

Figure 6. J-cycles and contribution rate in Germany (normalized cyclical components of trends of the annual growth rates), 1968–2016
Source: Federal Health Reporting Information System, 2016; Flegelskamp, 2008; Maddison, 2010; World Bank, 2017; own calculations.

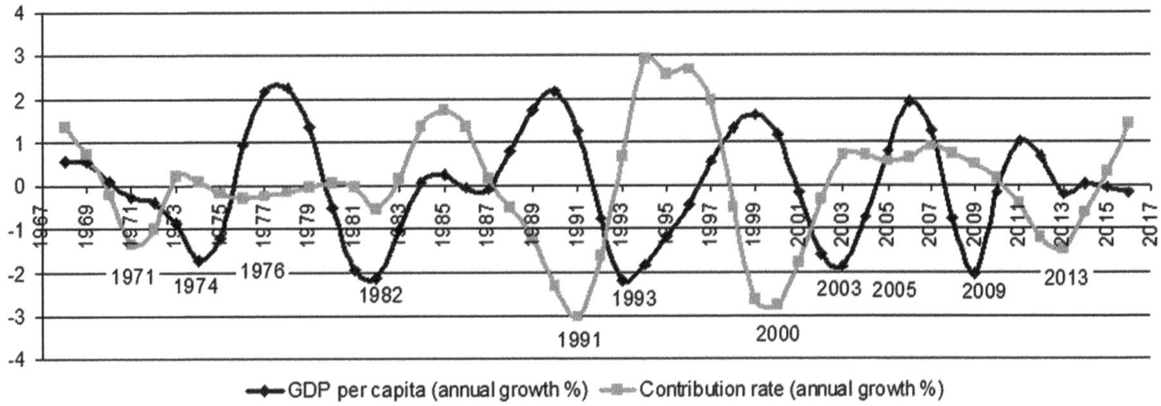

the upswing phase of the contribution rate in the periods 1978–1980, 1991–1993, 2000–2003 is strongly pronounced. There was also observed a concordance between a peak of the contribution rate growth and a trough of the national business cycle in 1993–1994 and an opposite concordance in 1990–1991 and in 1999–2000.

In regard to timing types of economic indicators the contribution rate growth is leading indicator of national and global business-cycles. This means that with a probability of .99 the contribution rate growth made to predict the turning points of business cycles (peak or trough). On the other hand, GDP per capita growth is a lagging indicator of the cyclical of the contribution rate growth. Hence GDP per capita can be used to confirm that the turning point contribution rate cycle already passed. A similar statement is true for world GDP.

In addition to the contribution rate, the parameters of the cyclicity of public pension spending as a percentage of GDP are analyzed. First of all, this concerns pension spending in the UK. Historical data are available for 1906–2016 (UK Public Spending, 2017). As above, when detecting K-cycles and J-cycles, annual growths (%) are used. The study is conducted on the basis of three-year averages to identify the relationship with K-cycles. Calculations have shown that pension spending in the UK are procyclical indicator (a significance level of .05) and leading indicator of K-cycles turning points significance level .01). Figure 7 shows that the turning points of pension spending occur before the K-cycles (1913–1916, 1922–1925, 1940–1943), but in the second half of the 20th century this pattern is violated (as in the case of Germany).

In view of this, special mention should be made of the relationship between the cycles of pension spending and medium-term J-cycles in this period. As for Germany, 1968–2016 is analyzed. The resulting maximum value of the cross-correlation coefficient –.7665 indicates a countercyclical and coincident behavior of the public pension spending UK with respect to the world business cycle (significance level .01). Figure 8 confirms the estimated behavior of the public pension spending: the peaks of the cycles of pension spending occur around the same time as the business cycle troughs and vice versa.

As for other countries, the methodology and assumptions should be common to the analysis of all countries, allowing the relationship of economic fluctuations and pension spending to be compared directly. But unfortunately, comparable data on public pension spending are available only for the period 1980–2011 and for some OECD countries in OECD data (OECD, 2017). Therefore, research is limited to these countries and the period. Table 2 shows that public pension spending growth are countercyclical and coincident indicator relative to the global business cycle (significance level is no more than .05) in 13 countries from 21 countries.

Countries that demonstrate the acyclicity of public pension spending and a multidirectional temporary relationship to fluctuations in the global economy can be divided into two groups. The first group includes the PIGS countries (Portugal, Italy, Greece, and Spain) and Turkey. The lack of correlations

Figure 7. K-cycles and public pension spending of the United Kingdom (normalized cyclical components of trends of the annual growth rates), 1906–2016
Source: Maddison, 2010; UK Public Spending, 2017; World Bank, 2017; own calculations

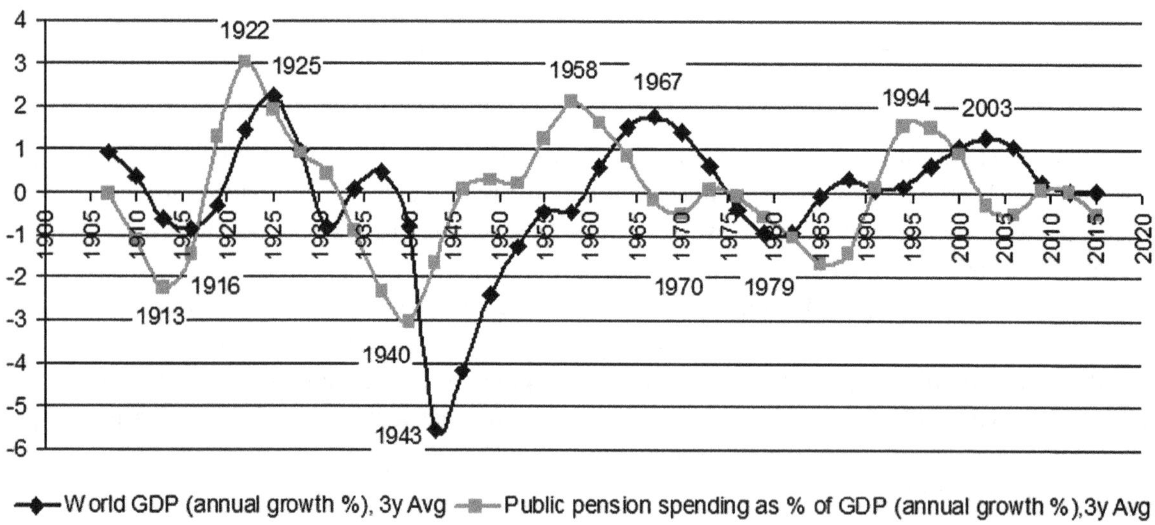

Figure 8. J-cycles and public pension spending of the United Kingdom (normalized cyclical components of trends of the annual growth rates), 1968-2016
Source: Maddison, 2010; UK Public Spending, 2017; World Bank, 2017; own calculations.

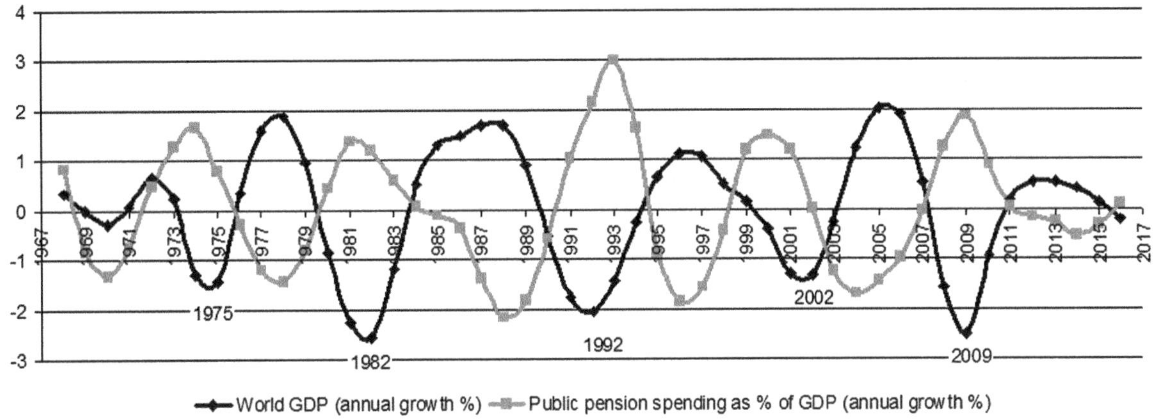

between public pension spending in PIGS countries and the "health" of the economy is explained by a long-term ineffective social policy that relies not on economic expediency, but on the populism of governments. This led to the fact that, due to the global financial crisis of 2008–2010, these countries were in fact unable to service their external obligations, including due to high pension spending. In addition, the small measures taken did not lead to a reduction in pension spending. The PIGS countries traditionally have the largest pension spending. The Pensions at a Glance (OECD, 2015) provides the following data on pension spending for 2014: 13.0% of GDP in Portugal, 15.8% in Italy, 14.5% in Greece, 10.5% in Spain (7.9% in OECD average). It should be noted that the public pension spending amount to 13.8% in France and 10.6% in Germany. These countries are united by the type of pension system (modern modification of the Bismarck model, i.e., mainly the PAYG system and undeveloped funded levels) and the problem of population aging.

The second group of countries includes Denmark and "significantly acyclic" Australia and New Zealand. These countries have a basic Beveridge model and pension spending below the OECD average (3.5% in Australia, 4.9% in New Zealand and 6.2% in Denmark). Countries with small PAYG-systems are usually considered to be under less financial stress (Finke & Sabatini, 2016). As can be seen from Table 5, the same countries are in the Top 5 of the Allianz Pension Sustainability Index (Finke & Sabatini, 2016) together with Sweden and the Netherlands, as opposed to the PIGS countries that are in the last five.

Thus, despite the convergence of pension systems in the process of evolution, the type of the basic model now determines the amount of public pension spending. Figure 9 shows that the average amount of public pension spending and the rate of their growth in the countries of the Bismarck model are higher.

The average amount of public pension spending in the Beveridge model countries increased from 5.14% in 1980 to 6.17% in 2011, while in the Bismarck model countries, from 7.33% to 11.30%. The amplitude of fluctuations in public pension spending also is also slightly higher in the Bismarck model countries, as Figure 10 shows.

Volatility of public pension spending is important in determining the impact of fluctuations of the economy and to assess the sustainability of pension systems and leverage decisions since volatility is synonymous with risk. The formal measure of absolute volatility is the standard deviation of the series

Table 2. Parameters of the cyclicity of public pension spending (annual growth %) and the Pension Sustainability Index in selected OECD countries

Country	Direction/Timing	Cross-Correlation Coefficient	Significance Level	Student t-Value	Volatility Absolute	Volatility Relative	PSI*
(A-S)**	countercyclical / coincident	–.8305	.01	–8.0286	1.3547	0.9988	6.24
(A-S)	countercyclical / coincident	–.8117	.01	–7.4842	1.3612	1.0035	7.05
Germany (c)***	countercyclical / coincident	–.7413	.01	–5.9486	1.5388	1.1345	6.49
(A-S)	countercyclical / coincident	–.6904	.01	–5.1389	1.3517	.9966	7.81
Belgium (c)	countercyclical / coincident	–.6615	.01	–4.7502	1.2195	.8991	6.43
United States (A-S)	countercyclical / coincident	–.6592	.01	–4.7208	1.2649	.9325	7.23
Switzerland (c)	countercyclical / coincident	–.5988	.01	–4.0267	1.2560	.9260	7.11
France (c)	countercyclical / coincident	–.5560	.01	–3.6022	1.4887	1.0976	6.28
Finland (c)	countercyclical / coincident	–.5444	.01	–3.4950	1.2788	.9428	6.93
Netherlands (A-S)	countercyclical / coincident	–.5421	.01	–3.4738	1.4497	1.0688	7.75
Luxembourg (c)	countercyclical / coincident	–.5169	.01	3.2513	1.2715	.9374	6.85
United Kingdom (A-S)	countercyclical / coincident	–.4679	.01	–2.8513	1.3117	.9671	7.20
(A-S)	countercyclical / coincident	–.3598	.05	–2.0769	1.2518	.9229	5.78
Italy (c)	acyclic / coincident	–.2992	.15	–1.6885	1.1951	0.8811	5.68
Portugal (c)	acyclic / coincident	–.2974	.15	–1.6777	1.2921	.9527	6.09
Denmark (A-S)	acyclic / coincident	–.2827	.15	1.5873	1.5467	1.1403	7.93
Greece (c)	acyclic / lagging $(\kappa = 1)$	–.2620 / .3657	.20 / .05	–1.4623 / 2.0794	1.3989	1.0313	5.49
Spain (c)	acyclic / coincident	–.2230	.25	–1.2320	1.2740	.9393	5.83
Turkey (c)	acyclic / coincident	–.1928	.30	1.582	1.1822	.8716	6.36
New Zealand (A-S)	acyclic / lagging $(\kappa = 1)$	–.1798 / .4049	.35 / .05	–.9841 / 2.3433	1.3252	.9771	7.51
Australia (A-S)	acyclic / leading $(\kappa = 1)$	.1294 / –.2667	.50 / .15	.7026 / –1.4904	1.5059	1.1103	8.08

Source: Finke & Sabatini, 2016; OECD, 2017; World Bank, 2017; own calculations.

* Pension Sustainability Index. ** Anglo-Saxon (Beveridge) pension system. *** Continental (Bismarck) pension system.

σ_y. To compare the volatility of pension spending cycles and economic cycles, relative volatility was calculated as the ratio of standard deviations of the corresponding series of cyclic components. Table 2 provides the results of estimating the volatility of cyclic components series of the public pension spending growth.

The calculations show that there is no significant relationship between the volatility of public pension spending growth and the type of basic pension model. On the other hand, it was found that the amount and volatility of public pension spending (not annual growth) depends on the basic pension model and have higher values in Bismarck model countries. The relative volatility of the average public pension

Figure 9. Comparison of the dynamics of public pension spending in the countries of the Bismarck and Beveridge model, 1980–2011
Source: OECD, 2017; own calculations.

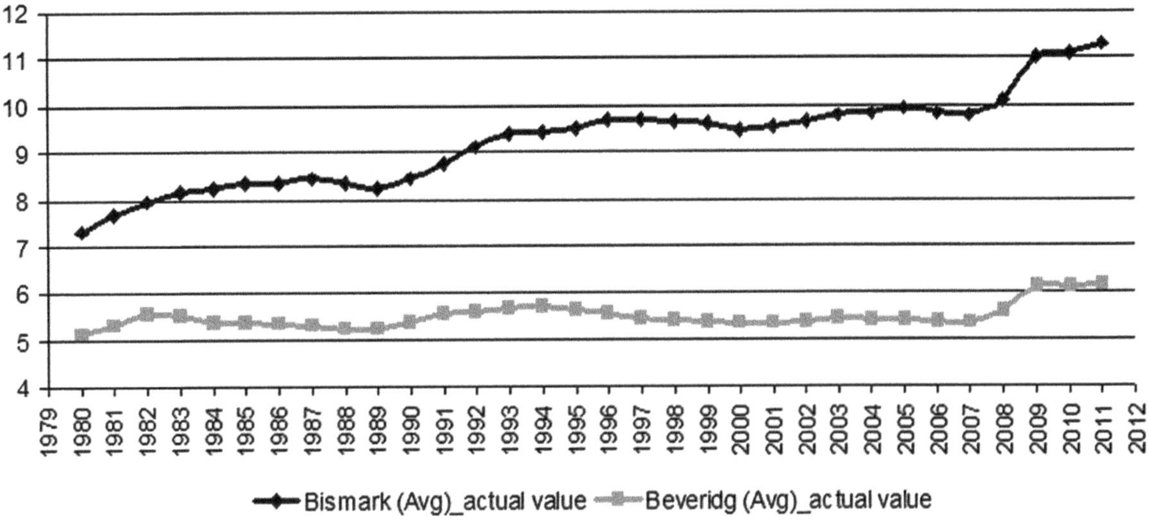

Figure 10. Comparison of fluctuations in public pension spending in the countries of the Bismarck and Beveridge model, 1980-2011
Source: OECD, 2017; own calculations.

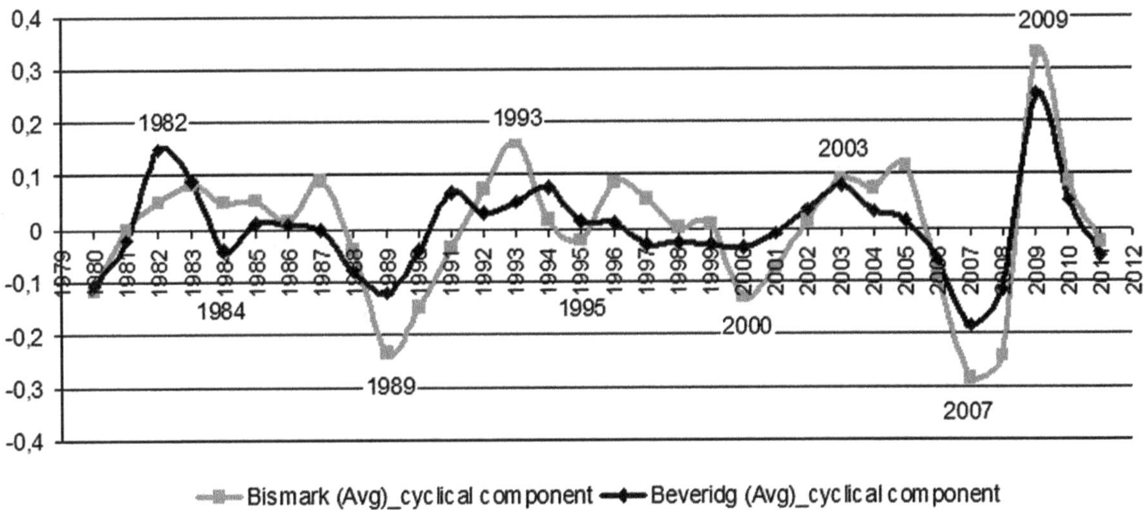

spending (series of cyclic components) in Bismarck model countries is 1.4908, and in Beveridge model countries is .9959. This indicates a greater risk of the Bismarck pension model. Figure 10 illustrates this conclusion.

Summing up the process of pension evolution, we can distinguish four global waves of pension reforms that led to the modern multi-pillar design of national pension systems. The first wave of reforms refers to 50s–60s of the 20th century and is caused by the rapid economical growth the post-war economies and the favorable demographical structure with small number of elderly people. So the reforms of the first wave may be characterized by the increase of rights for pension provision as well as the pension models convergence. For example, in Germany the assistance to those in need was introduced, in the United Kingdom lord Beveridge introduced a three-pillar type of pension system: the government secures the living minimum for the whole population, the employer–professional earnings of the workers, i.e., compensation of the lost earnings, while the worker himself provides the supplementary pension on a voluntary basis (Mayhew, 2001).

The second wave of reforms swept over pension systems in the early 1970s–late 1980s. Weakened by the energy crisis the economies don't withstand the load of the generous pension systems. Thus, first, the conditions of getting pensions became more rigid (e.g., the increase of pension age, length of service, and payments), second, the governments tried to transfer some part of their responsibility for pension provision on the shoulders of the citizens themselves; therefore private pension schemes are actively introduced.

The third wave of reforms (the end of the 20th and the beginning of the 21st centuries) is caused by the demographical crisis–the number of elderly people increased dramatically. The distributive pension systems which presented themselves in a good light in the young society failed in the elderly one. The further stimulation of private accumulating pension provision, the increase of pension age and insurance length of payments are observed.

But no measures taken by the governments in the period of the third wave of reforms solve the problem completely. The demographical structure of the post-industrial society is irreversibly changing, and the load of the pension system on the economics is growing. The problem is aggravated by the economic crisis of 2008, and the fourth wave of reforms directed to the reduction of the public pension spending is covered pension systems.

As a result of the evolution of the basic pension models, the national pension systems are transformed, as a rule, into three-pillar: the first pillar is mandatory and redistributive (basic), the second pillar is mandatory and funded (public or private), the third pillar is voluntary and funded (private). However, the structure of pension systems, the level of public pension spending and pension behavior of the population still depends on the basic pension model. In the countries which chose model of Bismarck, the load of pension provision mainly lies on the government. The introduction of private funded programs, even stimulated by the government, is coming into practice with much difficulty. For example, Rister-reform in Germany (the government stimulates by cash bonuses the voluntary payments in the pension funds of 1–4% annual gross income (Guardiancich, 2010)) does not actually find support among the public oriented towards the government paternalism. As for the public pension spending, one can observe the rapid growth in these countries (see Figure 9), because just PAYG-systems are extremely sensible to demographical risks.

So, forecasting the future dynamics of pension systems development, it is reasonable to state that the failure of the idea of the solidarity distributive principle of pension provision in the conditions of decreasing birth-rate, growing length of life and individualization of the post-industrial society is inevitable. The future is for multi-pillar systems in which the responsibility for pension provision is diversified between social partners: individuals, employers, government. At this, the government only provides social assistance to those who are in need. The able-bodied people provide their old years through corporative and individual funded programs themselves.

Thus, based on the results of the research conducted, the evolution of pension systems can be defined as the process of parametric and structural reorganization in time, as a result of which a form or structure appears that qualitatively differs from the previous one. The change of the parameters and structure of the national pension systems is caused by the change of the pension system load on the country's economics, crisis situations (economical, demographical, political), which can have both positive (rise) and negative (recession) sign, being the push towards revolutionary changes. So, it is possible to present the evolution of pension systems as several successive phases of stability and crises between them, i.e., to describe it by the model of the type of evolution stage models.

The construction of the evolutionary model of the stages of the evolution of national pension systems is based on a hypothesis. Each stage of the evolution of the pension system comprises a stability phase and crisis phase that follows it; each stability phase is characterized by dominate type of the pension system (structure and parameters), each crisis phase is characterized by dominant issue of determining the parameters and structure of pension systems to be determined to restore its stability. When building the model, six key parameters are defined: the stability phase, the crisis phase, the age of the pension system, the burden of the pension system on the national economy, the growth rate of the national economy and the demographic structure of society. Analysis of the evolution of the PAYG-systems and forecasting future trends of development allowed distinguish six evolutionary stages:

Stage I: Growth of PAYG-system;
Stage II: Enhancing the growth of PAYG-system;
Stage III: Changing the growth of PAYG-system;
Stage IV: Survival of the PAYG-system and the formation of a funded pension system;
Stage V: Survival of the PAYG-system and the development of a funded pension system;
Stage VI: Enhancing growth of a funded pension system.

The indicator of a pension system development is not the length of existence, but the burden of the pension system on the national economy. Crisis is the direct predecessor of pension reforms (in contrast to the Greiner model (Greiner, 1998)). The evolutionary model of stages of pension systems by the example of the evolution of the German pension system is illustrated in Figure 11.

It should be noted that, unlike the models of life cycle, in the stage models the system doesn't always go through all the stages and different successions of passing through the stages are possible. This thesis may be confirmed by the evolution of the pension systems of the former socialist countries (transitional economies). For example, the pension system of Ukraine evolved from the government-treasury redistributive only for officials at the beginning of the 18th century to the multi-pillar system with the private funded level for the whole population at the beginning of the 21st century returning to the centralized distributive type in the period of socialism. The basic pension model in Ukraine, as well as in other transitional economies, is model of Bismarck. Figure 12 shows the stages of the evolution of Ukraine's pension system.

FUTURE RESEARCH DIRECTIONS

The methodology for modeling the impact of economic cyclicality on the evolutionary dynamics of pension systems introduced in this chapter provides a natural guide to future research. The study proposed an HP filter to detect cycles, but it has some drawbacks (in particular, the problem of the "end point"). Hence, natural directions for future research include the using of alternative methods to identify cycles

Figure 11. Evolutionary model of the stages of pension systems (for example Germany)
Source: German Federal Pension Insurance, 2016; Heide, 2016; Recht, 2007

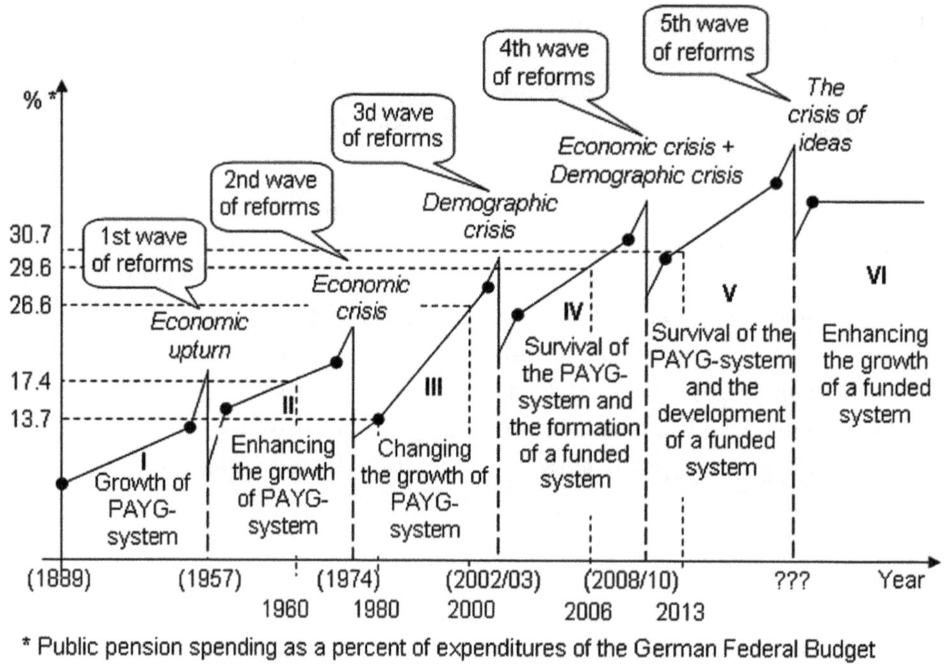

of pension dynamics. For example, it can be such known methods as Fourier transform, fractal (R / S) analysis or LOWESS algorithm. For an enhanced study of pension cycle parameters, it would be appropriate to extend the model tools. For example, the Bry-Boschan method (Bry & Boschan, 1971) can be used to dating of the turning points, and to study the persistence of time series can be used R/S analysis.

However, the most promising is probably the use of spline approximation methodology to the detection, visualization and analysis of cyclic structures in the pension dynamics. This can be both a phase spline-analysis (Vintizenko & Yakovenko, 2008), and the following tools for data mining: (i) Multivariate Adaptive Regression Splines (MARS) is a methodology suggested by J. Friedman (1991) for nonlinear regression modeling (due to the division of baseline data into areas by basic functions, MARS allows to determine emissions); (ii) Conic Multivariate Adaptive Regression Splines (CMARS) which has been developed as an alternative to the backward stepwise part of the MARS algorithm (Weber et al., 2012); (iii) CMARS extensions, e.g., CGPLM (Conic Generalized Partial Linear Model) or CMARS-MSOM (Mean-Shift Outlier Model with Conic Multivariate Adaptive Regression Splines) (Yerlikaya-Ozkurt, 2013). These new methods can be applied to high-dimensional data sets with a large number of predictors which exhibit nonlinear (complex) relationships in the pension dynamics.

The second direction of future research concerns the demographic context of pensions. This problem has two aspects: the rapid demographic aging of the population and the third demographic transition (immigration and ethnic change in low-fertility countries). These questions also require careful research.

The third direction of future research concerns the behavior patterns, as behavioral heuristics affect the implementation of pension reforms, the retirement planning and participation in individual pension programs. The use of behavioral economics will increase the explanatory properties of models.

Figure 12. Evolutionary model of the stages of pension systems (for example Ukraine)
Source: Leonov, 2011; National Bank of Ukraine, 2017; State Statistics Service of Ukraine, 2017.

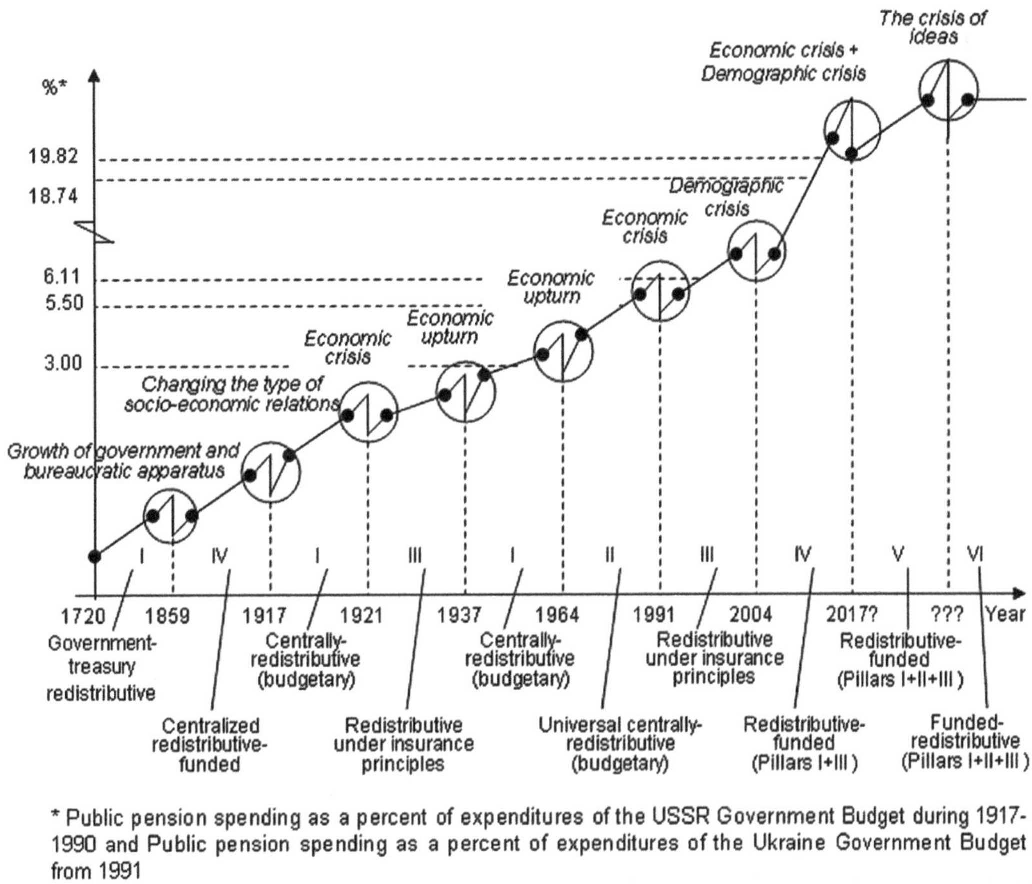

CONCLUSION

In this chapter, using the Hodrick-Prescott filter and statistical analysis of the direction and timing cyclical indicators, it is proved that the pension dynamics is rhythmic. Expansion of pension rights is replaced by an attempt to reduce them. The period of growing public pension expenditures is replaced by a period of their decline. The fluctuations of pension dynamics correlate with economic cycles. As a rule, structural pension reforms are carried out or planned in the troughs of the big K-cycles. Parametric pension reforms are carried out or planned in the troughs of the medium-term J-cycles.

Economic cyclicality and demographic trend have caused the convergence of the basic pension models of Bismarck and Beveridge. Modern pension systems generally have a multi-level structure that allows to solve a dual task, first, to diversify financial, economic and demographic risks, and, secondly, to diversify responsibility between social partners: individuals, employers, government. Nevertheless, as the results show, the type of the basic model still determines the public pension spending and behavioral reactions to the crisis.

The results show that, even with limited statistics, it is possible to establish meaningful economic and statistically significant correlations between pension evolutionary dynamics and economic cyclicality.

In conclusion, it should be noted that the results of the study can only be the first step to understand the long-term behavior of pension systems. Only an integrated system approach can help with forecasting, strategic decision making and long-term pension policy planning.

REFERENCES

Bry, G., & Boschan, C. (1971). *Cyclical Analysis of Time Series: Selected Procedures and Computer Programs.* Technical Paper 20. Retrieved from http://papers.nber.org/books/bry_71-1

Chlon-Dominczak, A. (2016). Impact of changes in multi-pillar pension systems in CEE countries on individual pension wealth. *Journal of Pension Economics and Finance*, 1–12. doi:10.1017/S1474747216000238

Chybalski, F. (2011). The resilience of pension systems in the CEE countries to financial and economic crisis: the need for higher diversification. In *Proceedings of the 13th International Conference of Finance and Banking* (pp. 258–267). Ostrava, Czech Republic: Silesian University, School of Business Administration. Retrieved from http://www.opf.slu.cz/kfi/icfb/proc2011/pdf/22_Chybalski.pdf

Federal Health Reporting Information System. (2016). *Entwicklung der beitragssatze in der sozialversicherung* [Development of contribution rates in social security]. Federal Health Reporting Information System of Germany. Retrieved from http://www.gbe-bund.de/gbe10/abrechnung.prc_abr_test_logon?p_uid=gast&p_aid=0&p_knoten=FID&p_sprache=D&p_suchstring=8862

Finke, R., & Sabatini, M. (Eds.). (2016). *2016 Pension Sustainability Index. International Pension Papers 1.* Munich, Germany: Allianz SE. Retrieved from http://projectm-online.com/app/uploads/Allianz_2016_Pension_Sustainability_Index.pdf

Flegelskamp, G. (2008). Ubersicht der Rentenbeitragssutze ab 1949 [Contribution rates to social security] [Data file]. Available from http://www.flegel-g.de/beitragssatz.html

Friedman, J. H. (1991). Multivariate Adaptive Regression Splines. *Annals of Statistics*, *19*(1), 1–67. doi:10.1214/aos/1176347963

German Federal Pension Insurance. (2016). *Rentenversicherung in Zahlen 2016* [Pension insurance in figures 2016]. Berlin, Germany: German Federal Pension Insurance. Retrieved from http://www.deutsche-rentenversicherung.de/cae/servlet/contentblob/238692/publicationFile/61815/01_rv_in_zahlen_2013.pdf

Greiner, L. E. (1998, May). Evolution and revolution as organizations grow. *Harvard Business Review*, 1–11. Retrieved from http://www.ceriumpartners.com/insights/Ev%20%20Rev%20as%20Orgs%20Grow-%20Greiner.pdf PMID:10179654

Grinin, L., Korotayev, A., & Tausch, A. (2016). *Economic cycles, crises, and the global periphery.* Springer. doi:10.1007/978-3-319-41262-7

Grinin, L. E., Devezas, T. C., & Korotayev, A. V. (2014). *Kondratieff Waves. Juglar – Kuznets – Kondratieff*. Volgograd, Russia: 'Uchitel' Publishing House.

Guardiancich, I. (2010). *Germany. Current pension system: First assessment of reform outcomes and output*. Country Report. European Social Observatory. Retrieved from http://www.ose.be/files/publication/2010/country_reports_pension/OSE_2010_CRpension_Germany.pdf

Heide, D. V. D. (Ed.). (2016). *Rentenversicherung in Zeitreihen* [Pension insurance in time series]. Berlin, Germany: German Federal Pension Insurance. Retrieved from http://www.deutsche-rentenversicherung.de/cae/servlet/contentblob/238700/publicationFile/50913/rv_in_zeitreihen.pdf

Hinz, R., Zviniene, A., Biletskyy, S., & Bogomolova, T. (2009, October). *The Long Term Impact of the Financial Crisis on Public Pension System Financing: Stress Testing Models of Mandatory Pension Systems in Middle Income and Developing Countries*. Discussion Paper provided for the Expert Group Meeting on Population Ageing, Intergenerational Transfers and Social Protection, Santiago, Chile. Retrieved from http://www.cepal.org/sites/default/files/events/files/hinz.pdf

Hodrick, R. J., & Prescott, E. C. (1997). Postwar U.S. Business Cycles: An Empirical Investigation. *Journal of Money, Credit and Banking*, *29*(1), 1–16. doi:10.2307/2953682

International Labour Organization. (2014). *World Social Protection Report 2014/15: Building economic recovery, inclusive development and social justice*. Geneva, Switzerland: International Labour Office. Retrieved from http://www.ilo.org/wcmsp5/groups/public/---dgreports/---dcomm/documents/publication/wcms_245201.pdf

Kondratev, N. D. (1922). *Mirovoye khozyaystvo i ego koniunktury vo vremia i posle voyny* [The world economy and its conjuncture during and after the war]. Vologda, Russia: Oblastnoe Otdelenie Gosudarstvennogo Izdatelstva.

Leonov, M. V. (2011). Osnovnyye problemy reformirovaniya pensionnykh sistem v stranakh byvshego SSSR [Main problems of pension systems reforming in former USSR countries]. *Ekonomika, predprinimatelstvo i pravo, 1*(11), 37–54.

Lindeman, D., Rutkowski, M., & Sluchynskyy, O. (2000). *The evolution of pension systems in Eastern Europe and Central Asia: opportunities, constraints, dilemmas and emerging best practices*. Washington, DC: The World Bank. Retrieved from www.oecd.org/daf/insurance-pensions/

Maddison, A. (2010). Statistics on World Population, GDP and Per Capita GDP, 1-2008 AD (Vertical file, copyright Angus Maddison, university of Groningen) [Data file]. Retrieved from http://www.ggdc.net/maddison/oriindex.htm

Mayhew, L. (2001). *A Comparative Analysis of the UK Pension System Including the Views of Ten Pension Experts*. SEERL – Birkbeck College, University of London. Retrieved from https://www.researchgate.net/publication/255608035_A_Comparative_Analysis_of_the_UK_Pension_System_Including_the_Views_of_Ten_Pension_Experts

Mladen, L. (2012). A comparative review over the pension systems performance in Central and Eastern European countries. *Journal of Knowledge Management, Economics and Information Technology, 2*(5), 1–11. Retrieved from http://www.scientificpapers.org/wp-content/files/1311_Luise_Mladen__A_COMPARATIVE_REVIEW_OVER_THE_PENSION_SYSTEMS_.pdf

Natali, D. (2011). *Pensions after the financial and economic crisis: a comparative analysis of recent reforms in Europe.* ETUI Working Paper 2011.07. doi:10.2139/ssrn.2221827

National Bank of Ukraine. (2017). *Macroeconomic indicators.* Retrieved April 28, 2017, https://bank.gov.ua/control/en/publish/article?art_id=24038763&cat_id=47389

OECD. (2009). *OECD recommendation on core principles of occupational pension regulation.* Retrieved April 24, 2017, from http://www.oecd.org/finance/private-pensions/33619987.pdf

OECD. (2013). *Pensions at a Glance 2013: OECD and G20 Indicators.* Paris: OECD Publishing. doi:10.1787/pension_glance-2013-

OECD. (2015). *Pensions at a Glance 2015: OECD and G20 Indicators.* Paris: OECD Publishing. doi:10.1787/pension_glance-2015-

OECD. (2017). Pension spending (indicator) [Data file]. Retrieved April 24, 2017, from https://data.oecd.org/socialexp/pension-spending.htm#indicator-chart

Orenstein, M. A. (2003). Mapping the diffusion of pension innovation. In R. Holzmann, M. A. Orenstein, & M. Rutkowski (Eds.), *Pension Reform in Europe: Process and Progress* (pp. 171–193). Washington, DC: The World Bank. Retrieved from http://www.tulane.edu/~dnelson/PolTransConv/Orenstein.pdf

Ravn, M., & Uhlig, H. (2002). On adjusting the Hodrick-Prescott filter for the frequency of observations. *The Review of Economics and Statistics, 84*(2), 371–380. doi:10.1162/003465302317411604

Recht, G. (2007). *Fortentwicklung der gesetzlichen Rentenversicherung* [Further development of the statutory pension insurance] [PowerPoint slides]. Retrieved from: https://www.f04.th-koeln.de/imperia/md/content/hp-f04/ivw/forschungundevents/symposium/vortrag_recht.pdf

Schumpeter, J. A. (1939). *Business cycles: A theoretical, historical and statistical analysis of the capitalist process.* New York: McGraw-Hill Book Company.

Schwarz, A. M., Arias, O. S., Zviniene, A., Rudolph, H. P., Eckardt, S., Koettl, J., ... Abels, M. (2014). *The Inverting Pyramid: Pension Systems Facing Demographic Challenges in Europe and Central Asia.* Washington, DC: World Bank. doi:10.1596/978-0-8213-9908-8

State Statistics Service of Ukraine. (2016). *Social protection: Statistical publication.* Retrieved April 24, 2017, from http://www.ukrstat.gov.ua/

Sun, S., & Hu, J. (2014). The Impact of Pension Systems on Financial Development: An Empirical Study. In M. Faure & N. Philipsen (Eds.), *The Role of Law and Regulation in Sustaining Financial Markets* (pp. 54–72). Oxon, UK: Routledge. Retrieved from https://papers.ssrn.com/sol3/papers.cfm?abstract_id=2481749

The Evolution of Public Pension Plans: Past, Present and Future. (2008). Washington, DC: National Conference on Public Employee Retirement Systems. Retrieved from http://www.ncpers.org/files/evolution_of_public_pensions_2d.pdf

Tuesta, D. (2011). A review of the pension systems in Latin America. *BBVA Research, 11*(15), 1–27. Retrieved from https://www.bbvaresearch.com/KETD/fbin/mult/WP_1115_tcm348-255095.pdf

UK Public Spending. (2017). Social Protection: Fiscal Years 1900 to 2020 [Data file]. Retrieved April 24, 2017, from http://www.ukpublicspending.co.uk/spending_chart_1900_2020UKp_16c1li011mcn_40t00t00t_Social_Protection

Vintizenko, I. G., & Yakovenko, V. S. (2008). Ekonomicheskaya tsiklomatika [Economic cyclomatics]. Moscow, Russia: Finance and Statistics. – Stavropol, Russia: AGRUS.

Weber, G. W., Batmaz, I., Koksal, G., Taylan, P., & Yerlikaya-Ozkurt, F. (2012). CMARS: A New Contribution to Nonparametric Regression with Multivariate Adaptive Regression Splines Supported by Continuous Optimisation. *Inverse Problems in Science and Engineering, 20*(3), 371–400. Retrieved from http://citeseerx.ist.psu.edu/viewdoc/download?doi=10.1.1.323.5864&rep=rep1&type=pdf

World Bank. (2009). *Pensions in crisis: Regional report for Europe and Central Asia*. Washington, DC: The World Bank. Retrieved from http://siteresources.worldbank.org/FINANCIALSECTOR/Resources/Pensions_in_Crisis_Nov09.pdf

World Bank. (2017). World Development Indicators [Data file]. Available from http://databank.worldbank.org/data/reports.aspx?source=2&series=NY.GDP.MKTP.KD.ZG&country=

Yerlikaya-Ozkurt, F. (2013). *Refinements, Extensions and Modern Applications of Conic Multivariate Adaptive Regression Splines.* (Doctoral dissertation). Retrieved from http://etd.lib.metu.edu.tr/upload/12616723/index.pdf

Yermo, J., & Severinson, C. (2010). The impact of the financial crisis on defined benefit plans and the need for counter-cyclical funding regulations. *OECD Working Papers on Finance, Insurance and Private Pensions, 3*. doi: 10.1787/5km91p3jszxw-en

KEY TERMS AND DEFINITIONS

Cycle Juglar: The cycle of medium-term fluctuations of business activity with a typical period of 7–11 years.

Cycle Kondratiev: The big cycle of economic conjuncture with a typical period of about 50 years.

Evolution of Pension Systems: The process of parametric and structural reorganization of pension systems in time, as a result of which a form or structure appears that qualitatively differs from the previous one.

Hodrick-Prescott Filter: The method of decomposition of time series into trend and cyclical components.

Indicator Acyclic: The pension indicator, whose cyclical dynamics is not related to the economic cycle.

Indicator Coincident: The pension indicator that moves simultaneously with the economic cycle.

Indicator Countercyclical: The pension indicator that moves in the opposite direction to the economic cycle.

Indicator Lagging: The pension indicator that moves behind the economic cycle.

Indicator Leading: The pension indicator that moves ahead of the economic cycle.

Indicator Procyclical: The pension indicator that moves in the same direction as the economic cycle.

Chapter 8
Financial and Economic Security of BRICS Countries

Elena V. Fomicheva
Southern Federal University, Russia

Marina S. Shikun
Southern Federal University, Russia

ABSTRACT

With the growing tension in the global community and the instability of the overall situation, problems with financial centers and regions have arisen, and as a result, the necessity of reforms have appeared in the global governance system and in the financial and economic system. BRICS countries are very interested in solving problems and developing relations within the bloc. Each country pursues its own interest, but the emerging dialogue of countries allows them to find a compromise in complex issues. The current economic situation in the world, where Western countries speak the language of sanctions, makes BRICS countries strengthen their interaction. It concerns the provision of financial and economic security. Increased financial risks in the capital market as well as the dominance of the dollar and the euro level, push BRICS countries to strengthen the financial mechanisms. BRICS countries support the creation of a global network of financial instruments.

INTRODUCTION

In the current geopolitical and geo-economics instability conditions, when the negative consequences of regularly recurring economic and financial crises exist, the instability of the global financial system have reached a level of real economic threat for each country, - the notion "Economic security" has got a new meaning and relevance.

Within the framework of interstate relations, it is customary to highlight the economic aspects of national and international security, and in Russian science this problem has been developed quite deeply and independently of Western researchers. The fundamental difference between the national and Western "schools" of the economic security is a narrower and more specific treatment by Western experts of the term "security" of the state. In the context of the bipolar confrontation of superpowers, the main

DOI: 10.4018/978-1-5225-3767-0.ch008

theoretical problem from the point of view of the "international economic aspects of security" is to find the optimal balance between the volumes and structure of defense spending, on the one hand, and the economic efficiency and competitiveness of the national economy, on the other.

The economic security of the regional association of countries (international macro-regional economic security) is viewed as the state and dynamics of the socio-economic system formed as a result of the economic integration of states, which allows to determine objectively and effectively implement regional economic goals and interests, eliminate threats in the economic sphere, ensure integrity, stability, economic efficiency and independence, competitiveness, of a sustainable growth and self-development of regional integration of countries as a subject of international relations and world economic relations in the conditions of globalization and instability of the world economy.

Assessing the economic security of the macroregion, it is necessary to note the following strengths. BRICS is a large integration grouping of a new type, with a large and constantly developing economic potential. BRICS countries account for 26% of the Earth's territory, 42% of the world's population and about 32.3% of world GDP. The largest GDP growth among the BRICS countries belongs to China, which brings 18% of world GDP. Russia is in the top five countries in defense and security (2nd place), China, India and Brazil are leading in the ranking on agriculture, India is the second in the field of information technology development, China is also among the five countries in the development of medicine, bio- and nanotechnologies, power engineering. All these characteristics indicate the ability of the integration group to provide economic sovereignty, sustainable economic development and the ability to grow, that is, to ensure the economic security of the subject of international economic relations.

In the 80-ies. XX century in the global economy chronologically coincided with two major processes - globalization and the beginning of modernization of Brazil, Russia, India and China. Despite the fact that Russia and China were moving away from the "command economy", while Brazil and India remained in the market economy system, all four countries began reforms towards a mixed economy. Reforms of the BRIC countries differed in content, forms, stages, and national specifics. But the vector of reforms was aimed at creating a competitive national economy and business, improving the quality of life, creating a new center of power in world geo-economics and geopolitics.

Since the reform strategies of the BRICS countries assumed the advantages of globalization, and the most developed part of the global economy is global finance, it is necessary to analyze the processes and forms of the entry of the national financial markets of the BRIC countries into the global finance system.

Under these conditions, the creation, formation and development of the group of BRICS countries is assessed as a worthy response to the existing imbalance of forces, the elimination of contradictions and the concentration of common efforts of countries primarily on the economic direction.

One of the factors of economic growth of BRICS was a strong monetary policy aimed at ensuring geo-economic and macroeconomic goals. The Central Bank of China supported the currency regime of the fixed exchange rate of the yuan for 30 years. The course was understated, but provided financial security of the country and allowed to create additional competitive advantages in the international commodity markets. The Central Bank of Russia within the framework of the currency corridor pursued a policy of stability of the course in the interests of export-oriented industries. This policy did not prevent the devaluation of the ruble in 1998 and did not create currency incentives for the processing industries, but ensured the formation of the state budget from the currency earnings of the commodity companies by 2/3. The Central Bank of Brazil and South Africa, pursuing the policy of free currency regimes, with the task of ensuring the stability of national currencies as a whole coped.

FINANCIAL AND ECONOMIC SECURITY

Economic security (including its financial aspects) is of particular interest in the light of integration processes as a characteristic feature of the development of modern international economic relations, as a close interconnection of economic, political, internal and external interests. All five BRICS countries unite not only the similarity of interests and common challenges, primarily the acceleration of modernization in all spheres of activity, but also the complementary nature of economic complexes, natural, technological, intellectual and financial resources of all five countries. The BRICS concept emerged not as a diplomatic but as a market initiative and, as a management from the point of view of a fashionable "soft power" the BRICS site allows discussing various and eclectic problems in essence, despite the existence of a political agenda in each country.

Financial issues are an important area of cooperation between the BRICS countries: most of these issues are considered during the meetings of finance ministers and central bank governors of the BRICS countries. Areas of cooperation include:

- The work of the New Development Bank to mobilize resources for infrastructure and sustainable development projects in the BRICS countries, in other emerging market countries and in developing countries, and to complement existing efforts by multilateral and regional financial institutions for Support of global growth and development;
- The work of the Currency Resources Fund to help strengthen the global financial security network and complement existing international monetary and financial mechanisms;
- Exchange of views and points of view on the main issues on the G20 agenda, including measures to minimize negative side effects in the world economy, as well as steps to ensure growth and job creation, stimulate investment and infrastructure development, strengthen the financial system and cooperation in tax matters, etc.;
- Promoting the reforms of international financial institutions (ifis), in particular, the International Monetary Fund (IMF) and the World Bank Group (WB);
- Discussion of new topics of mutual interest, which may include cooperation in the sphere of payment systems, in particular, supervision of payment systems and standards of financial messages. In addition, the BRICS development banks will cooperate in the framework of the Financial Forum.

Western countries, accustomed to the fact that most processes in the global economy are controlled only by them, are already worried that alternative structures appear and they are not controlled by the USA and its allies. The total outflow of capital from the economies of China, Brazil, Russia, India and South Africa for the past 10 years amounted to more than 3.5 trillion. Half of this amount was withdrawn for the last 4 years. Increasingly, international financial institutions are being used as an instrument of the West's influence on the BRICS countries. In this regard, it is urgent to work out measures to ensure the financial and economic security of the BRICS countries. Reducing dependence on global financial structures will help the BRICS countries to protect their economies from possible threats. The BRICS countries continue to follow the chosen course in the framework of large-scale negotiations on reforming the world monetary and financial system with a view to create "a more representative international financial architecture in which emerging and developing countries will have a greater voice and greater representativeness".

In the mainstream of this policy three main areas are identified:

- Modernization of global governance institutions, primarily the IMF and the World Bank;
- Formation of own credit and banking structures with the purpose of increasing the investment potential of the BRICS countries and weakening their dependence on the policy of the world financial centers;
- Creating a new system of reserve currencies and increasing the role of national monetary units in mutual settlements between the BRICS countries.

The problem of global security can be traced in the work of the countries at each meeting. The issues under consideration and summits summaries can be seen in Table 1.

A big attention is paid to the issue of organization of financial and economic security at the meetings of the BRICS member countries. First of all, they are talking about the restructuring of the work of the International Monetary Fund and the World Bank, aimed primarily at servicing the interests of developed countries holding leading positions in these institutions. In the practical terms this means changing the weight of developing countries in the international financial institutions, including the mechanism for calculating quotas and votes of the International Monetary Fund. By now, there is an obvious disproportion between the "weight" of the BRICS countries and other developing countries in the world economy and the proportion of their votes in the IMF. They have only 11% of the total share in the IMF, compared with 16.8% of the votes of the United States, which allows Washington to block any decisions of the IMF (for their adoption it is necessary to collect 85% of the votes). At the second BRICS summit in Brazil, a joint statement of the parties raised the issue of the need for large-scale reforms of the Bretton-Woods institutions, major changes in favor of emerging market countries and developing countries in order to bring their participation in the decision-making process. According with the decision to reform the old system taken in 2010 under pressure from developing countries, an increase in the total vote share of the BRICS countries must change from 11 to 14.13%. However, the process of reviewing quotas is delayed, first of all, at the initiative of the United States and in the Ufa declaration in 2015, disappointment was expressed in connection with the failure to implement the IMF reforms announced in 2010, which negatively affects the legitimacy, reputation and effectiveness of the fund. Therefore, BRICS called for the immediate 14th round of a general review of quotas, as well as a final agreement on a new quota calculation formula in conjunction with the 15th round. As a counteraction to economic pressure, the creation of the New BRICS Development Bank, an agreement on the creation of which was signed at the Fortaleza summit, to mobilize resources to finance infrastructure projects and projects in the field of sustainable development in the BRICS countries, other emerging market economies and developing countries. The creation of the Bank can be seen as a significant contribution to the formation of an alternative system aimed at weakening dependence on the policies of the IMF and the World Bank.

BRICS leaders called for the reform of international financial institutions to increase the weight of developing countries in them. So, it is necessary to improve the international monetary system, providing for the creation of an international system of reserve currencies, providing stability and reliability. At the same time, it is necessary to reform the International Monetary Fund (IMF) by strengthening the representation of the poorest members of the organization, including sub-Saharan Africa.

In May 2015, Russian President Vladimir Putin signed a law on the ratification of the treaty on the creation of a pool of foreign exchange reserves of the BRICS countries in the amount of $ 100 billion. It should help to strengthen the global network of financial security and complement existing international

Table 1. The main characteristics of the BRICS summits

N°	Summit, Date	Discussed Issues	The Main Outcomes of the Summit
1	Yekaterinburg, Russia 16.06.2009	Opportunities to overcome the consequences of the global economic crisis, investment policy, international trade, food security, interaction in the field of education and science, etc.	1. Joint statement of the leaders of the BRIC countries. 2. Joint statement by the BRIC countries on global food security.
2	Brasilia, Brazil 16.04.2010	Consequences of the crisis, the formation of a new financial order, international trade, poverty issues, climate change and the role of the BRIC in international organizations The IMF, the World Bank.	1. Memorandum on cooperation between state financial institutions for the development and support of exports of the BRIC member countries.
3	Sanya, China 14.04.2011	Measures to stabilize world prices for raw materials through the regulation of the commodity derivatives market, economic interaction with the use of national currencies.	1. A joint declaration was adopted on the results of the discussion.
4	New Delhi, India 29.03.2012	Prospects for establishing a development bank, mechanisms for convergence of stock exchanges, issues of world economic policy, reform of global governance institutions, international security, etc.	1. The Declaration "The Economy of the BRICS Countries: Brazil, Russia, India, China and South Africa" was adopted. 2. The General Agreement on the general procedure for opening credit lines in the national currencies of BRICS was signed. 3. A multilateral agreement on confirmation of letters of credit within the framework of the mechanism of interbank cooperation of the BRICS countries was signed.
5	Durban, South Africa 26-27.03.2013	The main theme of discussions is "BRICS and Africa: Partnership for Development, Integration and Industrialization"	1. A declaration and action plan was adopted following the discussion, which reflects the assessment of the current political and economic situation in the world, the main approaches of the BRICS countries on topical issues of multilateral cooperation. 2. The Declaration on the Establishment of the Consortium of Experts Centers of the BRICS countries was signed, on the publication of a joint statistical publication of the BRICS countries. 3. Agreements on co-financing of infrastructure projects in Africa and on cooperation in the field of "green economy" have been signed. 4. The Declaration on the Establishment of the BRICS Business Council was adopted.
6	Fortaleza, Brazil 13-17.06.2014	Establishment of the New Development Bank, which aims to finance infrastructure projects and projects for sustainable development in the BRICS countries and developing countries.	1. The Agreement on the Establishment of a New Development Bank and the Agreement on the Creation of a Pool of Conditional Foreign Exchange Reserves of the BRICS Countries were signed.
7	Ufa, Russia 8-9.07.2015	Creation of a New Development Bank and a pool of foreign exchange reserves	1. Continuation of the creation of the own structure of financial organizations.
8	Goa, India 15-16.10.2016	Strengthening the strategic partnership of the BRICS countries, the development of multi-sectoral cooperation, the fight against international terrorism, the settlement of relations in Syria, the stability of the world economy and finance.	

monetary and financial mechanisms. The document provides for the establishment of a self-governing pool of conditional foreign exchange reserves in order to counter possible pressure on the balance of payments of the BRICS countries and to provide mutual support. Formation of own financial structures will allow increasing the investment potential of the BRICS countries and reducing their dependence on the policy of the world financial centers. BRICS financial structures can become a prototype and catalyst for the development of a new financial world order, for the establishment of which not only developing, but also many developed states vote. In the conditions that do not stop sanctions against Russia and to protect themselves from possible pressure from Western financial institutions, the BRICS countries need to consider the possibility of creating their own rating agency, as well as an analogue of the SWIFT system. The creation of a similar system within the BRICS Group is necessary in connection with the creation of the New Development Bank of the BRICS countries, as well as an agreement that involves the mutual opening of credit lines in national currencies by the central banks of the BRICS countries. As a result, BRICS had a purely technological need for an analogue of SWIFT. The deliberate under-reporting of the ratings of companies and the sovereign rating of various countries causes significant damage to the economic interests of these countries. The creation of an own rating agency that will serve as an alternative to the "big western three" (Standard & Poor's, Moody's and Fitch), which dominates the credit risk analysis market, will be another important step towards the financial independence of the BRICS countries. The BRICS countries consider the internationalization of national currencies as one of the important directions for the reforming the international monetary and financial system. Speech in this case is not only about measures to reduce the risks of destabilization of their currency markets associated with massive cross-border movements of foreign capital, but also about the overall increase in the role of national currencies in mutual settlements between the members of the "group of the strongest five countries" and the development of cooperation in the financial sphere.

At the time of cooperation between countries, the settlement and payment infrastructure was developed only between China and Russia, so the results so far are only calculated in national currency between these countries. Trading in yuan / ruble currencies is carried out on only two exchanges - the Chinese trading system CFETS and the Moscow Stock Exchange.

Despite the fact that certain results have been achieved, the volumes of operations with a pair of currencies on the Russian market are not yet significant. China currently regards the ruble as not a stable currency and does not use national currencies in strategic and long-term contracts.

Countries should ensure that member countries exit to domestic capital markets. The development of stock markets would be facilitated by issuing bonds in national currencies. The liberalization of foreign exchange relations will facilitate the access of capital to the markets of the BRICS countries.

Table 2. Trading volumes of the yuan / ruble currency pair on the Moscow stock exchange

	2011	2012	2013	2014	2015	2016
Объем торгов, млрд. руб.	4,5	7,7	37	305	858	776
Прирост, %	–	72	381	724	181	- 9,5

Source: Moscow Stock Exchange.

Table 3. Trading volumes of the yuan / ruble currency pair on the Chinese trading system

	2014	2015	2016
Объем торгов, млн. долл.	3976	3532	1760
Прирост, %	–	- 11	- 50

Source: China Foreign Exchange Trade System.

In addition to the mechanism of interbank cooperation, there are such formats of currency interaction as the Exchange Alliance and BRICS Business Council.

Creation of the Exchange alliance falls on 2011, which included the major large stock exchanges of member countries. This alliance involves the creation of new products and services. In 2012, a cross-listing of derivative financial institutions was introduced into its own indices, easy access to which can be obtained by investors in trading on the sites of the alliance countries.

In 2013, the BRICS Business Council was created, whose activities include determining the deterrent factors for the development of economic ties between countries and their overcoming. In 2014, the Council approved a list of sectoral working groups on strategic areas of interaction.

Both Russia and other participating countries need to clearly understand and take into account and calculate the consequences of the fact that even allies and close partner countries will always not only put their country's interests in the dominant, but also try to use and act out "Conjunctural trump cards", and only then "collective interest" and "geopolitical nuance".

In the modern conditions much attention is paid to the harmonization of financial relations of participants in world economic relations, which is connected with the development of international economic ties and ensuring the security of banking activities. World experience has shown that globalization in the financial sphere is directly related to the internationalization of the economy that provides development. A vivid manifestation of financial globalization is the transnational activity of intermediary banks, which ensure the economic interests of economic entities. A special group of emerging market problems is formed by the mechanisms of the activity of credit institutions - subjects of national banking systems and participants in the world capital markets. Globalization covers the financial markets of most countries, regardless of their economic development. There is a process of merging the national capital markets, goods, services and coordination of foreign economic policies of states, as shown in the Table 4.

It is important to note that the hasty liberalization of financial markets is dangerous for a country with weak banking systems, since there is a threat of a fall in the rate of the national currency and the

Table 4. Aspects of the internationalization of financial ties in the context of globalization

Aspects of the Internationalization of Financial Ties	Economic Basis	Evolution of World Economic Relations
Financial Globalization	A new stage in the development of financial cooperation based on geo-economics	The high level of interdependence of national economies as a result of historical development
The driving forces of financial globalization	International competition and the use of new technologies	Modernization of international reproduction
The role of national governments and regulatory institutions	Reduction of the institutions of national regulation	Transformation and Unification
Consequences	Expansion of the possibilities of international economic relations	Perfection of world economic relations
Theoretical concept	Reorganization of all spheres of life on the new principles of the world economy	Internationalization at the global level, regionalization, cardinal reorganization (reengineering) of international financial and economic cooperation

Source: Toloraya, G.D., Yarygina, I.Z. (2016). Harmonization of financial relations as a factor in the development of the BRICS. Banking Services, 2-9 pp.

intensification of crisis phenomena that have a negative impact on the world economy. In conditions of high volatility of financial markets, it is advisable to strengthen regulation, to balance the inflow of foreign capital, and to improve it by regulation.

Harmonization of financial regulations contributes to counteracting the criminalization of economic relations, facilitates the improvement of relations with customers and strengthens the security of banking activities. The progressive character of the formation of the BRICS financial strategy is a solid basis for the transnationalization of the activities of credit institutions operating in partner countries.

In turn, it is advisable to take into account the contradictory impact of transnationalization on the economies of developing countries.

Positive consequences include improving the macroeconomics of countries (improving operations, introducing advanced technologies, offering new products and services of high quality, borrowing experience in business administration, increasing customer confidence, strengthening the local credit and financial market due to the high reliability of foreign investors, products and services on the local market); formation of the strategic alliances in the banking and corporate business.

It seems possible to attribute to the negative consequences of transnationalization, for example, the derivation of profits in various forms from the country of operations, which worsens the balance of payments of the host country and reduces the revenue of the state budget. In addition, it increases the import of equipment and materials, which worsens the local payment budget and reduces the gross domestic product. In general, thus, local business is suppressed by foreign ones. The formation of a roadmap for short-term and long-term coordinated financial and economic development of partner countries will contribute to the progressive development of the BRICS economies in order to improve the living standards of the population.

The expansion of foreign capital into the markets of developing countries and the desire to strengthen their business niche in the global economy are associated with the consolidation of economic capital, which determines the key universal trend in the development of national banking systems. Under the consolidation of economic capital, the authors mean the strengthening of the aggregate of tangible and intangible assets that support the activity and development of credit institutions in a competitive environment. A widespread form of consolidation of economic capital in modern conditions is the strategic partnership of the local and foreign banks. This partnership ensures the formation of conditions for the foreign bank to enter the local market, reduce the time limitation of development, replenish technological and administrative resources, expand the horizons of professional activity.

Foreign entities use the method of direct or portfolio investment when entering the local market, through opening a branch or acquiring a national bank, or buying a stake in a national bank.

Practice has proved a great relationship between the use of new technologies and the development of banking systems. It seems advisable within the BRICS to pay special attention to the growing role of cyber banking, which is based on the objective needs of geo-economics. A unified approach to solving technological problems and ensuring the security of financial relations of BRICS will create favorable conditions for cooperation.

The main source of reserve assets in the modern financial architecture is still the dollar. In the BRICS countries, which are the largest holders of international reserves (39% of the world ones and 69% of the reserves of the developing countries), it holds the largest share in foreign exchange reserves (58%). The instability of the main reserve currencies, above all the USA dollar, which has intensified in recent years, casts doubt on the stability of the current monetary system, and the unconventional monetary policy of the United States intensifies the volatility of world markets, which adversely affects developing countries, including the BRICS countries.

It is planned to create a single payment system within the BRICS, which can become an undoubted advantage for their independence and competitiveness. The question of the introduction of a single currency is not yet going on, it requires a deeper level of integration. However, countries tend to conduct trade transactions in national currencies. The necessary infrastructure has been created for a long time, but real deals and progress in this direction have appeared more recently in connection with the aggravation of the tension of the geopolitical situation around Russia; attempts of negative impact on it through various mechanisms of the dollar settlement system; a sharp weakening of the ruble.

Calculation in national currencies can ensure mutual benefit and minimization of risks (reduction of costs for currency purchase, no need to hedge currency risks), if subsequently there is no need to convert them. This situation arises if the mutual payments of partners are relatively similar in volume, which will allow for the conduct of clearing operations, or one of the parties is interested in investing in the country of the counterparty.

Another motivating factor that plays a significant role in this process may be political and geopolitical aggregate interest. This tool can be an effective mechanism for strengthening the macro-regional security of BRICS. Certain threats may arise in connection with currency and transaction risks, but with a stable and evolving system of international economic relations between the interacting countries, they will be avoided or minimized.

The activation of the process of using the national currency for trade in energy resources and other goods is recognized at the official level as one of the important tasks of developing international cooperation of Russia. China has a significant potential for development of this direction, especially taking into account the expansion of Russian-Chinese cooperation in the context of Western sanctions against Russia. The use of this system within the regional integration group and, in cooperation with external economic partners, will further strengthen both the economic security of each individual country and the entire association as a whole. This, undoubtedly, should positively affect the financial stability of the regional integration group.

Expanding interbank cooperation and reducing the use of the dollar and euro in its own settlement system will increase financial security for the BRICS countries. Within the framework of the BRICS interbank cooperation mechanism, the development banks of the five countries signed a general agreement on the provision of loans in national currencies and an agreement on the confirmation of credit letters. The document itself provides for reducing the dependence of local currencies on the dollar and euro. Foreign currency markets of the "most strongest five countries" will also develop on the basis of the exchange alliance of the BRICS countries. This became possible due to the agreements reached in October 2011 between the key BRICS exchanges. At present, the exchanges have started to carry out a cross-listing of futures contracts for the key stock indices of the BRICS countries.

A potentially profitable and promising "integration project" may be the embodiment of the idea of creating transport corridors "China" - "Europe". Its implementation entails the need to create and a real opportunity for the development of logistics centers, the construction of roads, gas stations, systems for processing and storing goods, the construction of hotels - a set of small derivative projects with the potential participation of private business and private investment from all BRICS member countries, places, will raise the level of incomes of a number of strata of the population.

To achieve stable development indicators that contribute to the financial and economic security of BRICS, it is necessary:

1. Creation of territories for advanced development or special economic zones. It will allow to test and develop a number of effective mechanisms to increase the investment attractiveness of countries, which can later be used on the scale of the entire BRICS grouping.
2. Development of "road maps" aimed at implementing measures to simplify, reduce the cost and accelerate the procedures on the territory of the BRICS related to doing business; to reduce administrative barriers and transaction costs.
3. Stimulating competition in the sphere of attracting foreign capital; active promotion of investment attractiveness of countries in general; development of the business environment (creation of a council for improving the investment climate, approval of the investment strategy, etc.).
4. Support for high-tech industries, science and education.
5. Support the development of fundamental science, language and business education to maintain a high level of qualification of human resources and build an innovative economy in the BRICS countries.
6. Ensuring environmental safety in the modernization of economies and in the process of innovative development.
7. Financing and developing the necessary infrastructure to support start-ups (newly established companies in the development or exploration of promising markets).
8. Improving the migration regime: lifting quotas for highly qualified specialists and granting them preferential treatment; control over compliance with the quota of unskilled labor.

Cooperation of BRICS partners on international financial and economic issues will remain for a long time one of the key directions of the site's work taking into account its great importance as a factor of ensuring financial stability and economic security.

Priority tasks in this area are:

- A qualitative strengthening of the coordination of the positions of the BRICS economies on the issues of the G20 agenda during the chairmanship of Turkey (2015) and the PRC (2016), based on the decisions of the St. Petersburg G20 summit (September 2013). The work should be conducted in the following areas: promotion of reforms of international financial institutions, primarily the IMF;
- Implementation of the Framework Agreement for ensuring a confident, sustainable and balanced growth, including the implementation of "comprehensive growth strategies" and minimizing the negative "side effects" from the normalization of monetary policy in developed countries;
- The development of a common position of the BRICS countries with respect to the consequences for their economies from the curtailment of unconventional monetary policy in the United States and other developed countries;
- The beginning of the full-format work of the New Development Bank and the Pool of Conditional Foreign Exchange Reserves of the BRICS countries;
- The adoption by BRICS participating States of measures to increase accessibility

- Financial services for the population, combined with increased protection of consumers' rights of such services and increased financial literacy of citizens;
- Strengthening coordination with BRICS partners as part of the implementation of the Action Plan to Combat the Dilution of the Tax Base and the Withdrawal of Tax Benefits (BEPS) and consideration of possible steps aimed at de-orphotizing national economies.

The next stage requires the development of new products that can be created on the basis of the currencies of the BRICS countries. With the transition to a new model of sustainable economic development that meets the requirements of the global economy of the 21st century, investment plays a special role. To deepen investment cooperation between the BRICS countries concrete measures must be taken. The most important of these could be the development of the Roadmap for Investment Cooperation, which will stimulate the activation of other forms of foreign economic interaction, primarily foreign trade, industrial and scientific and technological cooperation. Stimulation of mutual investments of the BRICS countries can be carried out within the framework of both general stimulation of foreign investments and additional measures for each specific cooperation project. In particular, through the provision of benefits for individual taxes, customs payments, the selective creation of special conditions for doing business, etc.

To increase the effectiveness of the dialogue and develop common positions in the sphere of the security, it is also necessary to intensify the cooperation of the BRICS expert centers and conduct joint research. According to the rating of the University of Pennsylvania, the TOP-150 of the world's leading centers include 25 research organizations from the BRICS countries, the Top 100 non-American expert centers of the world include 19 organizations from the BRICS countries, and 9 organizations from the BRICS countries are in the Top 70 of all the best centers peace in the sphere of international relations and security. The development of cooperation between the leading expert centers of the BRICS countries will help to develop common positions, develop a dialogue on foreign policy planning, a qualitative analysis of the policy pursued, and the formation of future joint projects.

CONCLUSION

Summing up the potential results of the measures examined, the authors can say that they reinforce the claims of the five states to their wider representation in international financial institutions and, in general, in the management of the world economy. The creation of a joint financial infrastructure within the BRICS Group will give an important impetus to the real sector of the economy. The use of national currencies in mutual settlements and lending will reduce the association's dependence on the conjuncture in traditional centers of the world economy. The development of measures to manage financial globalization in the face of increased threats and challenges is necessary to ensure the financial and economic security of the BRICS Group.

Looking at the near future, we have serious reasons to talk about the trend of further strengthening and stability of all BRICS currencies. They are connected with the strengthening of the positions of the economies within the framework of the closer economic alliance of the group, the creation of its own "financial security cushion" in the form of a bank and pool (possibly in the future the establishment of an analogue of SWIFT and a joint aggregate currency) and a more intensive shift to world trade in CHY and RUR. The forecast for the future of currencies has good reasons.

REFERENCES

BRICS Economic Partnership Strategy. (2015). Retrieved April 15, 2017, from http://static.kremlin.ru/media/events/files/ru/KT0SBHnIZjOpIuAj2AOXCnszNQA8u7HL.pdf

Delhi Declaration on the results of the 4th Summit on March 29, 2012. (n.d.). Retrieved March 27, 2017, from http: //www.brics.mid.ru/brics.nsf/WEBdocBric/1DB82FF3745BA8FE442579D5004778E6

Fomicheva, E. V. (2015). Geoeconomic positioning of Russia in the current conditions of international cooperation: New challenges and protection of national interests. South Federal University.

Glinkina, S. P. (Ed.). (2014). *New direction of the Russian foreign and foreign economic policy - interaction in BRICS*. Moscow: Institute of Economics, Russian Academy of Sciences.

Kheifets, B. A. (Ed.). (2014). Russia and the BRICS. New opportunities for mutual investment. Moscow: Publishing and Trading Corporation "Dashkov and K".

Samarukha, V. I. (n.d.). *On the role of the BRICS countries in the formation of a multipolar world economy*. Retrieved from https://library.ru/item.asp?id=26226239

Shavshukov, V. M. (2015). BRICS in the system of global finance: ten years later. World Economy, 6, 141-160.

The World Bank. (n.d.). *Indicators, 2000-2016*. Retrieved April 11, 2017, from http: //data.worldbank.org/indicator

Toloraya, G. D., & Yarygina, I. Z. (2016). Harmonization of financial relations as a factor in the development of the BRICS. Banking Services, 2-9.

Chapter 9
The Use of Fractal Analysis in Assessing the Reliability of Taxpayers Information

Galina Artemenko
Southern Federal University, Russia

Dmitry Artemenko
Southern Federal University, Russia

ABSTRACT

Tax authorities often have to make decisions under conditions of incomplete information about the actual situation. The incompleteness and uncertainty of information may arise from the actions of the two sides with distinct interests. One side is the tax authority (local, regional, or national level), and the other is the taxpayers (organizations or individuals). This largely explains the interest of tax authorities in the creation of software capable of providing quantitative and qualitative analysis of the declarations submitted by taxpayers to identify suspicious declarations for their additional inspection. The purpose of this chapter is to carry out a comparative pre-forecast analysis of the time series enterprise performance indicators and time series of its tax payments.

INTRODUCTION

Information provided to tax authorities by the taxpayer is the subject of research in this chapter.

The authors propose a model of controlling the reliability of the information provided to tax authorities by the taxpayer. Specificity of the model consists in using a fractal analysis of time series of tax payments. Also, when constructing the model, a phase analysis of the time series is used.

The study was carried out with the aim of developing methodical and information support of tax authorities, whose decision support systems are currently limited to the use of statistical methods.

DOI: 10.4018/978-1-5225-3767-0.ch009

METHODOLOGY AND BACKGROUND

The authors proposed the use of models, tools and methods of fractal analysis, methods of nonlinear dynamics, namely R / S analysis, phase analysis. In addition, the methods of comparison, statistical analysis and graphical interpretation were used in the work. The model proposed by the authors will allow the tax authorities to improve the quality of control over the reliability of the initial information of taxpayers.

The proposed approach can be used for all groups of taxpayers, especially for legal entities. As practice shows, employees of tax authorities almost always have to make decisions in conditions of incomplete information about the actual state of affairs of taxpayers. Due to inconsistencies between the interests of taxpayers (legal entities and individuals) on one hand and tax authorities (local, regional or state level) on the other, there is incompleteness and uncertainty of information coming to the tax authorities. This largely explains the interest of tax authorities in the creation of software products that can provide quantitative and qualitative analysis of declarations submitted by taxpayers for the purpose of identifying questionable declarations for their additional verification (D.G. Chernik et al., 2000).

As noted in D.G. Chernik et al. (2000), for today such programs are already actively being created and used. The basis of the algorithm for such a program is the recommendations of experts. Based on their rich experience and specific evidence they define the degree of suspiciousness granted by the declaration. The problem of tax evasion exists even in countries with well-developed, established taxation systems, and therefore it is widely discussed in the economic literature.

Cowell and Gordon (1995) compare the various available strategies of audit when collecting an indirect tax. They modeled tax evasion in the following way. The taxpayer company makes a choice between taxable activity in the legal market and activities in the shadow sector of the economy. During the inspection of a firm, if the fact of its activity in the shadow sector is revealed, then the missing tax and a fine is imposed on it. One possible strategy is to randomly check any firm with a certain fixed probability. An alternative policy is the decision-making, depending on the available information about each taxpayer. Cowell and Gordon consider a simple form of such a policy: the probability of verification depends on the declaration according to the "cut off" rule, i.e. firms declaring income less/no less than a certain amount, are always checked/never checked (Maksishko N.K. et al., 2008).

Chander and Wilde (1992) analyze the interaction of tax inspectors (auditors) and taxpayers, given the possibility of corruption emergence. In the model proposed by them (called the CW-model), the taxpayer has a high or low income with a certain probability and, depending on real income, he is obliged to pay taxes.

Vasin A.A. and Panova E.I. (1999) consider a model similar to the Chander and Wilde model and they solve a similar problem of comparative analysis. However, there are a number of differences between these approaches.

The CW-model determines the probability of audit check by Nash's equilibrium strategies of the players, while the probability of re-verification is fixed. In fact, the inspection leadership controls both indicators.

The main difference of the CW-model from the Nash's equilibrium strategies is the consideration of optimal rules of check instead of fixed probabilities, when conducting a comparative analysis of the net tax charge, depending on the size of fines and tax rates.

If take into account the fact that it is easier to change the inspection rules than the rates of taxes and penalties established by the legislation, the considered approach seems justified.

In this chapter, the authors propose to use methods of nonlinear dynamics as a supplement to the approach described above, applicable to background information in the form of time series (Z. Wang, A. Bovik, H. Sheikh, & E. Simoncelli, 2004).

On the basis tools of nonlinear dynamics, it is theoretically possible to construct a model that will allow forecasting the size of the tax payment at the end of the reporting period. In case of a significant discrepancy between the forecasted value and the actual amount of taxes paid by an enterprise (in the direction of decreasing the latter), the enterprise should be subjected to additional verification for the fact of reliability of information submitted to the tax authorities (Maksishko N.K. et al., 2008).

It is important to note that only under abidance adequacy of the constructed forecast model and the reliability of the information previously submitted to the tax authorities, it is possible to use the above approach.

In order to build an adequate forecast model, at a minimum, it is necessary to conduct a deep pre-forecast analysis for the research system. The object here can be time series of observations for a fairly long period of time (≥100 observations). Pre-forecast analysis is one of the most important stages by which the construction of an adequate predictive model becomes possible. Based on this analysis, a decision can be made whether it is possible to build a forecast for the system in general, and whether it is possible to obtain an accurate estimate of the forecast corresponding to a certain level.

Pre-predictive analysis involves studying the specifics of changes in a particular time series.

Modern approaches to the study of this subject are concluded, as a rule, in the search for an adequate answer to the next questions: in the researched time series, there is a trend and a cyclic component (Peters, E. E., 1994), in carrying out this type of analysis, there is a deterministic chaos (Sigel E., 2002) and whether there is a latent fractional quasi periodicity (Sergeeva L.N., 2002).

GOAL AND TASKS

The purpose of this study is carrying out a comparative pre-forecast analysis of the time series which are comprised of enterprise performance indicators (basic – the shipped products time series) and time series of its tax payments.

The tool for holding pre-predictive analysis are the so-called sequential R/S analysis algorithm (Maksishko N.K. et al., 2008; Perepelica V.A. et al., 2005), as well as the phase analysis of time series (Maksishko N.K, et al., 2008; Yanovsky L.P., 2000).

If using these methods, cyclicity, periodicity, and long-term memory are identified in the investigated system, then it makes sense to proceed to the forecasting (the second stage).

STATEMENT OF PROBLEM AND LEARNED RESULTS

For the pre-forecast analysis, the authors used the data of the accounting statements of the Russian Joint Stock Company «Rostselmash».[1] The study period is 5 years, from 2010 to 2015, the interval of the time series is one month. The main indicators of the enterprise for the construction of time series are finished products, shipped products, value-added tax, profit tax, personal income tax and unified social tax.

These indicators are represented by the following symbols:

Time series of the finished products

$$Z^1 = \{z_i^1\},\ i = 1,2,3,\ldots,n^1,\ n^1=223; \tag{1}$$

Time series of the shipped products

$$Z^2 = \{z_j^2\},\ j = 1,2,3,\ldots,n^2,\ n^2=234; \tag{2}$$

Time series of the value-added tax

$$Z^3 = \{z_k^3\},\ k = 1,2,3,\ldots,n^3,\ n^3=94; \tag{3}$$

Time series of the income tax

$$Z^4 = \{z_l^4\},\ l= 1,2,3,\ldots,n^4,\ n^4=80; \tag{4}$$

Time series of the personal income tax

$$Z^5 = \{z_m^5\},\ m = 1,2,3,\ldots,n^5,\ n^5=67 \tag{5}$$

Time series of the unified social tax

$$Z^6 = \{z_t^6\},\ t = 1,2,3,\ldots,n^6,\ n^6=228 \tag{6}$$

where

Z_r^v = the time series designation for each of the 6 variables;
r = the number of observations in the sample for each variable;
V = the sequence number of the time series for each variable (v = 1, ..., 6).

All time series are described by indicators measured in rubles, and the time series of finished products $Z^1(1)$ - at the cost of goods, the time series of shipped products $Z^2(2)$ - at the selling price of the products sold. The researchers also note that the time series of the finished products characterizes the volume of production of the enterprise, while the time series of the shipped products - the volumes of sales. The authors emphasize that the above taxes, namely value-added tax, profit tax, personal income tax and unified social tax, were selected from the entire list of taxes paid by enterprises not accidentally. These types of taxes are compulsory for all enterprises that are on the general tax regime (Gilmore C.G., 1993).

For all of the above taxes, there are accruals at fixed interest rates, all of them must be declared in the tax authorities, taking into account the above-described relationship between the series in question, suggests the emergence of common trends and similarities in the change of all-time series. It should be noted that finished products -are not always sold (shipped) products. Part of it consists of commodity stocks (remains unrealized in stock). Shipped products however, aren't always revenue, because sometimes it happens that business partners do not have the opportunity to pay contractual obligations on time. In addition, there are situations when the amount payable for the shipped goods (accounts receivable) hangs on the accounts of the enterprise for a long period, and sometimes it is not repaid at all. Risks related to the sales of products and the repayments of receivables, in the current period of time are most often found in the real sector of the economy. They have a very tangible effect on the dynamics development of the entire enterprise, which also affects the time series that are discussed in this chapter.

It should be noted that analysis, forecasting, management of the above risks - occupy a key position in the management of a modern enterprise, and in planning the effective functioning of tax authorities.

Figures 1 and 2 are diagrams of the segments of the shipped products time series and the value-added tax time series.

In these diagrams, the number of observations (or the length of the time series) is shown along the abscissa axis, and the amount of cash inflow - along the ordinate axis.

As a result of the statistical analysis of the time series were constructed graphics of empirical distribution functions of the random values. The study allowed the authors to obtain numerical values of mathematical expectations, variances, standard deviations and coefficients of the asymmetry and kurtosis for time series. All results are presented in Table 1 and in Figures 3 - 7. From the graphs of the empirical

Figure 1. Diagram of the shipped products time series Z^2 (2)
Source: Compiled by the authors according to the materials of the financial statements of the Russian Joint Stock Company "Rostselmash".

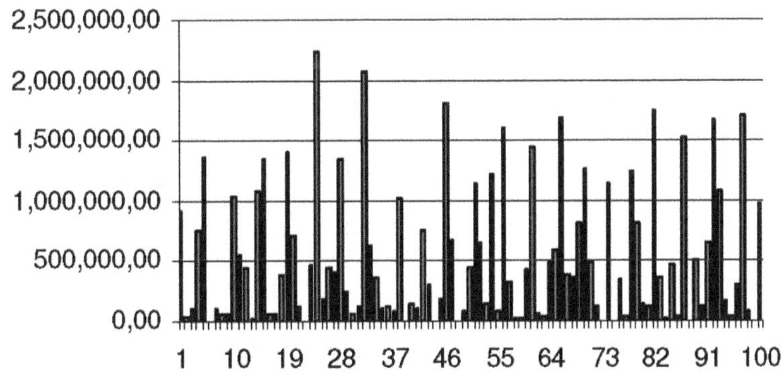

Figure 2. Diagram of the value-added tax time series Z^3 (3)
Source: Compiled by the authors according to the materials of the financial statements of the Russian Joint Stock Company "Rostselmash".

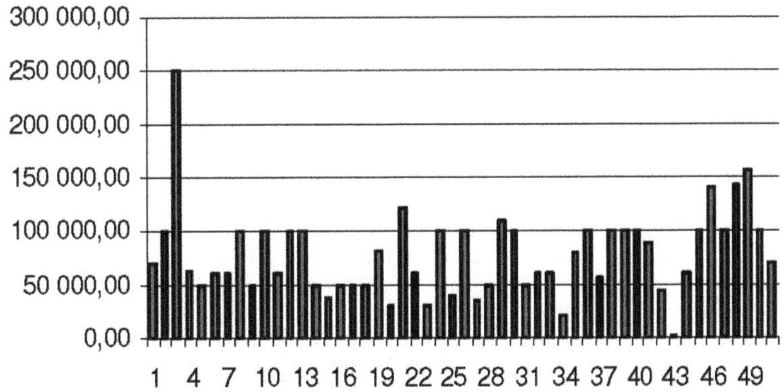

distribution functions of the time series Z^1, Z^2, ... Z^6 (Figures 3 - 7), as well as the obtained values of statistical indicators (Table 1), it becomes obvious that time series do not obey the normal distribution law. This is the reason for the low informativeness of the statistical estimates obtained, as well as for the low "predictability" of the time series by classical methods (Peters, E.E., 1991).

Figure 3. Histogram of the empirical distribution of finished products time series Z^1 (1)
Source: Compiled by the authors according to the materials of the financial statements of the Russian Joint Stock Company "Rostselmash".

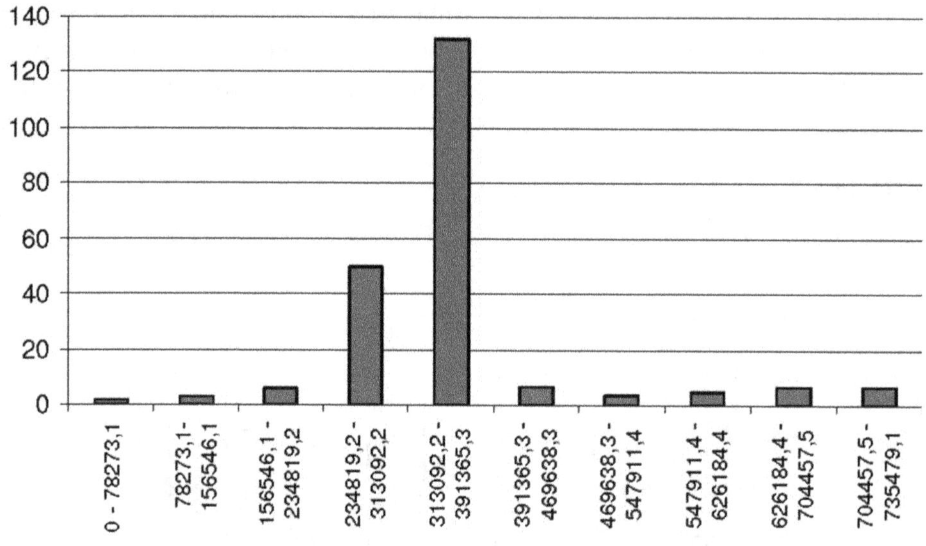

Figure 4. Histogram of the empirical distribution of shipped products time series Z^2 (2)
Source: Compiled by the authors according to the materials of the financial statements of the Russian Joint Stock Company "Rostselmash".

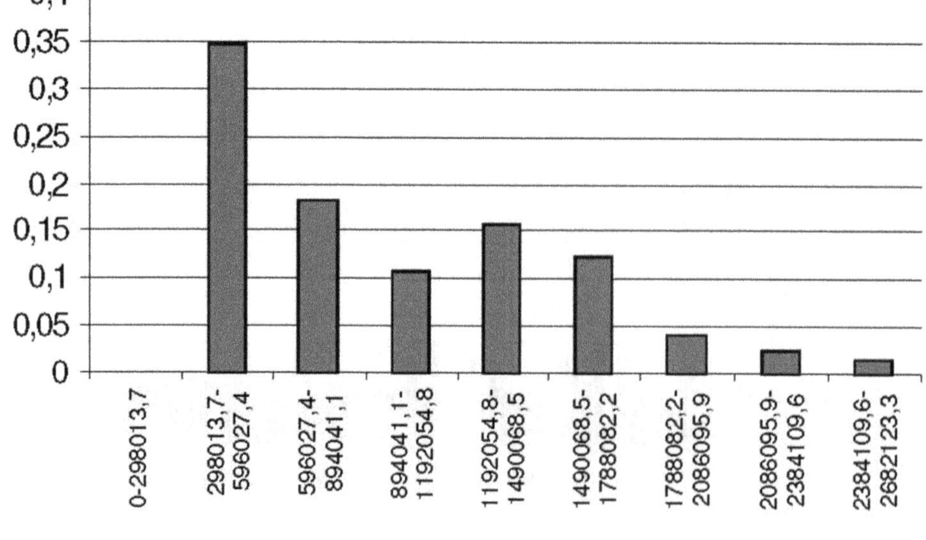

Table 1. Statistical indicators of time series risk $Z^1(1)$-$Z^6(6)$

Time Series Name	Mathematical Expected	Mean Square Deviation	The Coefficient of Variation	Asymmetry Coefficient	Kurtosis Coefficient
$Z^1(1)$	354166,875	114546,1719	0,323424	1,696644	4,0082252
$Z^2(2)$	551855,3125	613372,4375	1,111473	1,183048	0,430439
$Z^3(3)$	82254,73438	53852,33594	0,654702	2,034386	5,153335
$Z^4(4)$	44247,93359	36681,02344	0,828988	1,515809	3,262901
$Z^5(5)$	42310,04297	36596,21094	0,864953	1,709008	3,541281
$Z^6(6)$	35542,94531	38078,30859	1,071332	2,062252	5,687513

Source: Own calculations

Figure 5. Histogram of the empirical distribution of value-added tax time series Z^3 (3)
Source: Compiled by the authors according to the materials of the financial statements of the Russian Joint Stock Company "Rostselmash".

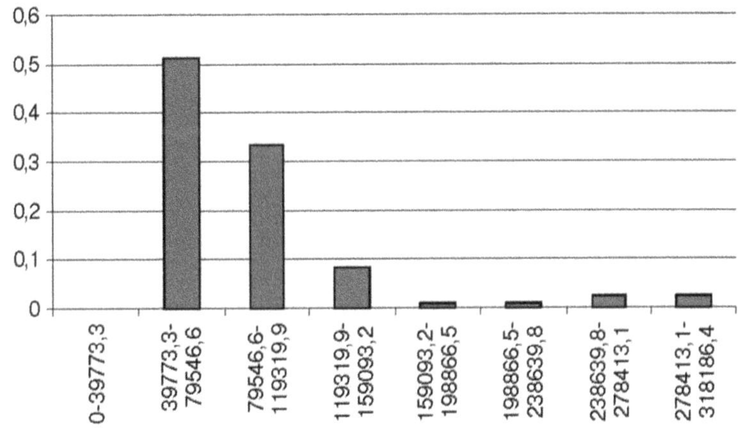

Figure 6. Histogram of the empirical distribution of income tax time series Z^4 (4)
Source: Compiled by the authors according to the materials of the financial statements of the Russian Joint Stock Company "Rostselmash".

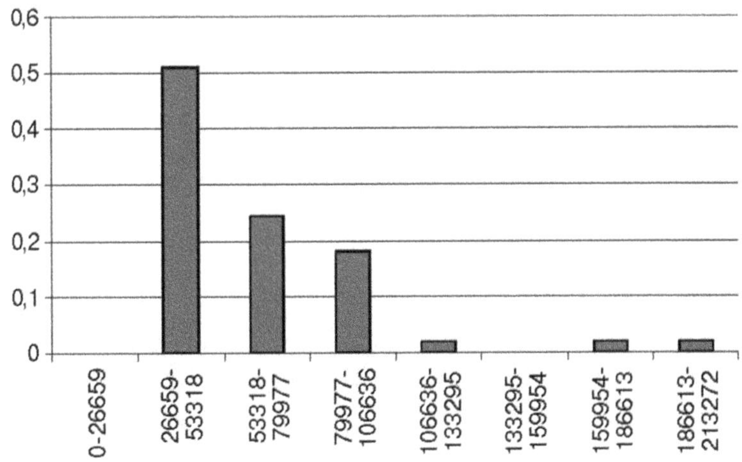

Figure 7. Histogram of the empirical distribution of personal income tax time series Z^5 (5)
Source: Compiled by the authors according to the materials of the financial statements of the Russian Joint Stock Company "Rostselmash".

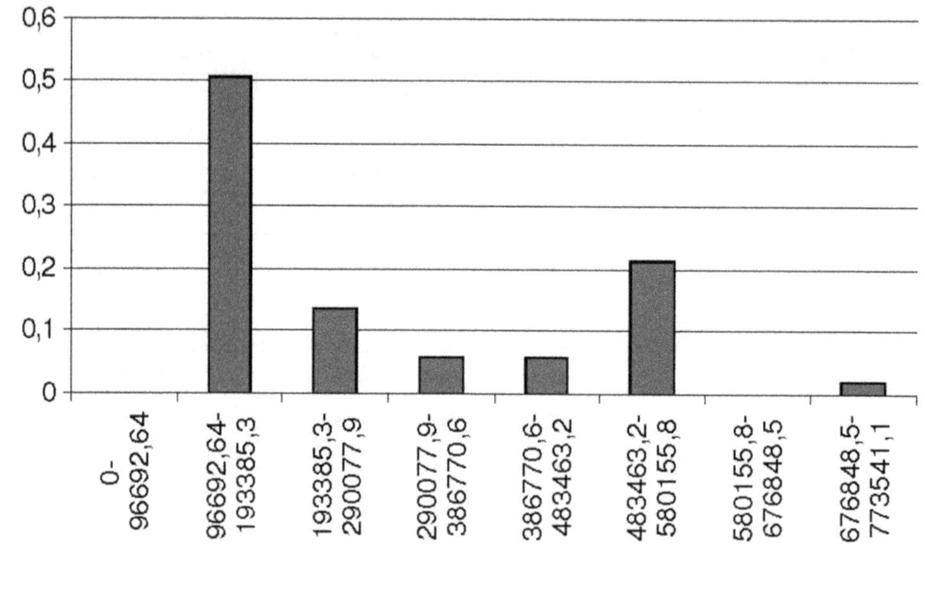

In order to eliminate this shortcoming (in the sense of choosing a more adequate forecasting method) the authors perform a pre-forecast analysis of one pair of time series of shipped products Z^2 (2) and value-added tax Z^3 (3) (Figures 1, 2). For this authors use other approaches. This is a set of methods for nonlinear dynamics: sequential R/S analysis and phase analysis. It should be taken into account that the value-added tax has the most stable taxable base. This tax is most proportional to the taxable base. The key factor in the taxable base for calculating the value-added tax is the time series of the shipped products.

As a result of applying the phase analysis method (Janovskij L.P., 2000) for the time series of products shipped $Z^2(2)$ and the value-added tax $Z^3(3)$, phase trajectories are constructed. And then each phase trajectory is decomposed into quasi-cycles (Figures 10, 11).

In the phase trajectories of time series of products shipped and value-added tax 46 and 19 quasi cycles were identified. An example of typical quasi-cycles in this decomposition is shown in Figures 10 and 12. To the same time series the algorithm of sequential R/S-analysis was applied (Stanislavchik E.N., 2002), in conclusion sequences of H-trajectories and R/S-trajectories were obtained. Exhaustion of a quasi-cycle in the corresponding time series (Figures 9, 11) is determined by the point number of the trend change of a particular R/S-trajectory. It should be specially noted that the lengths of almost all quasi-cycles obtained by phase analysis coincided exactly with the lengths of the corresponding quasi-cycles obtained by means of a serial R/S analysis (Figures 9-12).

Figures 3-8 show histograms of the empirical distribution functions of time series for the six analyzed variables (1 - 6).

Figures 9, 10 show the phase trajectories of time series of shipped products and value-added tax.

Further the authors can observe examples of a quasi-cycle of the time series of shipped products $Z^2(2)$, achieved on the basis of R/S analysis and phase analysis. The beginning of the quasi-cycle corresponds to the level $z^2_{156} \in Z^2$.

Figure 8. Histogram of the empirical distribution of the unified social tax time series Z^6 (6)
Source: Compiled by the authors according to the materials of the financial statements of the Russian Joint Stock Company "Rostselmash".

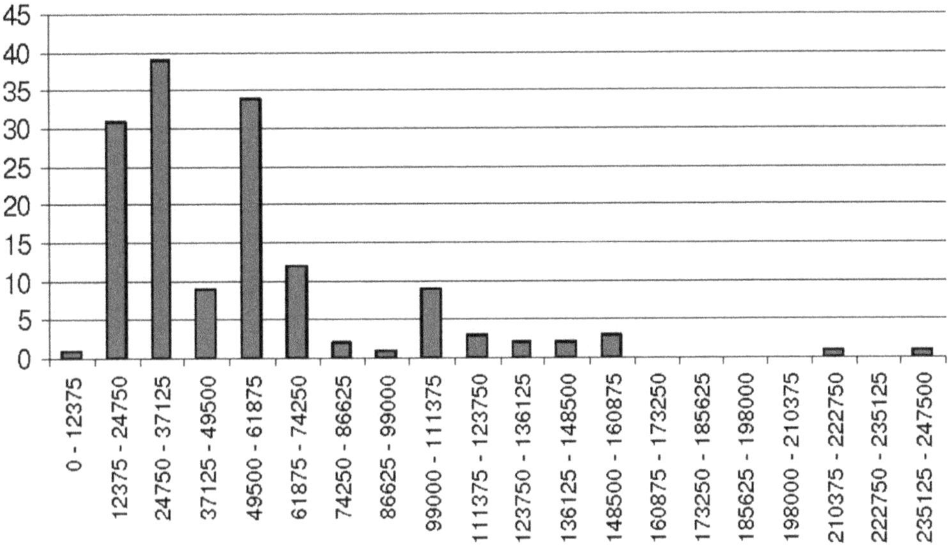

Figure 9. Phase trajectory of the time series of shipped products
Source: The official site of the corporate information disclosure center // http://www.e-disclosure.ru/portal/ files.aspx?id=421&type=3, own calculations.

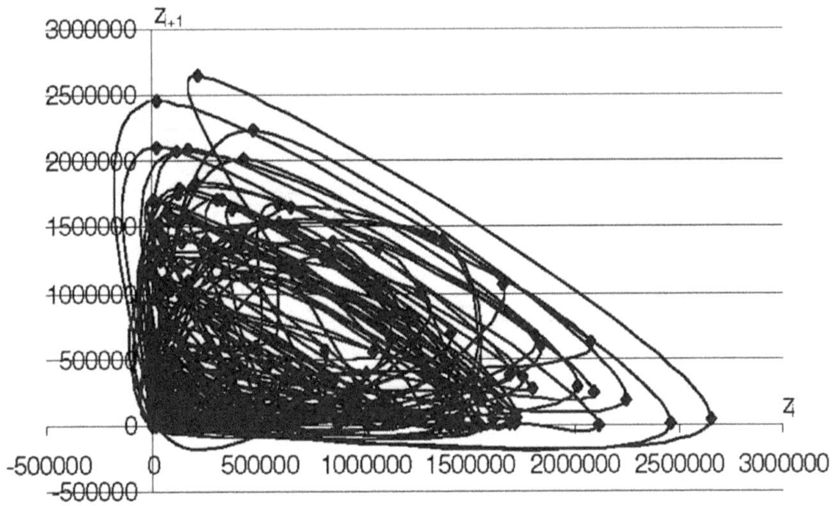

On the abscissa axis, a log (observation number t / 2) is plotted, and log (R/S) ; H (t/2) – is along the ordinate axis.

Then the authors considered examples of a quasi-cycle of the time series of value-added tax Z3(3), obtained on the basis of R/S analysis and phase analysis. Beginning of the quasi-cycle corresponds to the level - z326 € Z3

Figure 10. Phase trajectory of the time series of value-added tax
Source: The official site of the corporate information disclosure center // http://www.e-disclosure.ru/portal/ files.aspx?id=421&type=3, own calculations.

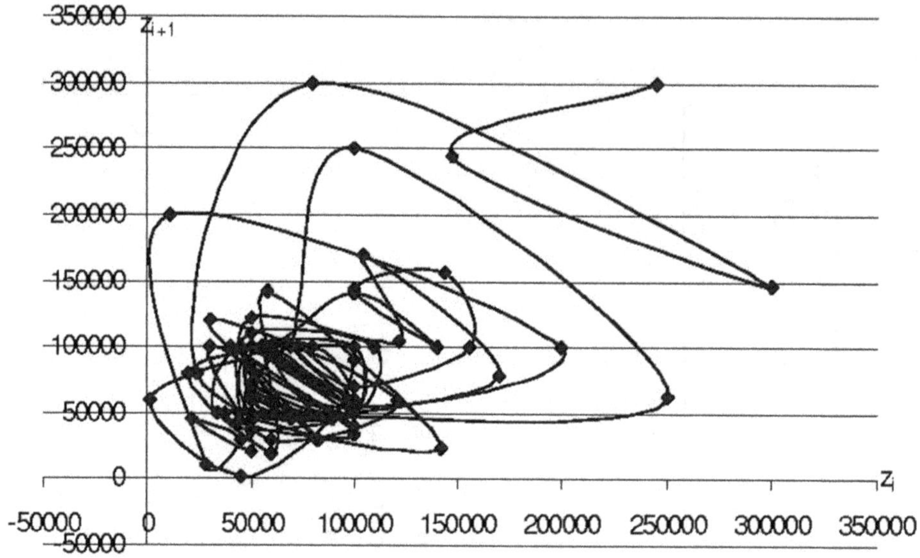

Figure 11. Quasi-cycle of the time series of shipped products Z2(2), achieved on the basis of R/S analysis
Source: The official site of the corporate information disclosure center // http://www.e-disclosure.ru/portal, own calculations

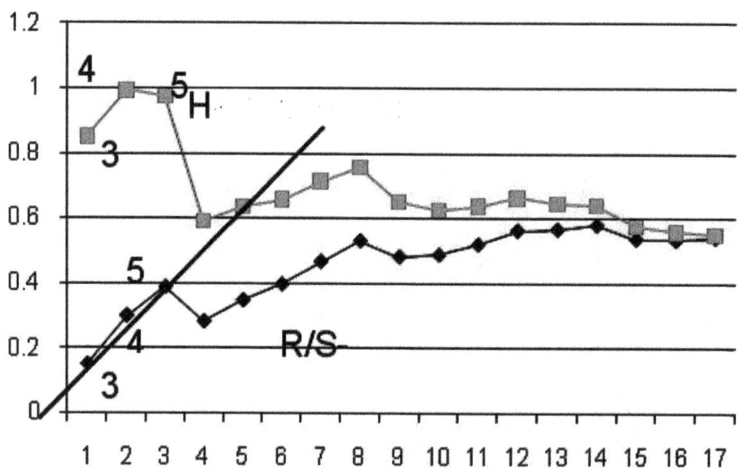

For further presentation of the comparative fractal analysis of time series of shipped products and time series of value-added tax, the authors introduced the following notation:

- p is the length of the quasi-cycle;
- h^r_p is the frequency of appearance of quasi-cycles of length p in the series Z^r, $r = 1, 6$ (at %) for the conducted sequential R / S-analysis;

Figure 12. Quasi-cycle of the time series of shipped products Z2(2), achieved on the basis of phase analysis
Source: The official site of the corporate information disclosure center // http://www.e-disclosure.ru/portal, own calculations

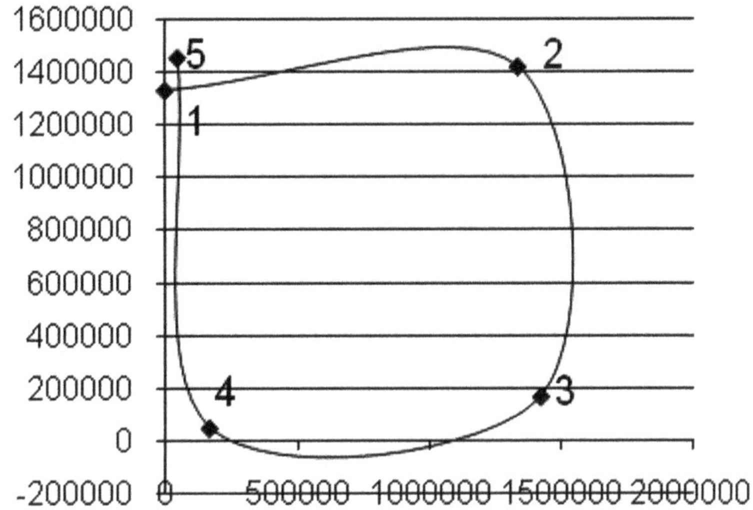

Figure 13. Quasi-cycle of the time series of value-added tax Z3(3), obtained on the basis of R/S analysis
Source: The official site of the corporate information disclosure center // http://www.e-disclosure.ru/portal, own calculations

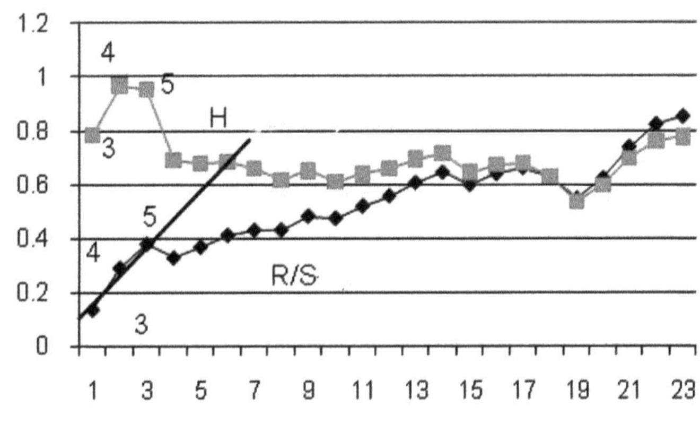

Figure 14. Quasi-cycle of the time series of value-added tax Z3(3), obtained on the basis of phase analysis
Source: The official site of the corporate information disclosure center // http://www.e-disclosure.ru/portal, own calculations

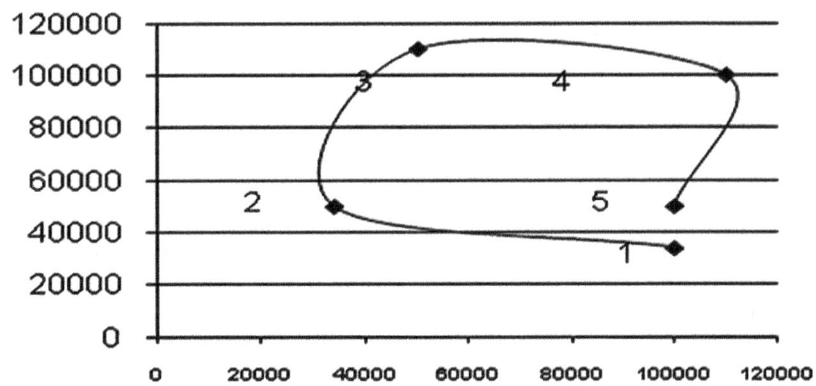

- q^r_p is the frequency of appearance of quasi-cycles of length p in the series Z^r, $r = 1, 6$ (R / S analysis) for the conducted phase analysis.

For time series of shipped products and value-added tax, consider the statistics of pairs (p, hp2) and (p, hp3) obtained from the R / S analysis, and pairs (p, qp2) and (p, qp3), obtained from the results of phase analysis.

By analyzing the direction of quasi-cyclic element rotation on figures 9-10 (clockwise or anticlockwise), it is clear that the majority of elements have a clockwise rotation, which is fundamentally important for carrying out the prediction analysis. However, there are some quasi-cycles which include elements that rotate anti-clockwise, although their part is 5/48, which means not more than 10,5%.

Pre-forecast information of time series phase portrait can be divided into 3 groups.

1. Pre-forecast information can be represented by decomposing the phase portrait of the time series into quasi-cycles.
2. Pre-forecast information is a trajectory of the drift of the quasi-cycles centers. Numbers of dots on such drifting trajectory match to the numbers of corresponding quasi-cycles, and the coordinates of these dots represent the coordinates of the centers of corresponding quasi-cycles.
3. Pre-forecast information is the drift trajectory of semi perimeters of the dimensional rectangles of quasi-cycles, which are attained as a result of phase portrait decomposition, and also by the phase portrait itself of this trajectory.

Based on the proposed phase analysis tools for time series Z2 (2) and Z3 (3), the following information can be formulated for further forecasting of the indicated time series:

- According to the drift trajectory of the centers, time series have the following clearly expressed regularity: the alternation of the increase and decrease of the coordinates of the points of the centers;
- The overwhelming majority of links of the previous quasi-cycles have a clockwise rotation direction;
- In Figures 12, 14 the decomposition of the phase portrait (Figures 9-10) on quasi-cycles there are no unfinished quasi-cycles, in other words, the last quasi-cycles are complete.

Based on the foregoing, the following recommendations for predicting the Z j + 1 level in the time series Z2 (2) and Z3 (3) under consideration can be proposed. The predicted value of Z j + 1 determines the predicted link Zj (Zj + 1), which belongs to a new clockwise rotating quasi-cycle. It leaves the ordinate of the predicted point on the phase portrait without any significant changes. In other words, in its essence, the forecast reduces to the fact that the predicted value of Zj + 1 will differ little from the value of the previous level Zj.

The final representation of these pairs is presented in Tables 2 and 3, as well as in Figures 15, 16.

In Figures 15 and 16 it is shown that the greatest number of quasi cycles for time series of shipped products corresponds to quasi-cycles of length 6, and for time series of value added tax – length 4.

Graphical representation of the quasi-cycles percentage ratio for time series of value added tax and shipped products was achieved by applying parallel R/S analysis and phase analysis.

Table 2. Percentages of the lengths of the allocated quasi-cycles of time series of shipped products and value added tax, received through the R/S analysis (at %)

Length of the Quasi Cycle (p)	**1**	**2**	**3**	**4**	**5**	**6**	**7**	**8**	**9**	**10**
(h^2_p)	0	0	12	17	25	28	14	3	0	0
(h^3_p)	0	0	21	39	18	5	7	4	5	1

Source: Own calculations.

Table 3. Percentages of the lengths of the allocated quasi-cycles of time series of shipped products and value added tax, received through the phase analysis (at %)

Length of the Quasi Cycle (p)	**1**	**2**	**3**	**4**	**5**	**6**	**7**	**8**	**9**	**10**
(q^2_p)	0	0	17	15	20	41	2	4	0	0
(q^3_p)	0	0	26	37	11	0	11	11	5	0

Source: Own calculations.

In Figures 15 and 16, p is the length of a quasi-cycle, hp is the frequency of appearance of quasi-cycles of the specified length (in percent).

In Figure 15, pink color indicates a time series of value-added tax, and blue - the time series of products shipped.

In Figure 16, the pink color indicates the time series of the value-added tax, and blue - the time series of the shipped products.

Figure 15. Chart of the quasi-cycles percentage ratio for time series of value-added tax and shipped products achieved by applying R/S analysis

Source: The official site of the corporate information disclosure center // http://www.e-disclosure.ru/portal, own calculations.

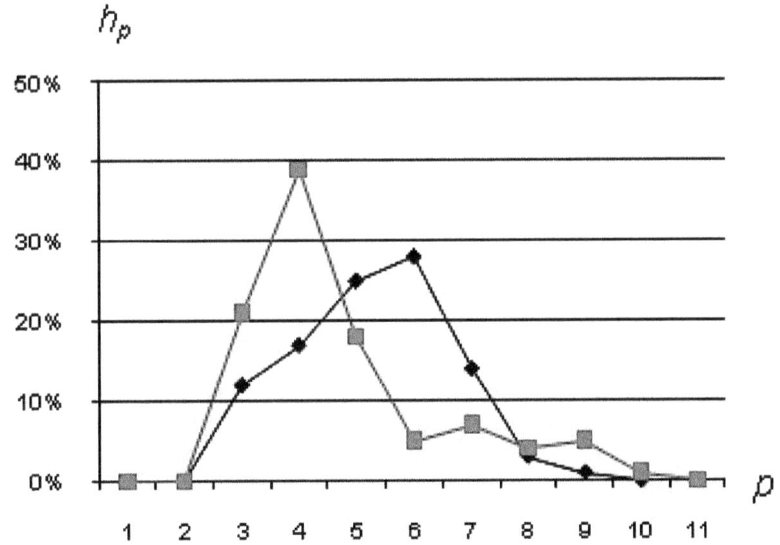

Figure 16. Chart of the quasi-cycles percentage ratio for time series of value-added tax and shipped products achieved by applying phase analysis
Source: The official site of the corporate information disclosure center // http://www.e-disclosure.ru/portal, own calculations.

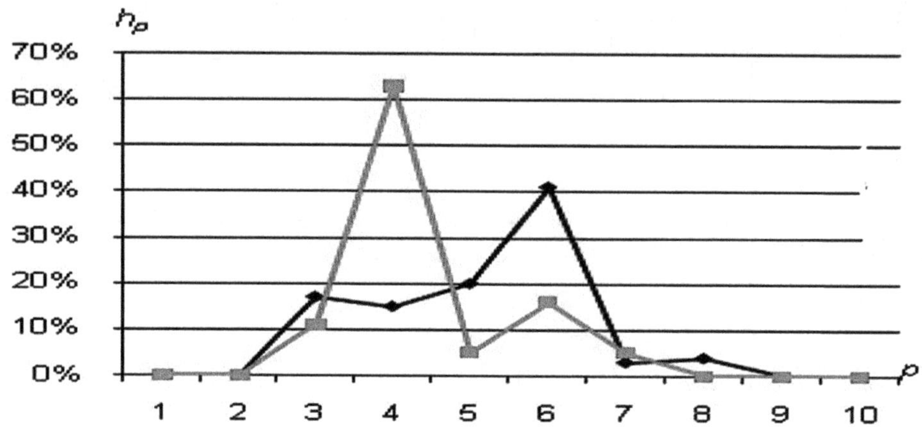

ELIMINATION OF ROW LENGTH MISMATCHES

The presented graphs demonstrate significant discrepancies between the dynamic properties of the series considered. This is explained as follows. The length of the time series under consideration is different. According to formulas 2, 3 the time series of shipped products contains 234 values, and the time series of the value added tax is only 94 values. As at the enterprise under consideration, tax deductions for shipped products are made in total for a certain period of time and correspond to several shipments.

In order to eliminate this discrepancy, all values z^2_j, corresponding to the sum of the added value tax z^3_k for a certain period of time were aggregated.

As a result, the authors obtained a time series derived from the time series of shipped products, aggregated by the value-added tax, which the authors denote by the following expression:

$$\check{Z}^2 = \{\check{z}^2_j\}, j=1,2,3,\ldots,\check{n}^2, \check{n}^2=94 \quad (7)$$

R/S analysis and phase analysis were also applied to the above time series.

The obtained results were compared with the results presented above according to the time series for the value-added tax (see Tables 4 and 5 and Figures 17, 18). Figures 17, 18 show, that the maximum number of quasi-cycles of the time series of value-added tax and the time series of products shipped

Table 4. The percentage of the lengths of the allocated quasi-cycles of the time series of value-added tax $Z^3(3)$ and the time series of the shipped products aggregated by the value-added tax $\check{Z}^2$, obtained by R/S analysis

The Length of the Quasi-Cycle (p)	1	2	3	4	5	6	7	8	9	10	11
Value-added tax	0	0	21	39	18	5	7	4	5	1	0
Shipped products	0	0	18	30	16	17	8	5	3	1	1

Source: Own calculations.

Table 5. The percentage of the lengths of the allocated quasi-cycles for the time series of the value-added tax and the time series of the shipped products aggregated by the value-added tax obtained by the phase analysis

The Length of the Quasi-Cycle (p)	1	2	3	4	5	6	7	8
Value-added tax	0	0	11	63	5	16	5	0
Shipped products	0	0	12	53	6	12	12	6

Source: Own calculations.

Figure 17. Chart of the percentage ratio of the lengths of quasi-cycles for Z3(3) and Ž2 obtained with the help of R/S analysis
Source: The official site of the corporate information disclosure center // http://www.e-disclosure.ru/portal, own calculations.

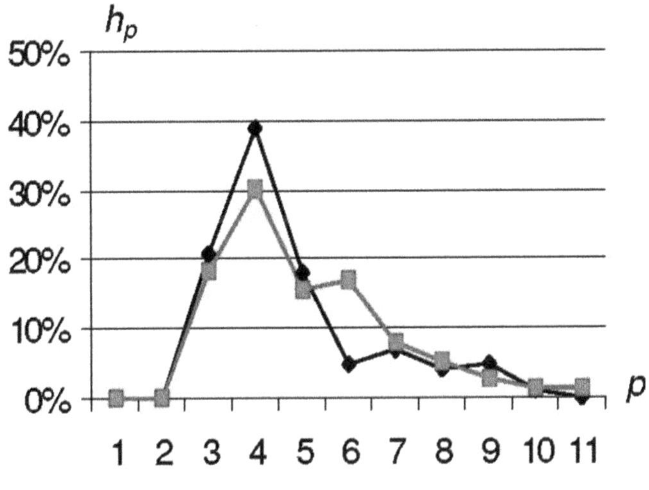

Figure 18. Chart of the percentage ratio of the lengths of quasi-cycles for Z3(3) and Ž2 obtained with the help of phase analysis
Source: The official site of the corporate information disclosure center // http://www.e-disclosure.ru/portal, own calculations.

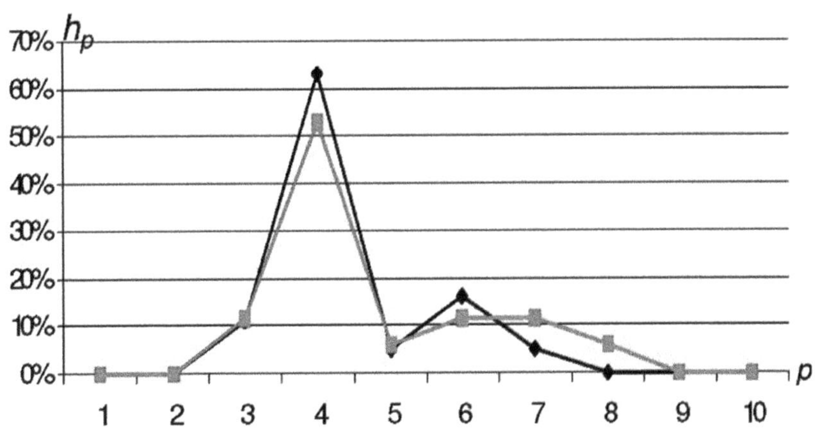

aggregated to the value-added tax corresponds to quasi-cycles of length 4 for both series. Graphical depiction of the percentage ratio for quasi-cycles, for a time series of value-added tax and time series of shipped products, aggregated on the value-added tax, also demonstrated the similarity of the dynamic properties of the considered series obtained with the help of R/S analysis and phase analysis.

In Figure 17, blue color indicates the time series of value-added tax, pink - the time series of shipped products, aggregated by the value-added tax.

In Figure 18 blue color indicates the time series of value-added tax, pink - the time series of shipped products, aggregated by the value-added tax.

It should be noted that in order to define the term "similarity" more strictly, in the context of the present work, it is necessary to select and present criteria for the degree of "similarity". As such criteria, the following dynamic characteristics of time series are offered: cyclicity (stable coincidence of the lengths of quasi-cycles, periodicity, as well as local maxima and local minima of the graphs of percentages of quasi cycle lengths under consideration *(Holden et al., 1990)*. By the "similarity" degree of two time series Z^r_1, Z^r_2, we mean the following vector:

$$\mu(Z^r_1, Z^r_2) = (\mu^h (Z^r_1, Z^r_2), \mu^q (Z^r_1, Z^r_2)), \tag{8}$$

where

$$\mu^h\left(Z_1^r, Z_2^r\right) = \frac{\sum_{p=1}^{\min\left(b_{r1}, b_{r2}\right)} h_p^{r1} - h_p^{r2}}{100\%},$$

$$\mu^q\left(Z_1^r, Z_2^r\right) = \frac{\sum_{p=1}^{\min\left(b_{r1}, b_{r2}\right)} q_p^{r1} - q_p^{r2}}{100\%}, \tag{9}$$

$$b_{r1} = \max\left(p\left(Z^{r1}\right)\right), b_{r2} = \max\left(p\left(Z^{r2}\right)\right),$$

Differences |hr1p - hr2p| and |qr1p - qr2p| from formula 9 reflect the distances between points p in the corresponding graphs (Figures 17-20).

The time series Z_r^1 and Z^r_2 are considered "similar" if the degree of "similarity" of these series μ^d (Z^r_1, Z^r_2), d ∈ {h,q} calculated according to formulas (8), (9) does not exceed 1/3, i.e. $\mu^d (Z^r_1, Z^r_2) \leq 1/3$, d ∈ {h,q}.

For the particular time series of shipped products $Z^2(2)$ and the time series of value-added tax $Z^3(3)$, based on formulas (8) and (9), the authors can calculate the degree of their similarity:

$$\mu(Z^2, Z^3) = (\mu^h(Z^2, Z^3), \mu^q(Z^2, Z^3)) = (0{,}75;\ 1{,}02) \tag{10}$$

where the values of each of the indicators have the following form:

$\mu^h(Z^2, Z^3) = 0{,}75 \gg 1/3$

$\mu^q(Z^2,Z^3) = 1{,}02 \gg 1/3$

Similar calculations were carried out for the time series of the value-added tax Z^3 and for the time series of products shipped aggregated by the value-added tax $\check{Z}^2$ (formula (7)), resulting in the following:

$$\mu(Z^2,Z^3) = (\mu^h(Z^2,Z^3),\ \mu^q(Z^2,Z^3)) = (0{,}31;\ 0{,}29), \qquad (11)$$

where the values of each of the indicators have the following form:

$\mu^h(Z^3,\check{Z}^2) = 0{,}31 \leq 1/3$

$\mu^q(Z^2,Z^3) = 0{,}29 \leq 1/3$

Further, for the comparative analysis of different time series of tax deductions, the authors aggregated the time series of shipped products Z (2) not only for a time series of value-added tax Z (3), but also for the time series of income tax Z (4).

Aggregation of the time series of shipped products according to the time series of the profit tax was carried out in the same way as aggregation on formula (7) (the values $Z2_j$ were summed corresponding to the time interval from the J-th to the (J+1) -th days of income tax deductions). Eventually, the authors obtained a time series of products shipped, aggregated by the income tax, which was denoted by the following expression:

$$\check{Z}^2 = z_j^2,\ j = 1,2,3,\ldots,\check{n}^2,\ \check{n}^2 = 80 \qquad (12)$$

As a result of applying R/S analysis and phase analysis to the time series of income tax Z^4(4) and aggregated on its time series of shipped products $\check{Z}^2$ (formula 12), the results presented in Tables 6 and 7 were obtained, as well as in Figures 19, 20 (by analogy with tables 4 and 5, and also Figures 17, 18). According to the results obtained, the time series of the profit tax (formula (4)) and aggregated on its time series of shipped products (formula (12)) demonstrated a complete discrepancy of the characteristics of analyzed time series dynamics (at the visualization level in Figure 20, a low degree of similarity is demonstrated).

In Figure 19 the time series of the profit tax is marked blue, and the time series of the shipped products aggregated under the profit tax is pink.

Table 6. A summary table of the percentage of the lengths of the allocated quasi-cycles of time series of profit tax and shipped products (formulas (4) and (12)), obtained by R/S analysis, at %

The Length of the Quasi Cycle (p)	1	2	3	4	5	6	7	8	9	10
h^4_p	0	0	20	17	21	20	14	4	4	0
h^{42}_p	0	0	10	30	23	23	3	0	7	3

Source: Own calculations.

Table 7. A summary table of the percentage of the lengths of the allocated quasi-cycles of time series of profit tax and shipped products (formulas (4) and (12)), obtained by phase analysis, at %

The Length of the Quasi Cycle (p)	1	2	3	4	5	6	7	8	9
q^4_p	0	0	21	14	21	21	14	0	7
$\check{g}^2_p$	0	0	0	38	63	0	0	0	0

Source: Own calculations.

Figure 19. Chart of the percentage ratio of the quasi-cycles lengths for the time series of the profit tax and time series of the shipped products aggregated upon this tax with the help of R / S analysis
Source: The official site of the corporate information disclosure center // http://www.e-disclosure.ru/portal, own calculations.

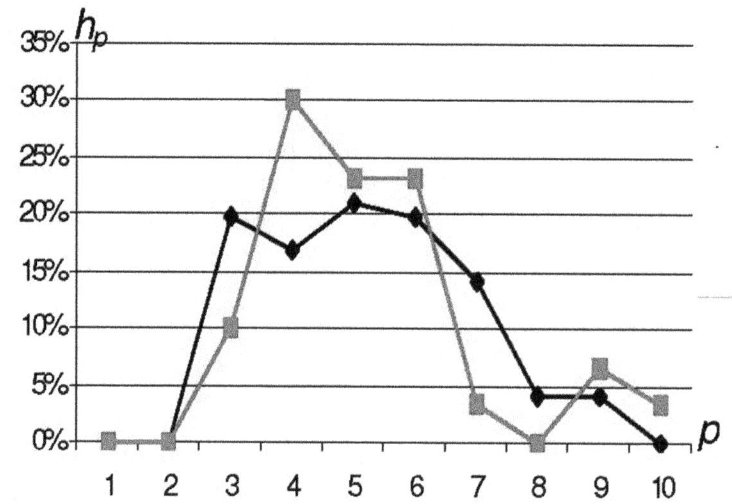

In Figure 20 the time series of the profit tax is marked blue, and the time series of the shipped products aggregated under the profit tax is pink.

The fact that the dynamic characteristics of $Z^4(4)$ и $\check{Z}^2$ do not match is confirmed by the similarity degree parameters calculated for them:

$$\mu(Z^4,\check{Z}^2) = (\mu^h(Z^4,\check{Z}^2), \mu^q(Z^4,\check{Z}^2)) = (0,45;1,3)$$

$$\mu^h(Z^4,\check{Z}^2) = 0,45 > 1/3$$

$$\mu^q(Z^4,\check{Z}^2) = 1,3 \gg 1/3$$

Thus, the time series Z^3 and $\check{Z}^2$ (presented in Figure 20) according to the selected criteria and the numerical value of the similarity degree introduced by the authors, can be qualified as similar. The criteria selected by the authors for comparing time series are as follows: stable coincidence of the quasi cycles lengths, periodicity, local maxima and minima of the quasi-cycle lengths percentages charts.

Figure 20. Chart of the percentage ratio of the quasi-cycles lengths for the time series of the profit tax and time series of the shipped products aggregated upon this tax with the help of phase analysis
Source: The official site of the corporate information disclosure center // http://www.e-disclosure.ru/portal, own calculations.

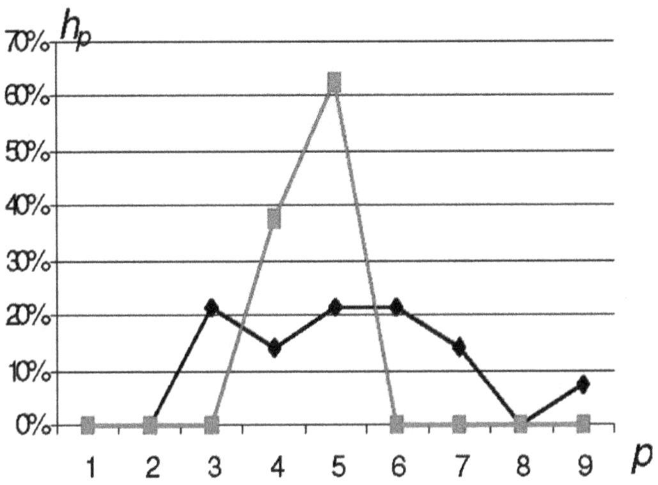

While, the remaining pairs of time series (the time series of goods shipped $Z^2(2)$ and value-added tax $Z^3(3)$, the time series of income tax $Z^4(4)$ and a time series of products shipped aggregated according to the income tax $\check{Z}^2$) impossible to qualify as similar pairs under the same criteria.

The latter is an indirect basis for posing the question of an additional analysis of the correctness of the deduction of tax payments.

Estimates of the similarity degree of the basic time series of goods shipped $Z^2(2)$ and the corresponding aggregate "tax time series" on the value-added tax (formula (7)), on the income tax (formula (12)) are based only on the characteristics of the cyclic component of these time series. At the same time, the trend component of these time series is not less important information. Represent the characteristics of this component for comparative analysis at the level of visualization can be done by constructing moving averages for time series that are comparable among themselves.

FUTURE RESEARCH DIRECTIONS

In this study, dynamic characteristics of time series the method of R/S analysis and H-analysis was not considered a problem not clearly expressed quasicycles. In the future it would be useful to distinguish classes of time series with fuzzy expressed quasicycle to attempt the procedure improve the accuracy of detecting the magnitude of a fuzzy set memory depth time series at the first level of two-level cellular-automata forecast models. This approach will give the opportunity to form a methodological knowledge base, which allows to computerize information processing in the part of the two-level cellular automata model, which until recently was provided for the assessment process cyclical components.

It is also necessary to conduct a study of the reference time series, allowing to reveal features of the dynamics of quasicycles with uncertainty and to clarify the concept of "incidence and trend", "memory loss in the early time series" and "memory depth".

CONCLUSION AND RECOMMENDATIONS

The paper analyzed economic and mathematical tools aimed at creating information systems for decision support in complex socio-economic systems of interaction of tax authorities with taxpayers. The given tools make the following conclusions and offers possible.

For purposes of strategic planning (forecasting) of economic indicators of taxpayers, the authors recommend to use the pre-forecasting model. The pre-forecast analysis was carried out using the R/S analysis algorithm and phase analysis.

The authors carried out a comparative analysis of the series of dynamics of tax revenues with the corresponding series of dynamics of the tax base.

The principle of self-similarity, characteristic for fractal and phase analysis, allowed the authors of the study to identify patterns and similar trends in the dynamics of time series of tax and accounting reporting indicators of taxpayers.

Based on the principle of self-similarity, it is possible to carry out pre-audit tax payments analysis in order to select the less reliable taxpayers-legal entities in order to carry out a field tax audit on their territory.

The authors identified one of the key problems of forecasting in the system of work of tax authorities - the problem of insufficient information (the problem of short time series). The solution to this problem is using the analysis of the dynamics of "short" time series which is based on the visualization method.

The practical significance of the research is that all the proposed recommendations can be widely used in the automated information systems of any department. However, according to the authors opinion, tax control has the greatest need for the proposed measures. The standard interaction procedures used between tax authorities and taxpayers, in the boundaries of current legislature, require significant material costs on the side of the government.

The sensible solution to the current issue is to use the automated data analysis, which is carried out on the basis of modern economic and mathematical methods.

The proposed approach can be used to develop mathematical and informational support for monitoring systems and decision support in tax administrations.

REFERENCES

Chander, P. (1992). Corruption in tax administration. *Journal of Public Economics, 49*, 333-349.

Cowell, F. (1995). Auditing with "ghosts." In F. Cowell & G. F Gordon (Eds.), *The Economics of Organized Crime* (pp. 184-198). Academic Press.

Gilmore, C. G. (1993). A new test for chaos. *Journal of Economic Behavior and Organization, 22*, 209-237.

Holden, K., Peel, D. A., & Thomson, J. L. (1990). *Economic forecasting: an introduction*. Press Syndicate of the University of Cambridge.

Maksishko, N. K., Perepelica, V. A., & Zahovalko, T. V. (2008). Ispol'zovanie metodov nelinejnoj dinamiki dlja ocenki stepeni dostovernosti informacii, predostavljaemoj nalogoplatel'shhikom. *Vestnik nacional'nogo Zaporozhskogo universiteta, 1*(3).

Perepelica, V. A., Tebueva, F. B., & Temirova, L. G. (2005). *Strukturirovanie dannyh metodami nelinejnoj dinamiki dlja dvuhurovnevogo modelirovanija*. Stavropol: Stavropol'skoe knizhnoe izdatel'stvo.

Peters, E. E. (1991). *Chaos and Order in the Capital Markets*. New York: John Wiley and Sons.

Peters, E. E. (1994). *Fractal Market Analysis: Applying Chaos Theory to Investment and Economics*. New York: John Wiley and Sons.

Sergeeva, L. N. (2002). Modelirovanie povedenija jekonomicheskih sistem metodami nelinejnoj dinamiki (teorii haosa). In *Ljudmila Nil'sovna Sergeeva*. Zaporozh'e: ZGU.

Stanislavchik, E. N. (2002). Risk-menedzhment na predprijatii. *Teorija i praktika*, 89.

Wang, Z., Bovik, A., Sheikh, H., & Simoncelli, E. (2004). Image quality assessment: From error visibility to structural similarity. *Image Processing, IEEE Transactions on*, *13*(4), 600–612. doi:10.1109/TIP.2003.819861 PMID:15376593

ENDNOTE

1 The official site of the corporate information disclosure center: http://www.e-disclosure.ru/portal/files. aspx?id=421&type=3

Chapter 10
Applying Neural Networks for Modeling of Financial Assets

Dmitry Averchenko
Southern Federal University, Russia

Artem Aldyrev
Southern Federal University, Russia

ABSTRACT

The purpose of this chapter is to develop an analytical system for forecasting prices of financial assets with the use of artificial neural networks technology. Proposed by the authors, the analytical system consists of several neural networks, each of which makes the forecast of financial assets prices. The system includes recurrence (with feedback) neural networks with sigmoidal activation formula. This allows the networks to "remember" a sequence of reactions to the same stimulus. The learning process of neural networks is performed using an algorithm of back propagation of error. The key parameters of forecast for this analytical system are the indicators presented by the terminal MetaTrader 4-broker Forex Club: Average Directional u Movement Index; Bollinger Bands; Envelopes; Ichimoku Kinko Hyo; Moving Average; Parabolic SAR; Standard Deviation; Average True Range; and others.

INTRODUCTION

The main task of the investor is to buy cheaper and to sell more expensively. The higher the asset price variability, the more opportunities to conduct winning trading strategies, even when considering transaction costs. Unfortunately, what appears to be a simple and obvious hindsight is not at all obvious. All forecasting methods and techniques of financial assets prices can be divided on two large groups: fundamental and technical. The concept of fundamental analysis is based primarily on prediction of price behavior as a result of the influence of those or other events in the world economy. The basic methods of fundamental analysis include: benchmarking method, deduction and induction, correlation, grouping and generalization, macroeconomic analysis, industry method, as well as the method of financial coefficients (Schwager & Turner, 1995; Kalmykova, 2007; Chirkova, 2014).

DOI: 10.4018/978-1-5225-3767-0.ch010

Technical analysis is based on three postulates: "the market price takes everything into account", "the market follows the trend", and "the market is the pattern". Technical analysis methods include technical indicators, wave and candle analysis, price charts, and methods of artificial intelligence (neural networks). (Murphy, 1999; Schwager, 1997; Elder, 2007; Kovel, 2007; Nison, 2003). We focus our attention in this chapter on the last method of artificial neural networks.

Neural networks are a class of powerful machine learning algorithms based on statistical methods of analysis. They have been successfully applied for many years in the development of trading strategies and financial models: tasks are determined for developing their own approaches to forecasting the financial market and designing profitable trading systems (Laletina, 2015). New methods and models are being explored to improve the accuracy of forecasts and solve the problems of approximation and classification (Vladimirova, 2014). Despite this situation, neural networks do not have a very good reputation due to frequent failures of their practical application. In most cases, the reasons for failure lie in inadequate design decisions and a general lack of understanding of how they work.

The novelty and relevance of this study is due to the combination of classical approaches of technical analysis and automatic optimization of this model through a neuron network. This excludes cases of inadequacy of the model, and not allowing for the situation of the "black box". The study describes the construction of a neuron network cluster using four networks through which the authors of the study will determine the current phase of the market - trend / flat. Subsequently, the model will use those indicators and approaches that are used for this phase of the market - trend / flat indicators.

The practical significance of the study is to develop a new approach for forecasting prices of financial assets and adopting appropriate investment decisions by combining the classical approaches of the technical analysis and the artificial intelligence. The main target audience of this research can be traders, investors with different investment horizons, and scientists with a similar sphere of scientific interests. We believe the fractal (the neuron analog) is the price of a financial asset, and neural networks are one of the most promising methods of fractal analysis. The key question here is whether it is possible to determine the direction, magnitude, and volatility of future asset price changes by extrapolating the available past data. Adaptive nonlinear systems can be trained to perform technical analysis with minimum possible assumptions.

Lately, the hypothesis of an effective market (EMH) developed by Eugene Fama is seriously criticized as unscientific literature. In its weak form, this hypothesis asserts that the investor cannot receive abnormal rate of return via technical analysis rules, which are based solely on past prices. In other words, information about past prices cannot be useful for extracting abnormal income (Baestaens, Van Den Berg, & Wood D., 1997).

Simultaneously, EMH does not specify the nature and manner of obtaining such information through past prices. Should the simple time-series autocorrelation, methods of Box-Jenkins or Fourier analysis, or many others filtering methods be used? Moreover, EMH is a combined hypothesis.

This means that EMH verification requires preliminary pricing model formation, which in turn depends on the degree of predictability. This circumstance makes the matter even more complicated. The main reason that does not allow the possibility to reject EMH is the presence of the market Index funds (the following strategies of passive management), which are very popular among pension funds. However, recently in the scientific literature there are assumptions that financial markets have some features of predictability (Peters, 1994). However, it should be remembered that the success in applying technical analysis depends entirely on the quality of the optimization method.

Thus, there is a growing need for information systems that can conduct approximation, classify and recognize different types of data over a wide range of their changes and are capable of flexible learning, depending on the changes, goals, and objectives set before them. In such systems, artificial neural networks (NN) are widely used. Currently, there is a large number of various software packages for working with NN, but the implementation of methods and algorithms in most software solutions (for example, algorithms for teaching NN or normalizing input training data) is hidden from the user.

In most cases, the user has no way of knowing which algorithms with which properties the developers of the package used to achieve the best result in teaching NN. Therefore, it is difficult to determine at what expense the properties of the algorithm different software packages demonstrate different results on the same training samples (TS). Additionally, this causes difficulties in the application of the package in own development and in the conduct of research. Consequently, it is impossible to select the optimal software package to work with NN. Modern artificial NNs are often used not as a separate tool, but are instead integrated into information systems that require flexible training depending on the changes of certain tasks. Therefore, to conduct a large amount of computational experiments, NN should be integrated into software complexes, with the use of which these experiments will be conducted.

The software package MatLab provides for working with the NN library Neural Network Toolbox, which allows one to model neural networks of different architectures. During the modeling when trying to scale the network or change for its own purposes, it is necessary to consider the following: The NN topology, weighting coefficients of neural connections, and NN activation functions that cannot be changed without changing the results of NN functioning. This can lead to a violation of the reproduction of results on the same training sample for networks having different structures. The larger the neural network (the greater the number of layers and neurons), the more difficult the evaluation (Turovskiy, 2014, 2015).

THEORETICAL FRAMEWORK

The idea of creating systems for the processing of intellectual information by the image of the device of the nervous system arose in the 1940s. McCulloch and Pitts (1943) created a simplified model of the neural cell – the neuron. NNs are computational structures that, for signal processing, use phenomena similar to those occurring in human brain neurons. The most important feature of the network, providing its broad capabilities, is the parallel processing of information by all links of the network. A large number of inter-neural connections permit speeding up the process of information processing significantly, which provides the possibility of processing signals in real time. In addition, the network becomes resistant to errors that occur (Yahyaeva, 2016).

Another no less important property of the NN is the ability to learn and generalize the accumulated knowledge. The NN has the properties of artificial intelligence. Trained on a limited set of data, the network can summarize the information received and display the results on new data not used in the learning process. The human nervous system is built from elements called neurons. The human brain contains up to 10^{11} neurons, which have approximately 10^{15} compounds among themselves. Each neuron has many properties in common with other organs of the body, but it has unique abilities: to receive, to process and transmit electrochemical signals along the nerve pathways that form the communication system of the brain.

Figure 1 shows the structure of a pair of typical biological neurons. Dendrites go from the body of the nerve cell to other neurons where they take signals at the junction points called synapses. Adopted

Figure 1. Structure of the biological neuron

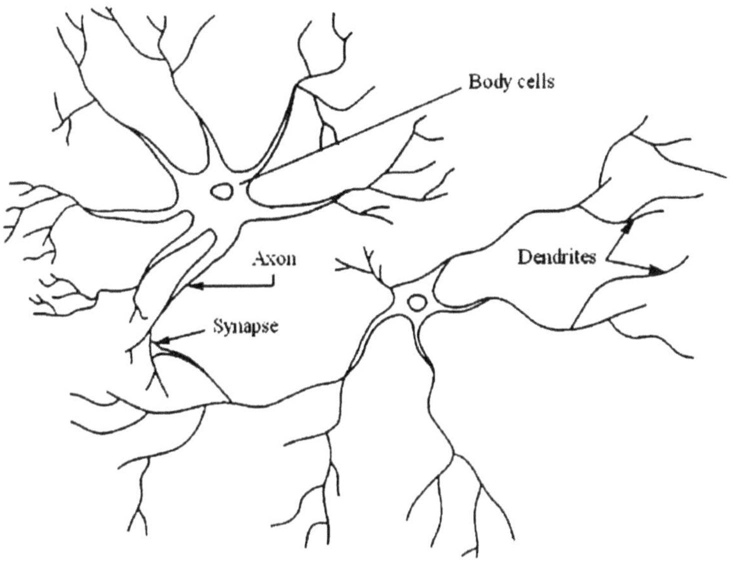

by the synapses, input signals are transmitted to the neuron's body. Here they are summed and some inputs aim to excite a neuron while others prevent its arousal. When the total excitation in the body of the neuron exceeds a certain threshold, the neuron is excited and sends signal to other neurons through the axon. This basic functional scheme has many complications and exceptions; nevertheless, most of the artificial neural networks simulate only these simple properties.

The artificial neuron imitates, in the first approximation, the properties of a biological neuron. At the input of an artificial neuron comes a set of signals, each of which is the output of another neuron. Each input is multiplied by the corresponding weight, similar to the synaptic force, and all products are summed, determining the activation level of the neuron (Bushuev, 2000).

Figure 2 presents a model that realizes this idea. The set of input signals, designated as $X_1, X_2, \ldots, X_n$, goes to the artificial neuron. These input signals, denoted by the vector X, correspond to the signals arriving at the synapses of the biological neuron. Each signal is multiplied by the corresponding weight - $\omega_1, \omega_2, \ldots, \omega_n$, and supplied on the summing block, designated as Σ each weight corresponds to the "strength" of one biological synaptic connection (the set of weights is denoted by the vector W). The

Figure 2. Model of the simplest artificial neural network

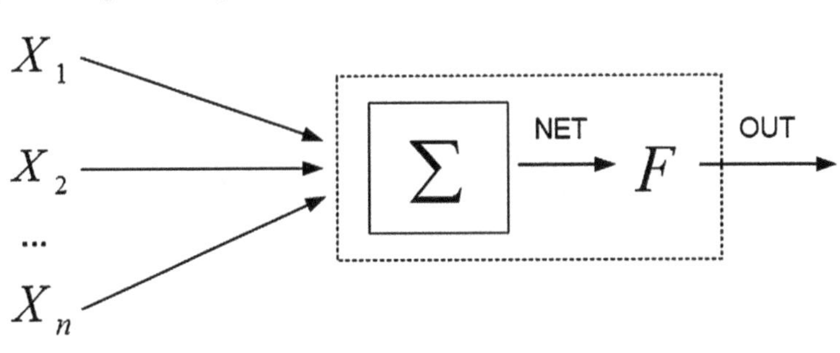

summing block corresponding to the body of the biological element sums the weighted inputs algebraically, thus creating an output which we will call NET. In vector notation this can be compactly written as follows:

$$NET = XW \tag{1}$$

The NET signal is converted by the activation function F and provides an output neural signal OUT. The activation function can be an ordinary linear function

$$OUT = F\left(NET\right) \tag{2}$$

where, F - is a constant of threshold function.

$$OUT = \begin{cases} 1, & if\ NET > T \\ 0, & if\ NET \leq T \end{cases} \tag{3}$$

where, T is some constant threshold value, or a function that is more accurate simulates the nonlinear transfer characteristic of a biological neuron and provides to the neural network great opportunities.

In Figure 2, the block designated F receives the signal NET and issues an OUT signal. If the block F narrows the range changes of quantity NET so that for any values of NET the OUT values belong to some finite interval, then F is called the "compressive" function. As a "compressive" function, the logistic or "sigmoidal" (S-shaped) function is often used (shown in Figure 3). This function is mathematically expressed as:

$$F\left(x\right) = \frac{1}{1 + e^{-x}} \tag{4}$$

Thus, using equations (3) and (4), the authors obtain:

Figure 3. Graph of the sigmoid function

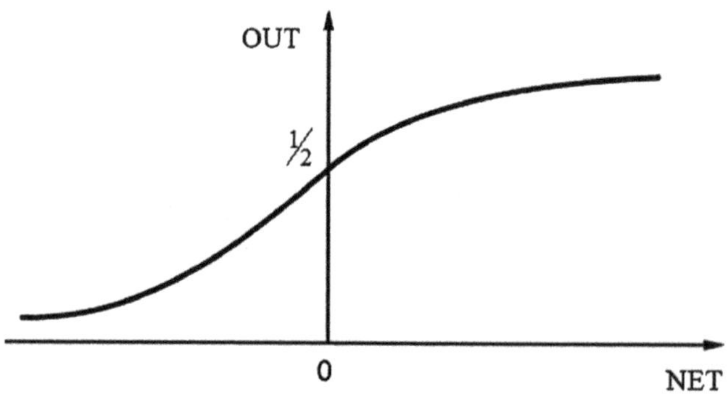

$$OUT = \frac{1}{1 + e^{-NET}} \tag{5}$$

By analogy with electronic systems, the activation function can be considered a nonlinear amplification characteristic of an artificial neuron. The increment factor is calculated as the ratio of the increment of the value of OUT that causes it to become small increment of the NET. It is expressed by the slope of the curve at a certain level of excitation and varies from small values at large negative excitations (the curve is almost horizontal) to the maximum value at zero excitation and again decreases when the excitation becomes large positive.

Grossberg (1973) found that such a nonlinear characteristic solves the dilemma of noising saturation set by him (Wasserman, 1992). How can the same network handle both weak and strong signals? Weak signals require a large network strengthening to give a usable output signal. However, amplifying cascades with high amplification coefficients can lead to saturation of the output by the noise of the amplifiers (random fluctuations), which are present in any physically realized network.

Strong input signals, in turn, will also lead to saturation of amplifying stages, excluding the possibility of a useful utilization of the output. The central area of the logistic function, which has a high gain coefficient, solves the problem of processing weak signals, while areas with a falling gain at the positive and negative ends are suitable for large excitations.

Thus, the neuron operates with a high amplification in a wide range level of input signal.

$$OUT = \frac{1}{1 + e^{-NET}} = F\left(NET\right) \tag{6}$$

Another widely used activation function is the hyperbolic tangent. In form it is similar to the logistic function and is often used by biologists as a mathematical model of nerve cell activation. As an activation function of an artificial neural network, it is written as follows:

$$OUT = th\left(x\right) \tag{7}$$

Similar to a logistic function, the hyperbolic tangent is an S-shaped function; however, it is symmetric about the origin and at the point NET = 0, the value of the output signal OUT is zero (see Figure 4). Unlike the logistic function, the hyperbolic tangent takes values of different signs and this property is applied for full number of networks series. The considered simple model of an artificial neuron ignores many properties of its biological twin. For example, it does not take into account the time delays that affect the dynamics of the system. Input signals immediately generate an output signal. And, more importantly, it does not consider the effects of the frequency modulation function or the synchronizing function of the biological neuron, which a number of researchers consider decisive in the nervous activity of the natural brain.

Despite these limitations, networks constructed from such neurons exhibit properties that strongly resemble a biological system. Only time and research will be able to answer the question of whether such coincidences are accidental or they are a consequence of the fact that the most important features of the biological neuron are correctly grasped by the model.

Figure 4. Chart of the hyperbolic tangent

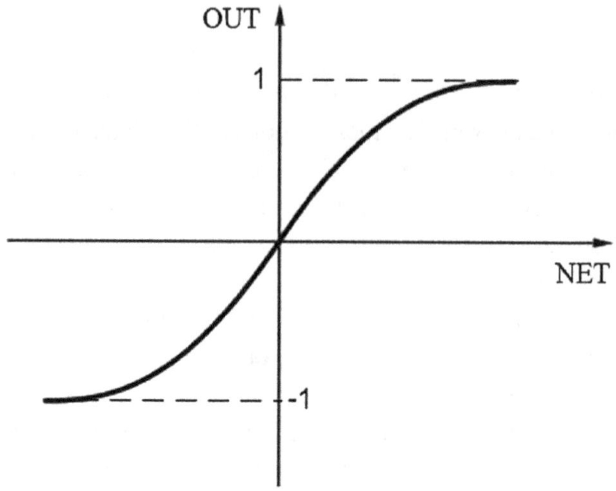

Single-Layer and Multi-Layer Artificial Neural Networks

Although one neuron is capable of performing the simplest recognition procedures, for serious calculations it is necessary to combine neurons in the network. The simplest network is composed of groups of neurons which form a layer, as shown in the right part of Figure 5.

Note that the vertex-circles on the left serve only for the distribution of the input signals. They do not perform any calculations and therefore will not be considered a layer. For greater clarity, let's designate them as circles to distinguish them from computational neurons, denoted by squares. Each element from the set of inputs X by a separate weight is connected to each artificial neuron. And each neuron gives a weighted sum of inputs to the network. There may also be connections between the outputs and the inputs of the elements in the layer.

It is convenient to consider the weights as elements of the matrix W. The matrix has m rows and n columns, where m is the number of inputs, and n is the number of neurons. For example, ω_{23} is the weight connecting the second input to the third neuron.

Figure 5. Model of a single layer artificial neural network

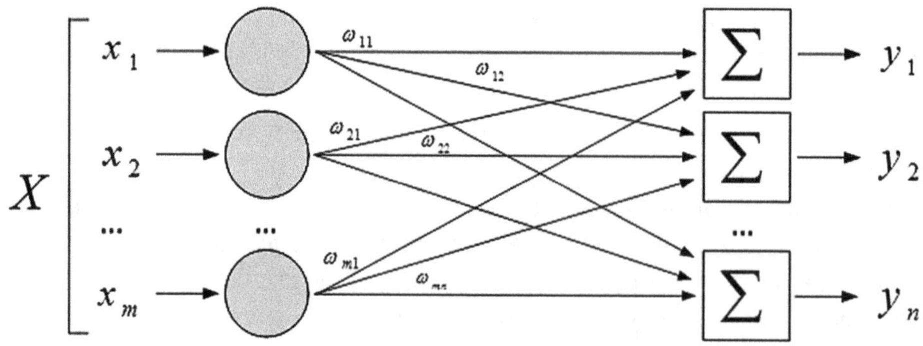

Thus, the calculation of the output vector N, components of which are outputs of OUT-neurons, is reduced to matrix multiplication:

$$N = XW \tag{8}$$

where, N and X are row-vectors.

Larger and more complex neural networks have, as a rule, greater computing power. Although there are networks of all configurations, layered organization of neurons copies the layered structures of certain parts of the brain. It turns out that such multi-layer networks have more possibilities than single-layer networks and in recent years, algorithms have been developed for their training.

Multi-layer networks can be built from cascades of layers. The output of one layer is the input for the subsequent layer. A similar network is shown in Figure 6 and it is displayed with all connections. Multi-layer networks cannot lead to an increase in processing power compared to a single-layer network if the activation function between the layers is linear. The calculation of the layer output consists in multiplying the input vector by the first weight matrix with the subsequent multiplication of the result vector by the second weight matrix (if there is no nonlinear activation function).

$$OUT = \left(XW_1\right)W_2 \tag{9}$$

Since the multiplication of matrices is associative, then:

$$\left(XW_1\right)W_2 = X\left(W_1W_2\right) \tag{10}$$

Figure 6 shows that a two-layer linear network is equivalent to a single-layer matrix, equal to the product of two weight matrices. Therefore, any multi-layer linear network can be replaced by an equivalent single-layer network. However, single-layer networks are very limited in their computing capabilities. Thus, for expansion of the capacity of networks, in comparison with a single-layer network, a nonlinear activation function is needed.

Figure 6. Model of multi-layer artificial neural network

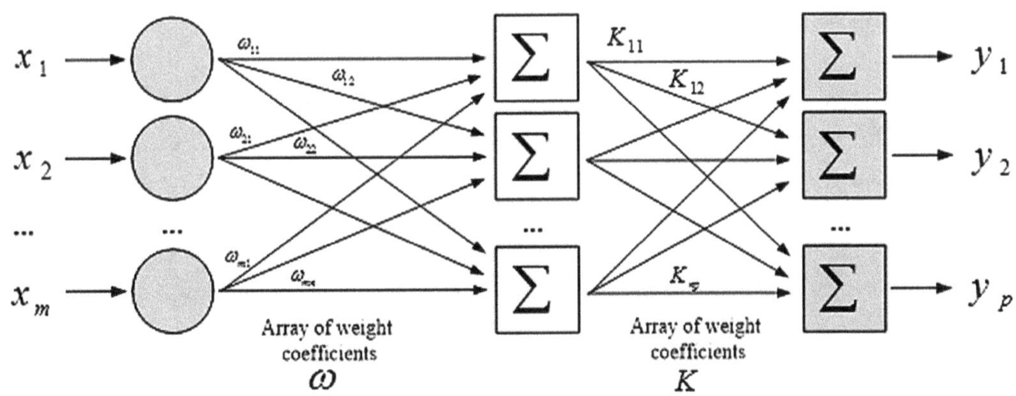

The networks considered so far had no feedbacks (i.e., connections) which are going from the exiting outputs of a certain layer to the inputs of the same layer or previous layers. This special class of networks, called networks without feedbacks or networks of direct distribution, arouse great interest and is widely used. Networks of a more general kind, having connections from output to input, are called networks with feedbacks. Networks without feedbacks have no memory as their output is completely determined by the current inputs and the values of the weights. In some configurations of networks with feedback, the previous output values are returned to the inputs; the output is determined by the current input and the previous outputs. Therefore, networks with feedbacks can have properties similar to short-term human memory, where network outputs also partly depend on previous inputs.

Unfortunately, there is no generally accepted method for counting the number of layers in a network. The multi-layer network consists, as shown in Figure 6, from alternating sets of neurons and weights. In Figure 5, it was already mentioned that the input layer does not perform the summation. These neurons serve only as ramifications for the first set of weights and do not affect to the computing capabilities of the network. Therefore, the first layer is not taken into account when counting layers and the network similar to that depicted in Figure 6 is considered to be two-layer, since only two layers perform calculations. The weights of the layer are connected with the neurons following them. Consequently, the layer consists of a set of weights with the neurons summing the weighted signals following after them.

Training of Artificial Neural Networks

Among all the interesting properties of artificial neural networks, not one captures imagination as much as their ability to train. Their training resembles the process of intellectual development of a human person to such an extent that it might seem that we have achieved a deep understanding of this process. But it is not so. The possibilities of learning artificial neural networks are limited and many difficult tasks need to be solved to determine whether the researchers are on the right track.

The Purpose of Training

The network is trained to give a desired set of outputs (or, at least, equivalent to it) for a set of inputs. Each such input set or output set is a vector. Training is carried out by sequential presentation of input vectors with simultaneous adjustment of weights in accordance with a certain procedure. In the process of learning, weights of the network gradually become such that each input vector produces an output vector.

Teaching With the Teacher

There are learning algorithms with and without a teacher. Training with the teacher assumes that for each input vector there exists a target vector representing the required output. Together they are called a learning pair. Usually the network is trained on a certain number of such training pairs. Presented an output vector, the network output is calculated and compared with the corresponding target vector. The difference (error) is fed into the network with the help of feedback and weights are changed in accordance with the algorithm that seeks to minimize the error. The vectors of the training set are presented sequentially, errors are calculated and weights are adjusted for each vector until the error throughout the training array does not reach an acceptable low level.

Teaching Without a Teacher

Despite numerous applied achievements, training with the teacher was criticized for its biological implausibility. It is difficult to imagine a learning mechanism in the brain, which would compare the desired and actual values of the outputs and performing feedback correction. Teaching without a teacher is a much more plausible model of learning for a biological system. Developed by Kohonen and many others, it does not need a target vector for outputs and consequently, does not require comparison with predefined ideal answers. The training set consists only of input vectors.

The training algorithm adjusts the network weights so that the agreed output vectors are obtained, i.e. the presentation of sufficiently close input vectors will give identical outputs. The learning process, therefore, highlights the statistical properties of the learning set and groups similar vectors into classes. Delivery of a vector to the input from this class will give a certain output vector, but before learning it is impossible to predict which output will be produced by this class of input vectors. Consequently, the outputs of such network must be transformed into some understandable form, conditioned by the learning process. This is not such a serious problem. It is usually not difficult to identify the connection between the input and the output set by the network.

Learning Algorithms

Most modern learning algorithms have grown from the Hebb concept. He proposed a model of teaching without a teacher in which the synaptic force (weight) increases, if both neurons (source and receiver) are activated. In this manner, frequently used paths in the network are amplified and the phenomena of habit and learning through the repetition are explained. In an artificial neural network using the training on the basis of Hebb, the build-up of weights is determined by the product of excitation levels of the transmitting and receiving neurons. This can be written as:

$$\omega_{ij} \cdot \left(n+1\right) = \omega\left(n\right) + \alpha \cdot OUT_i \cdot OUT_j \tag{11}$$

where,

$\omega_{ij}\left(n\right)$ = the weight value from the neuron i to the neuron j before the tuning,
$\omega_{ij}\left(n+1\right)$ = weight value from neuron i to neuron j after tuning,
α = coefficient of training speed,
OUT_i = output of the neuron i and the input of the neuron j,
OUT_j = output of neuron j.

The networks using the training by the Hebb have developed constructively; however, over the past 20 years more effective learning algorithms have been developed (Haykin, 2006). Therefore, in this investigation the backpropagation algorithm was used to which the authors devoted the following section.

Back Error Propagation Algorithm of Artificial Neural Network

To illustrate this process, we use a neural network consisting of three layers and having two inputs and one output of (Figure 7).

Each neuron consists of two elements. The first element (dendrites) adds weighting coefficient to the input signals. The second element (the body) realizes a non-linear function, the so-called neuron activation function. The signal ε - is the weighted sum of the input signals:

$$y = f(\varepsilon)$$
$$\varepsilon = x_1 \cdot \omega_1 + x_2 \cdot \omega_2 \tag{12}$$

To train a neural network, we must prepare the data for training (examples). In our case, the training data consists of the input signals (and) and the desired result. Training is a sequence of iterations (repetitions). In each iteration the weights coefficients of the neurons are adjusted with use of new data from training examples. The change in weight coefficients is the essence of the algorithm described below (Melihova, Veprintseva, & Chumichev, 2016).

Each step of training begins with the impact of input signals from the training examples. Subsequently, we can determine the values of the output signals for all neurons in each layer of the network. Figure 8 shows how the input signals, acting on the first neuron, are transmitted into the hidden layer. After that, how $f_6(\varepsilon)$ deliver the input signals from $f_4(\varepsilon)$ and $f_5(\varepsilon)$, output signal from $f_6(\varepsilon)$, (i.e., y) is compared with the desired output signal z, which is stored in the training data. The difference between these two signals is called the output network layer error δ.

$$y_i = f_i\left(x_1 \cdot \omega_{x_1 i} + x_2 \cdot \omega_{x_2 i}\right), i = \overline{1,3}$$
$$y_k = f_k\left(y_1 \cdot \omega_{1k} + y_2 \cdot \omega_{2k} + y_3 \cdot \omega_{3k}\right), k = \overline{4,5} \tag{13}$$
$$y = f_6\left(y_4 \cdot \omega_{46} + y_5 \cdot \omega_{56}\right)$$

Figure 7. Three-layer neural network

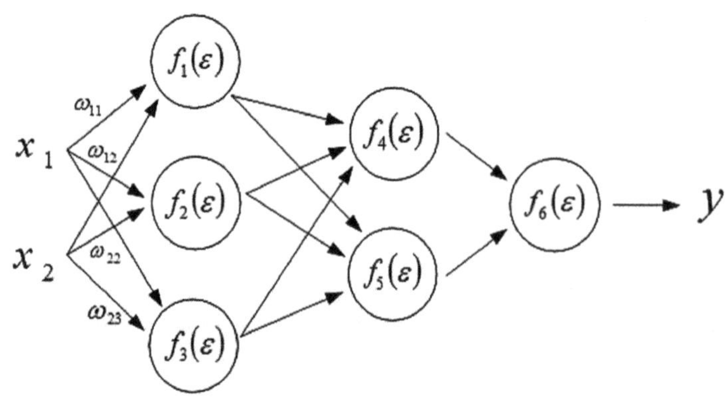

Figure 8. Distribution in the network of input signals

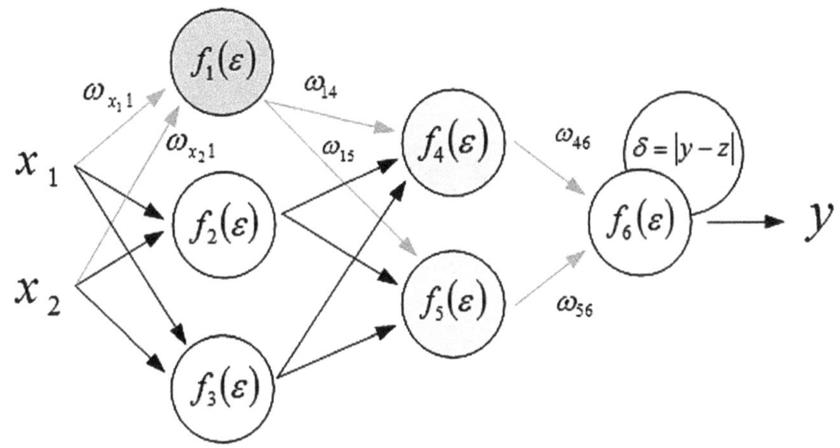

It is impossible to directly calculate the error signal for internal neurons, because the output values of these neurons are unknown. For many years, an effective method for learning a multi-layer network was unknown. Only in the mid-1980s was the algorithm for backpropagation of the error was developed. The idea is to propagate the error signal δ back to all neurons whose output signals were input signals for the last neuron (calculated in the learning step, Figure 9).

The weight coefficients ω_{mn}, used for backpropagation of the error, are equal to the same coefficients, which were used during the calculation of the output signal. Only the direction of the data flow changes (signals are transmitted from the output to the input). This process is repeated for all layers of the network. If the error has come from several neurons, it is summed up:

$$\begin{aligned} &\delta_i = \omega_{i4} \cdot \delta_4 + \omega_{i5} \cdot \delta_5, i = \overline{1,3} \\ &\delta_k = \omega_{k6} \cdot \delta, k = \overline{4,5} \\ &\delta = \left| z - y \right| \end{aligned} \tag{14}$$

When the value of the signal error is calculated for each neuron, it is possible to correct the coefficient weights of each neuron input (dendrite) node. The authors use the activation function of the sigmoid type (Figure 4):

$$S\left(x\right) = \frac{1}{1 + e^{-x}} \tag{15}$$

Then the derivative of the function $S\left(x\right)$ is as follows:

$$S'\left(x\right) = S\left(x\right) \cdot \left(1 - S\left(x\right)\right) \tag{16}$$

Figure 9. Back propagation of the error signal

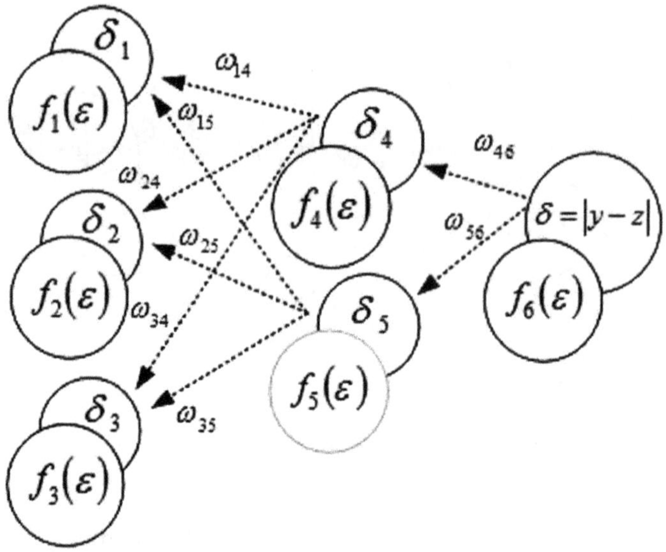

The computation of the derivative is necessary, because to adjust the weighting coefficients when learning NN using the backpropagation algorithm, the gradient descent method is used:

$$\begin{cases} \omega'_{x_1 i} = \omega_{x_1 i} + \mu \cdot \delta_i \dfrac{df_i(e)}{de} x_1 \\ \omega'_{x_2 i} = \omega_{x_2 i} + \mu \cdot \delta_i \dfrac{df_i(e)}{de} x_2 \end{cases}, i = \overline{1,3} \tag{17}$$

$$\begin{cases} \omega'_{1k} = \omega_{1k} + \mu \cdot \delta_k \dfrac{df_k(e)}{de} y_1 \\ \omega'_{2k} = \omega_{2k} + \mu \cdot \delta_k \dfrac{df_k(e)}{de} y_2, k = \overline{4,5} \\ \omega'_{3k} = \omega_{3k} + \mu \cdot \delta_k \dfrac{df_k(e)}{de} y_3 \end{cases} \tag{18}$$

$$\omega'_{j6} = \omega_{j6} + \mu \cdot \delta \frac{df_6(e)}{de} y_j, j = \overline{4,5} \tag{19}$$

The coefficient μ affects the learning rate of the network. There are several methods for selecting this parameter. The first way is to start the learning process with a large value of the parameter μ. During correction of weight coefficients, the parameter is gradually reduced.

The second is a more complex method of training begins with a small value of the parameter μ. In the learning process, the parameter increases, and then decreases again at the final stage of training. Beginning of the educational process with a low value of the parameter μ makes it possible to determine the sign of the weighting coefficients.

Forecasting the NASDAQ-100 Index Through a Neural Network

In the framework of this study the system for forecasting the rates of financial assets was implemented (based on the NASDAQ-100 index). It is based on four recurrent neural networks of Elman (1990). For calculation, the interval was taken from April 24, 2013 to March 21, 2017. The model was trained on this interval. The 4-hour and daily timeframes were used as a time period. Recurrent neural networks are one of the most complex. In them, in contrast to direct distribution networks, there are feedbacks. Feedbacks allow to provide adaptive memorization of past time events. Recurrent networks are a flexible tool for constructing nonlinear models (Bugorskiy, 2008).

At the heart of the Elman artificial neural network is a multilayer perceptron. The feedback is realized in the Elman network via the transmission on the input layer of baseline data and outputs neurons of the hidden layer. To the original data in this model, the authors use the next set of indicators submitted by the terminal MetaTrader 4, broker Alpari:

Moving Average the simplest and most common indicator of the technical analysis based on which most of the tools for algorithmic trading are built. This indicator allows researchers to determine the average price value of the instrument for the specified time interval. There are several options for calculating a moving average, such as Simple Moving Average (SMA), Exponential Moving Average (EMA), Smoothed Moving Average (SMMA), and Linear Weighted Moving Average (LWMA). In this case, the following period values were used to calculate the EMA: $N = 12, 24, 48, 120$;

Ichimoku Kinko Hyo, which can also call an independent trading system, including several different indicators with their time intervals and length:

Tenkan-sen shows the average value of the price for the first period of time, defined as the sum of the maximum and minimum for this time and divided by two. This method allows one to determine the strength of the trend, support and resistance levels. As values of the parameters was used an author's values of constants: $ShPer = 9$, $MidPer = 26$, $LonPer = 52$;

Average Directional Movement Index allows identifying the presence or absence of a price trend in the market, these indicators were first presented by Wells Wilder. For ADX, the standard period $N = 14$ was used;

Momentum can be used as the leading indicator for forecasting the turn of the market movement, by assumption, that most of the tendencies end with a co-directional sharp price impulse, when most of the participants hastily leave the market ($N = 14$).

Parabolic SAR is like a moving average, but has greater acceleration mainly used to manage the position; namely, its fixation. Where is an index of acceleration, step of price change when closing a position, in our case, has a value of 0.02;

HeikenAshi - is a special type candle, which are somewhat different from Japanese candles. The opening price of such candles is a smoothened quotation period in the range between the opening and the closing of previous HA candle. In other words, the opening and the closing prices are summed up and divided by two. It is important to highlight that in order to calculate the current candles, the values of the previous HA candle are taken. The closing price of the HA – is the average price of the opening, closing, maximum and minimum values of the previous period. Maximum HA candles are defined, as comparison of maximum parameters of ordinary Japanese candle with the price of opening and closing candles HA. Minimum HA are selected by comparing the minimum of an ordinary Japanese candle with the opening and closing of HA candles.

Bollinger Bands are two moving averages, equidistant lines from the price in the distance of some standard inclinations. Indicator lines react to the changes of volatility on the market, thus helping to identify the period of rapid upswing or flat. In the neural network, the following parameters were used: m = 1, N = 20.

Commodity Channel Index provides a numerical characteristic of the deviation of the instrument's price from its average value, thus introducing the levels of extreme values. It is possible to identify overbought or oversold instruments. To calculate the given index classical value of quantity of considered periods $N = 14$ were used.

Relative Strength Index, by the design of the founder, is capable of identifying price peaks and base while the indicator is entering the zones of overbuying and overselling. For the parameter of considered periods value = 9 was taken,

Stochastic Oscillator compares the current price with some price range. The given indicator can be related to the main oscillators which are used in technical analysis, in the base of which underlies the principle of overbuying and overselling of price value of the tool.

The system is presented as a neural network cluster and has the following structure:

Input: Input parameters;

Input for Artificial Neural Network 1: Input parameters, which enter the first neural network (indicators: EMA12, EMA24, EMA48, EMA120, Ichimoku_Span, ADX14_Main);

Input for Artificial Neural Network 2: Input parameters, which enter the second neural network (indicators: ADX14_Di, Momentum14, Ichimoku_tk, EMA12_cl, SAR_0.02_0.2, Ichimoku_Span2);

Input for Artificial Neural Network 3: Input parameters, which enter the third neural network (indicators: Heiken_Ashi, Bollinger_20_1, 38parots, CCI14, RSI9, Stochastic);

Artificial Neural Network 1: Neural network adviser specialized on the forecast of the trend and flat;

Artificial Neural Network 2: Neural network adviser specialized on the forecast of growth and fall of currency, within the conditions of flat;

Artificial Neural Network 3: Neural network adviser specialized on the forecast of growth and fall of currency, within the conditions of the trend;

Artificial Neural Network 3: Neural network, which generalizes the results of previous networks (Artificialneuralnetwork1, Artificialneuralnetwork 2, and Artificialneuralnetwork 3) and general input parameters (Figure 10).

Output: Results of the forecast, expressed in growth and fall of security (Giles & Lawrence, 2006).

Building a neural network system in a production environment of Matlab:

1. Upload the data from Excel with the name of "Data", from the lists TrainData and TestData (Code block number 1)
2. Creating arrays of training samples for the first three networks (Artificial neural network 1 - neural network adviser specializing in the forecast of a trend or flat. Artificial neural network 2 - a neural

Figure 10. Scheme of the neural system

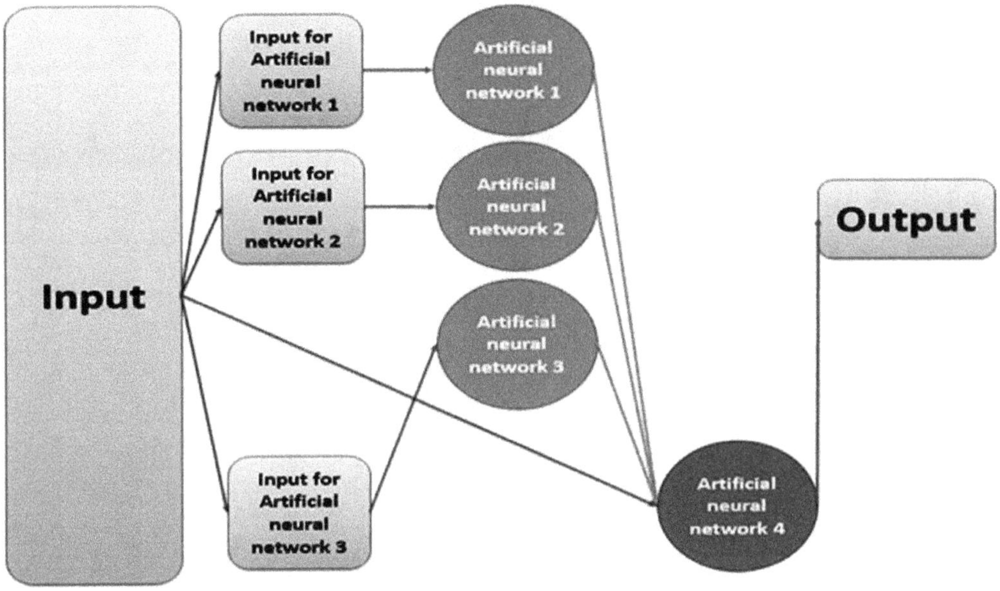

network adviser specializing in forecasting the growth or fall of a course in a flat environment. Artificial neural network 3 - a neural network adviser specializing in forecasting the growth or fall course in a trend) (Code block number 2)

3. Creating arrays of testing samples for the first three networks (Artificial neural network 1 - neural network adviser specializing in the forecast of a trend or flat. Artificial neural network 2 - a neural network adviser specializing in forecasting the growth or fall of a course in a flat environment. Artificial neural network 3 - a neural network adviser specializing in forecasting the growth or fall course in a trend.) (Code block number 3)
4. Transpose the arrays for filing into the neural network (Code block number 4)
5. Creating the arrays of outputs for all networks: Artificialneuralnetwork 1, Artificialneuralnetwork 2, Artificialneuralnetwork 3. Output of the fourth network is the output of all the system (Code block number 5)
6. Form an artificial neural network N° 1. (Artificial neural network 1 - a neural network adviser specializing in the forecast of a trend or a flat). (Code block number 6). Elman's neural network has 42 inputs, because transmitted on 6 converted indicators to the input for the last seven days. Output 1. The network has 3 hidden layers, each of which uses a sigmoid activation function; the method of learning is the method of backpropagation of the error. The network stops training either after 10,000 learning epochs (cycles), or on condition that the size of the error is zero. The execution function is the mean absolute error.
7. Teaching the artificial neural network N° 1 and carry out a test on the teaching sampling of data and on the testing sampling of data. (Figure 11, Figure 12) (Code block number 7)Best training performance is 0.37486 at epoch 9996.
8. Creating the artificial neural network No. 2 (Artificial neural network 2 - a neural network adviser specializing in forecasting the growth or fall of the course in a flat environment) (Code block

Figure 11. Window of the process of learning the artificial neural network No. 1

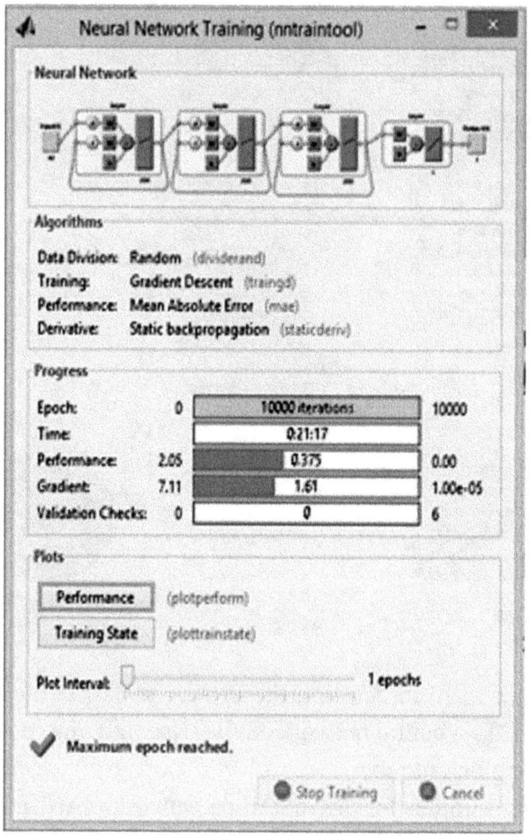

Figure 12. The graph of the learning error of artificial neural network No. 1

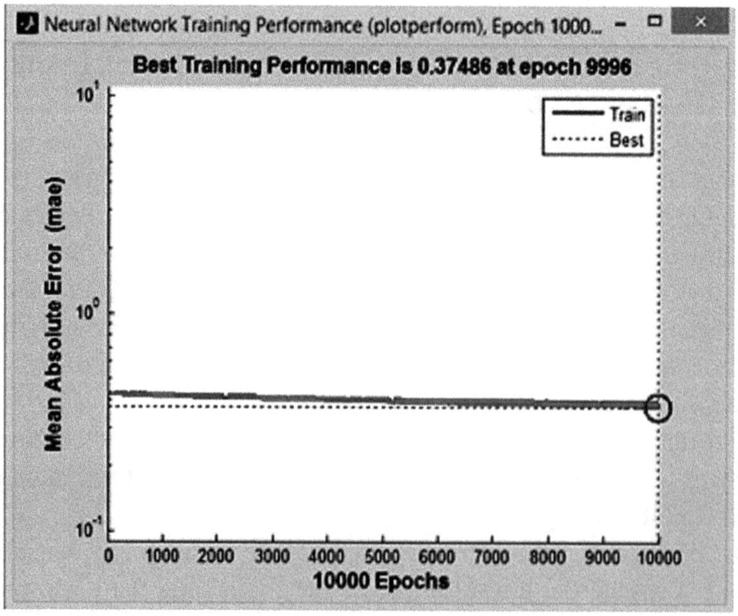

number 8). Elman's neural network has 42 inputs, because transmitted on 6 converted indicators to the input for the last seven days. Output 1. The network has three hidden layers, each of which uses a sigmoid activation function; the method of learning is the method of backpropagation of the error. The network stops training either after 10,000 learning epochs (cycles), or on condition that the size of the error is zero. The execution function is the mean absolute error.

9. Training of artificial neural network N° 2 and testing on the learning sample of data and on the testing sample of data(Figure 13, Figure 14) (Code block number 9)Best training performance is 0.40839 at epoch 9996.
10. Forming an artificial neural network No. 3 (Artificial neural network 3 - a neural network adviser specializing in forecasting the growth or fall of the course in a trend). Elman's neural network has 42 inputs, because transmitted on 6 converted indicators to the input for the last seven days. Output 1. The network has 3 hidden layers, each of which uses a sigmoid activation function; the method of learning is the method of backpropagation of the error. The network stops training either after 10,000 learning epochs (cycles), or on condition that the size of the error is zero. The execution function is the mean absolute error. (Code block number 10).

Figure 13. Window of the learning process of artificial neural network No. 2

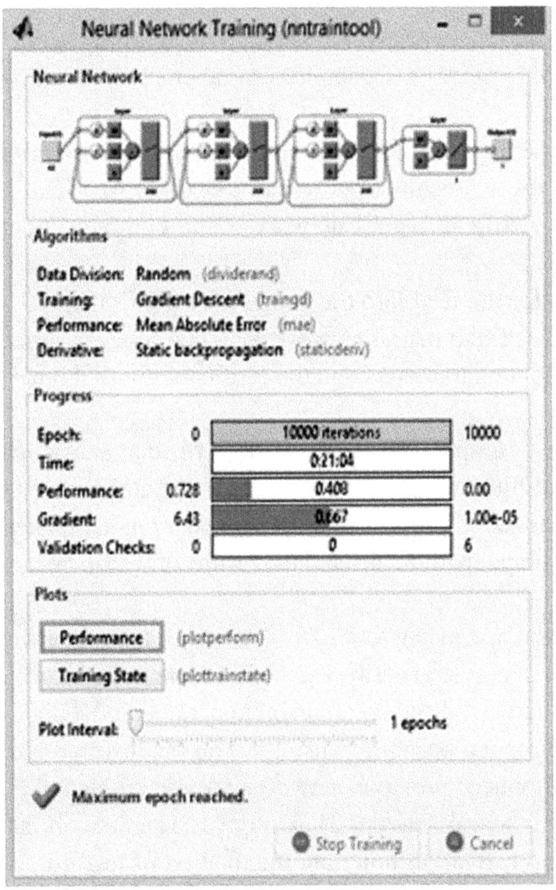

Figure 14. The graph of the learning error of the artificial neural network No. 2

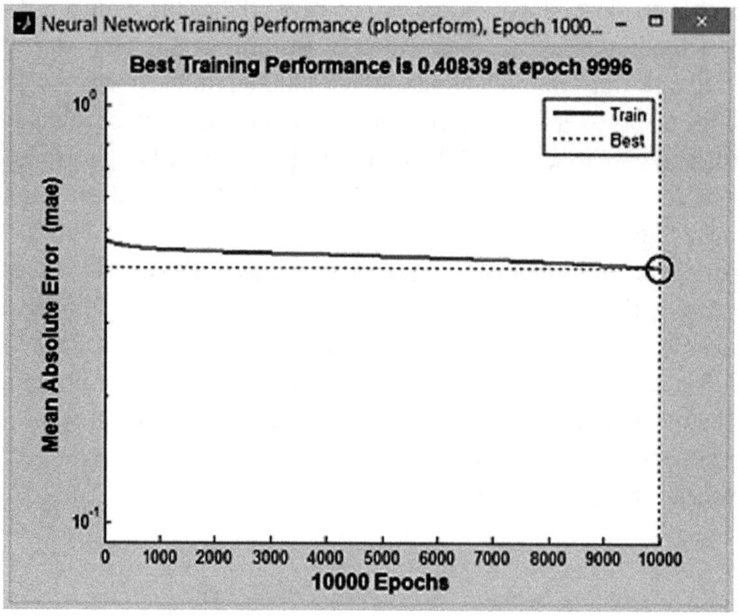

11. Training of artificial neural network N° 3 testing on the learning sample of data and on the testing data sample (Figure 15, Figure 16) (Code block number 11).Best training performance is 0.38879 at epoch 9976.
12. Forming an array of training samples for the 4th network (Artificial neural network 4 - a neural network that summarizes the results of previous networks: Artificial neural network 1, Artificial neural network 2 and Artificial neural network 3 and common input parameters) (Code block number 12)
13. Transpose the massifs for the feed into the neural network (Artificial neural network 1, Artificial neural network 2 and Artificial neural network 3) (Code block number 13)
14. Combine the original input parameters and the results of the first 3 networks (Code block number 14)
15. Forming an array of test samples for 4th network (Artificial neural network 4 - a neural network that summarizes the results of previous networks: Artificial neural network 1, Artificial neural network 2 and Artificial neural network 3 and common input parameters) (Code block number 15)
16. Integration the original input parameters and the results of the first 3 networks (Code block number 16)
17. Formation of an artificial neural network No. 4 (Artificial neural network 4 - a neural network that summarizes the results of previous networks: Artificial neural network 1, Artificial neural network 2 and Artificial neural network 3 and common input parameters). (Code block number 17). Elman's neural network has 45 inputs, because transmitted on 6 converted indicators to the input for last 7 days and also the outputs of previous networks (Artificial Neural Network 1, Artificial neural network 2 and Artificial neural network 3).. Output 1. The network has three hidden layers, each of which uses a sigmoid activation function; the method of learning is the method of backpropa-

Figure 15. The window of the process of learning the artificial neural network Nº 3

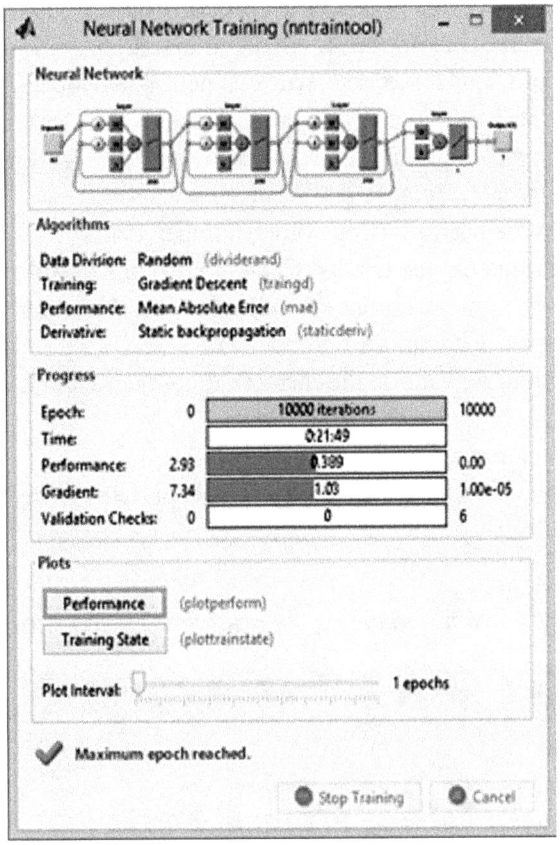

Figure 16. Chart of learning error of artificial neural network No. 3

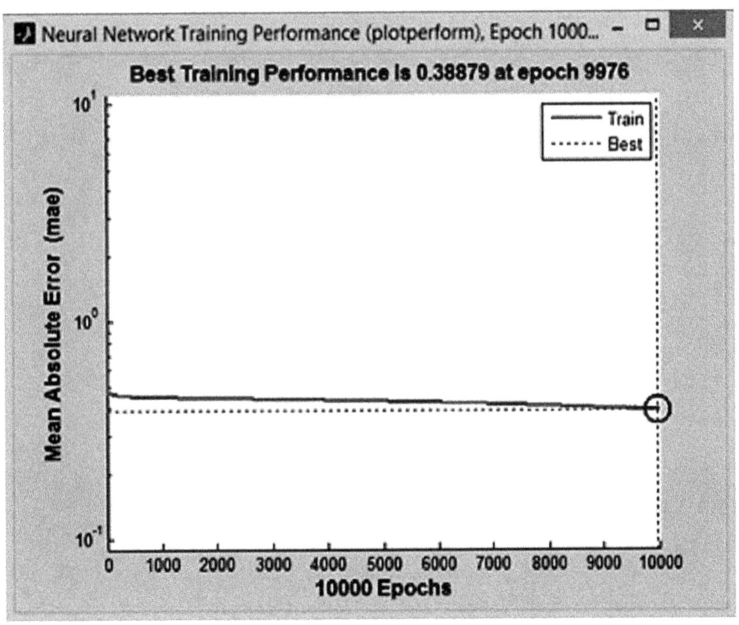

gation of the error. The network stops training either after 10,000 learning epochs (cycles), or on condition that the size of the error is zero. The execution function is the mean absolute error.

18. Training of artificial neural network N° 4. Artificial neural network 4 - a neural network which summarizes results of previous networks: Artificial neural network1, Artificial neural network 2, Artificial neural network 3 and common input parameters.Then testing artificial neural network 4on the learning data sample and on the testing data sample (Figure 17, Figure 18) (Code block number 18). Best training performance is 0.2284 at epoch 9941.
19. Transpose the output of the network (Code block number 19)
20. Forming a massif for evaluating the results (Code block number 20)
21. Convert an answer of the system response as follows: If on the output the result will be less than 0.5, then 0, if more, then 1 (Code block number 21)
22. If the answers of the system are equal to the correct values, then a unit is placed (Code block number 22)
23. Summation of all correct answers (Code block number 23)
24. The proportion of correct answers in the testing sample of data (Table 1) (Code block number 24)

Figure 17. The window of the learning process of an artificial neural network N° 4

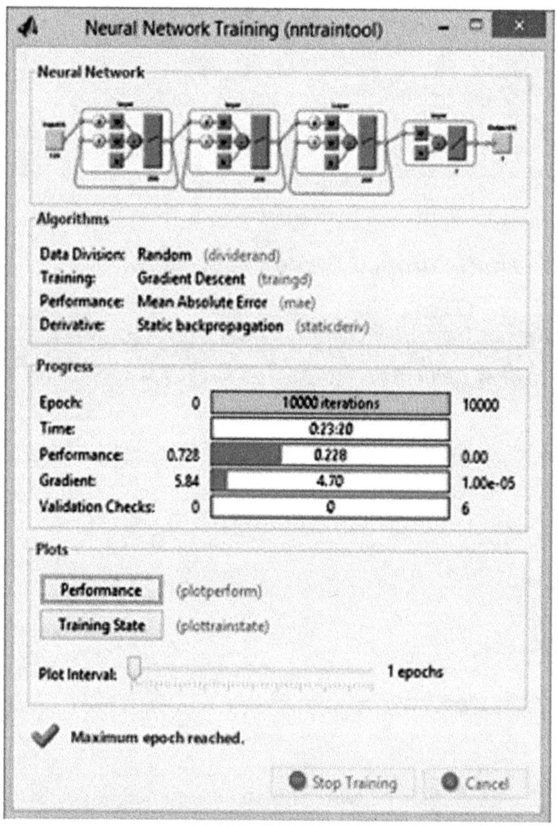

Figure 18. Graph of learning error of artificial neural network No. 4

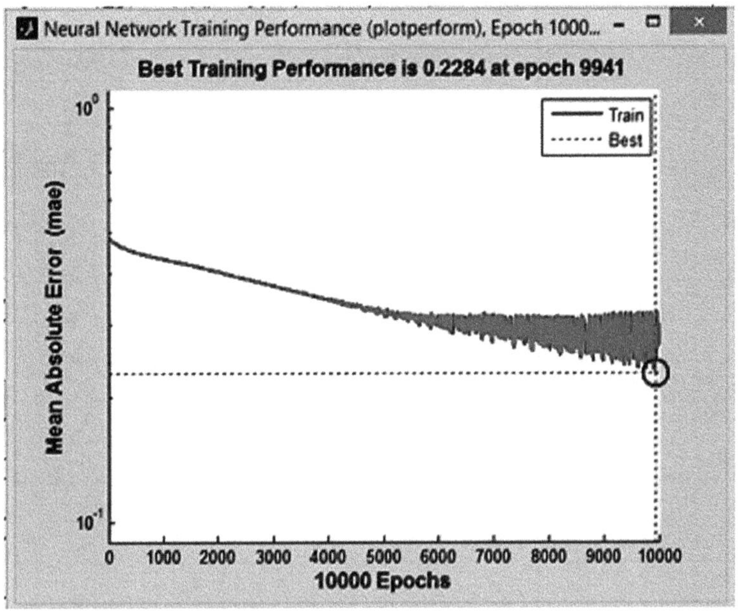

Table 1 presents the results of testing the model for the period March 3, 2017 to May 3, 2017. For convenience, the column names correspond to the names of the variables that were used in the code. The first column characterizes the output values of the model; the third column characterizes the result obtained. In the second and third columns, the value 0 means opening a short sale, and the value 1 means opening the purchase, within the same bar. When the values of the second and third columns coincide, in the fourth column recorded 1 - true, under mismatch of 0 - false. Thus, the constructed model allows to predict the direction of price movement for the next day with a probability of more than 86%.

CONCLUSION

Based on the results of the study the following conclusion can be drawn. Neural networks, unlike the other statistical methods of multidimensional classification analysis, are based on the parallel processing of information and possess the ability to conduct self-learning. This means that neural networks make it possible to obtain a valid result based on data that was not encountered in the learning process.

The ability to model nonlinear processes, adaptability and work with noisy data make it possible to apply neural networks to solve a wide class of financial problems. The main advantages of neural networks, in comparison with traditional computational methods, consist of the following advantages.

Problem Solving in Uncertain Conditions

The ability of a neural network to learning allows them to solve the tasks with unknown variables and dependencies between input and output data. Neural networks can work with incomplete data. In addi-

Table 1. Performance indicators of the trained neural network (learning on the interval 24.04.13 - 21.03.17) for the period 22.03.17 - 03.05.17

OutputTestNet4	TestDataTargetForNet4	OutputTestNet4v2	True / False
0,768384871	0	1	0
0,919348023	1	1	1
0,965955128	1	1	1
0,956145818	1	1	1
0,786567834	1	1	1
1,024603642	1	1	1
-0,06389309	0	0	1
1,138041594	1	1	1
0,945113273	1	1	1
0,09343831	0	0	1
0,8953719	1	1	1
0,187533021	0	0	1
0,910185852	1	1	1
-0,015994155	0	0	1
0,760703532	0	1	0
0,22351667	0	0	1
0,99790483	1	1	1
0,323678698	0	0	1
0,959294256	1	1	1
0,989780381	1	1	1
0,052524776	0	0	1
1,030220979	1	1	1
1,005773521	1	1	1
0,900863843	0	1	0
0,902306425	1	1	1
9,73E-05	0	0	1
1,025063458	1	1	1
0,704575764	1	1	1
0,451072272	1	0	0
0,726591368	1	1	1
Number of correct answers			26
Share of correct answers			0.8667

tion, the relationship between the data is not established in advance since the method involves studying the existing relationships on the finished models.

Resistance to Noises in the Input Parameters

The neural network can independently determine non-informative parameters for their analysis and screening.

Adaptation to the Environmental Changes

Neural networks can be retrained in new environmental conditions, described by slight fluctuations of the environment parameters. That is to state it is possible to retrain the neural networks on the basis of minor fluctuations of environment parameters. The possibility of neural network methods for modeling the behavior of prices of the American index NASDAQ-100 is substantiated. It is demonstrated that the model based on four recurrent neural networks of Elman allows predicting the future behavior of prices quite accurately.

As trading signals, the model generated a fractional numerical value. Furthermore, the obtained data were considered already in a discrete representation, where 0 was considered as a sell signal, and 1 - signal for a purchase. A trading solution was generated at the time of the opening of the day bar. Upon its closure, it was possible to compare the predicted results of the model and the actual market data. After training between April 24, 2013 to March 21, 2017, the model showed excellent results during the period from March 22 to May 3, 2017, with a forecast accuracy of 86.7%.

To test this model on other financial instruments, the authors prepared in the application the source code in the MQL4 (Appendix 1) language and applied package MatLab (Appendix 2), through which it is possible to obtain similar trading signals and conduct more detailed studies. It can be concluded that neural networks are a tool that allows solving problems with an unknown beforehand algorithm in advance, initially possessing a strong degree of oscillation and therefore it is difficult to formalize. However, the use of this tool requires a sensible selection of explanatory factors, effective work with data preprocessing. In general, the construction of such models can be used to conduct algorithmic trading (Vladimirova & Koksharova, 2014).

REFERENCES

Baestaens, D.E., Van Den Berg, W.M., & Wood, D. (1997). Neural Network solutions for trading in financial markets. *Pitman Bublished*, 56-77.

Bugorskiy, V.N., & Sergienko, A.G. (2008). Ispolzovanie neyronnyh setey dlya modelirovaniya prognoza kotirovok tsennyh bumag. *Prikladnaya Informatika, 3*.

Bushuev, K. (2000). Primeneniene yrosetevyh metodov pri prognozirovanii dinamiki rynka aktsiy. Modeli ekonomicheskih system I informatsionnye tehnologii. *Finansovay aakademiya pri Pravitelstve RF*, 30-35.

Chirkova, E. (2014). *How to evaluate business by analogy*. Alpina Pablisher.

Elder, A. (2007). *How to play and win on the stock exchange*. Alpina Pablisher.

Elman, J. L. (1990). Finding structure in time. *Cognitive Science*, *14*(2), 179–211. doi:10.120715516709cog1402_1

Giles, C.L., Lawrence, S., & Tsoi, A.C. (2006). *Rule inference for financial prediction using recurrent neural networks*. Academic Press.

Haykin, S. (2006). *Neural networks a comprehensive foundations*. Hamilton, Canada: McMaser University.

Kalmykova, L. (2007). *Fundamental analysis of financial markets*. Peter.

Kovel, M. (2007). *Trend Following: How Great Traders Make Millions in Up or Down Markets*. FT Press.

Laletina, A. (2015). Prognozirovanie finansovyh ryncov s ispol'zovaniem iskusstvennyh neyronnyh setey. Elektronnye sredstva I sistemy upravleniya. *Tomskiy gosudarstvennyj universitet system upravleniyai radioelektroniki*, 214-217.

MATLAB. (n.d.). Retrieved from http://www.mathworks.com/products/matlab/

Melihova, O. A., Veprintseva, O. V., & Chumichev, V. S. (2015). *Rezhimyobucheniya v iskusstvennyh neyronnyh setyah*. YUFU.

Murphy, J. (1999). *Technical Analysis of the Financial Markets: A Comprehensive Guide to Trading Methods and Applications*. New York Institute of Finance.

Nison, S. (2003). *The Candlestick Course*. Wiley.

Peters, E. E. (1994). *Fractal Market Analysis*. Wiley.

Schwager, J. (1997). *Technical Analysis*. Wiley.

Schwager, J., & Turner, S. (1995). Futures: Fundamental Analysis. Wiley.

StatSoft. (n.d.). Retrieved from http://www.statsoft.ru/products/STATISTICA_Base/

Turovskiy, Y. A. (2014). O pytmodelirovaniya alternativnyh cheloveko-mashinnyh interfeysov. *Aktualnye napravleniya nauchnyh issledovaniy XXI veka: teoriya I praktika, 5*, 249–252.

Turovskiy, Y. A. (2015). Otsenka skorosti raboty neyrokompyuternogo interfeysa, realizovannogo s ispolzovaniem gibridnogo intellekta. *Biomeditsinskaya radioelektronika, 3*, 61–70.

Vladimirova, D. (2014). Prognozirovanie finansovyh ryncov iskusstvennymi neiyronnymi setyami. Naukaibiznes: Putirazvitiya. *Fond razvitiya nauki I kultury*, 26-42.

Vladimirova, D. B., & Koksharova, A. A. (2014). Prognozirovanie finansovyh rynkov iskusstvennymi neyronnymi setyami. *FGBUVPO Permskiy natsionalnyy issledovatelskiy politehnicheskiy universitet, 3*, 45.

Wasserman, F. (1992). *Neurocomputer Techniques: Theory and Practice*. Moscow: Mir Publ.

Yahyaeva, G. E. (2016). *Osnovy teorii neyronnyh setey*. Natsionalnyy Otkrytyy Universitet.

APPENDIX 1

Preliminary Data in MQL4

```
#propertystrict    #propertyshow_inputs
externintN = 1000; // Numberofbars
voidOnStart()
{ int Handle; string  File_Name="Данные-"+Symbol()+".csv"; datetime Data; int
trend, H, result; double Sup, Inf;
Handle=FileOpen(File_Name,FILE_CSV|FILE_WRITE,";");// Open file
 if(Handle<0) { if(GetLastError()==4103) Alert("file not existence ",File_
Name); else Alert("Error open file ",File_Name); return; }
FileWrite(Handle, "Data", "EMA12", "P", "EMA24", "P", "EMA48", "P", "EMA120",
"P", "Ichimoku_Span", "P1", "P2", "ADX14_Main", "P", "TREND", "", "ADX14_Di",
"P1", "P2", "Momentum14", "P", "Ichimoku_tk", "P1", "P2", "EMA12_cl", "P",
"SAR_0.02_0.2", "P", "Ichimoku_Span2", "Result", "", "Heiken_Ashi", "P1",
"P2", "Bollinger_20_1", "P1", "P2", "38parots", "CCI14", "P", "RSI9", "P",
"Stochastic", "P", "Result", "", "Volume", "High", "Low", "Close", "Open");
FileWrite(Handle, "", "X_11", "Pr","X_12", "Pr", "X_13", "Pr", "X_14", "Pr",
"X_15", "Pr1", "Pr2", "X_16", "Pr", "Y_1", "", "X_21", "Pr1", "Pr2", "X_22",
"Pr", "X_23", "Pr1", "Pr2", "X_24", "Pr", "X_25", "Pr", "X_26", "Y_21", "",
"X_27", "Pr1", "Pr2", "X_28", "Pr1", "Pr2", "X_29", "X_210", "Pr", "X_211",
"Pr", "X_212", "Pr", "Y_22", "", "Volume", "High", "Low", "Close", "Open");
for (inti=2; i<=N; i++)
{ Data = iTime(Symbol(), 0, i);  // DateTime
   // find the bar number on H1 for the opening time of the current bar on D1
   H = iBarShift(Symbol(), PERIOD_H4, Data);  // search bar on H1 by time
   Sup = High[iHighest(Symbol(), 0, MODE_HIGH, 3, i)]; Inf =
Low[iLowest(Symbol(), 0, MODE_LOW, 3, i)];  // go beyond the range of 3 can-
dles before
if (Close[i-1] > Sup || Close[i-1] <Inf) trend = 1; else trend = -1;
if (Close[i-1] > Open[i-1]) result = 1; else if (Close[i-1] < Open[i-1]) re-
sult = -1; else result = 0; // the result of closing the candle
FileWrite(Handle, Data, Neiron_1_1MA(H), pNeiron_1_1MA(H), Neiron_1_2MA(H),
pNeiron_1_2MA(H), Neiron_1_3MA(H), pNeiron_1_3MA(H), Neiron_1_4MA(H),
pNeiron_1_4MA(H), Neiron_1_5Ishim(H), pNeiron_1_5Ishim(H),
p2Neiron_1_5Ishim(H), Neiron_1_6ADX(H), pNeiron_1_6ADX(H),trend, " ",
Neiron_2_1ADX(H), pNeiron_2_1ADX(H), p2Neiron_2_1ADX(H), Neiron_2_2Momentum(H),
pNeiron_2_2Momentum(H), Neiron_2_3IshimT(H), pNeiron_2_3IshimT(H),
p2Neiron_2_3IshimT(H), Neiron_2_4MA(H), pNeiron_2_4MA(H), Neiron_2_5SAR(H),
pNeiron_2_5SAR(H), Neiron_2_6Ishim(H), result, " ", Neiron_2_7Heiken_Ashi(H),
pNeiron_2_7Heiken_Ashi(H), p2Neiron_2_7Heiken_Ashi(H), Neiron_2_8Bollinger(H),
pNeiron_2_8Bollinger(H),  p2Neiron_2_8Bollinger(H), Neiron_2_9_38parrots(H),
```

```
Neiron_2_10CCI(H), pNeiron_2_10CCI(H), Neiron_2_11RSI(H), pNeiron_2_11RSI(H),
Neiron_2_12Stoch(H), pNeiron_2_12Stoch(H), result, "", Volume[H], High[H],
Low[H], Close[H], Open[H]); }   FileClose(Handle); // Close and save the re-
cord file return; }
//The first neuron of the first network
int Neiron_1_1MA(inti){intNeir; double MA1=iMA(NULL,PERIOD_H4,12,0,MODE_
EMA,PRICE_CLOSE,i); if(MA1<iHigh(Symbol(), PERIOD_H4, i) && MA1>iLow(Symbol(),
PERIOD_H4, i)) Neir=-1;  else Neir=1;  return(Neir);}
double pNeiron_1_1MA(inti){  double MA1=iMA(NULL,PERIOD_H4,12,0,MODE_EMA,PRICE_
CLOSE,i);  return(MA1); }
//The second neuron of the first network
int Neiron_1_2MA(inti){intNeir; double MA1=iMA(NULL,PERIOD_H4,24,0,MODE_
EMA,PRICE_CLOSE,i); if(MA1<iHigh(Symbol(), PERIOD_H4, i) && MA1>iLow(Symbol(),
PERIOD_H4, i)) Neir=-1; else Neir=1; return(Neir);}
double pNeiron_1_2MA(inti){double MA1=iMA(NULL,PERIOD_H4,24,0,MODE_EMA,PRICE_
CLOSE,i); return(MA1);}
//The third neuron of the first network
int Neiron_1_3MA(inti){intNeir; double MA1=iMA(NULL,PERIOD_H4,48,0,MODE_
EMA,PRICE_CLOSE,i); if(MA1<iHigh(Symbol(), PERIOD_H4, i) && MA1>iLow(Symbol(),
PERIOD_H4, i)) Neir=-1; else Neir=1; return(Neir);}
double pNeiron_1_3MA(inti){double MA1=iMA(NULL,PERIOD_H4,48,0,MODE_EMA,PRICE_
CLOSE,i); return(MA1);}
//The fourth neuron of the first network
int Neiron_1_4MA(inti){intNeir; double MA1=iMA(NULL,PERIOD_H4,120,0,MODE_
EMA,PRICE_CLOSE,i); if(MA1<iHigh(Symbol(), PERIOD_H4, i) && MA1>iLow(Symbol(),
PERIOD_H4, i)) Neir=-1; else Neir=1; return(Neir);}
double pNeiron_1_4MA(inti){double MA1=iMA(NULL,PERIOD_H4,120,0,MODE_EMA,PRICE_
CLOSE,i); return(MA1);}
//The fifth neuron of the first network
int Neiron_1_5Ishim(inti){ intNeir; double SpanA=iIchimoku(NULL,PERIOD_
H4,9,26,52,MODE_SENKOUSPANA,i); double SpanB=iIchimoku(NULL,PERIOD_
H4,9,26,52,MODE_SENKOUSPANB,i); double Supremum=MathMax(SpanA,SpanB); double
Infinum =MathMin(SpanA,SpanB); if(iClose(Symbol(), PERIOD_H4, i)<Supremum
&&iClose(Symbol(), PERIOD_H4, i)>Infinum) Neir=-1; else Neir=1; return(Neir);}
double pNeiron_1_5Ishim(inti){ double SpanA=iIchimoku(NULL,PERIOD_
H4,9,26,52,MODE_SENKOUSPANA,i); return(SpanA);}
double p2Neiron_1_5Ishim(inti){ double SpanB=iIchimoku(NULL,PERIOD_
H4,9,26,52,MODE_SENKOUSPANB,i); return(SpanB);}
//The sixth neuron of the first network
int Neiron_1_6ADX(inti){ intNeir; double ADX=iADX(NULL,PERIOD_H4,14,PRICE_
CLOSE,MODE_MAIN,i); if(ADX<25) Neir=-1; else Neir=1; return(Neir);}
double pNeiron_1_6ADX(inti){ double ADX=iADX(NULL,PERIOD_H4,14,PRICE_
CLOSE,MODE_MAIN,i); return(ADX);}
```

```
//The first neuron of the second network
int Neiron_2_1ADX(inti){intNeir=0; double DI_Plus =iADX(NULL,PERIOD_
H4,14,PRICE_CLOSE,MODE_PLUSDI,i); double DI_Minus=iADX(NULL,PERIOD_H4,14,PRICE_
CLOSE,MODE_MINUSDI,i); if(DI_Plus>DI_Minus) Neir= 1; if(DI_Plus<DI_Minus)
Neir=-1; return(Neir);}
double pNeiron_2_1ADX(inti){ double DI_Plus =iADX(NULL,PERIOD_H4,14,PRICE_
CLOSE,MODE_PLUSDI,i); return(DI_Plus);}
double p2Neiron_2_1ADX(inti){ double DI_Minus=iADX(NULL,PERIOD_H4,14,PRICE_
CLOSE,MODE_MINUSDI,i); return(DI_Minus);}
//The second neuron of the second network
int Neiron_2_2Momentum(inti){ intNeir=0; double Momentum=iMomentum(NULL,PERIOD_
H4,14,PRICE_CLOSE,i); if(Momentum>100) Neir= 1; if(Momentum<100) Neir=-1;
return(Neir);}
double pNeiron_2_2Momentum(inti){ double Momentum=iMomentum(NULL,PERIOD_
H4,14,PRICE_CLOSE,i); return(Momentum);}
//The third neuron of the second network
int Neiron_2_3IshimT(inti){ intNeir=0; double Fast=iIchimoku(NULL,PERIOD_
H4,9,26,52,MODE_TENKANSEN,i); double Slow=iIchimoku(NULL,PERIOD_
H4,9,26,52,MODE_KIJUNSEN,i); if(Fast>Slow) Neir= 1; if(Fast<Slow) Neir=-1;
return(Neir);}
double pNeiron_2_3IshimT(inti){double Fast=iIchimoku(NULL,PERIOD_
H4,9,26,52,MODE_TENKANSEN,i); return(Fast);}
double p2Neiron_2_3IshimT(inti){ double Slow=iIchimoku(NULL,PERIOD_
H4,9,26,52,MODE_KIJUNSEN,i); return(Slow);}
//The fourth neuron of the second network
int Neiron_2_4MA(inti){ intNeir=0; double MA1=iMA(NULL,PERIOD_H4,12,0,MODE_
EMA,PRICE_CLOSE,i); if(iClose(Symbol(), PERIOD_H4, i)>MA1) Neir= 1;
if(iClose(Symbol(), PERIOD_H4, i)<MA1) Neir=-1; return(Neir);}
double pNeiron_2_4MA(inti){double MA1=iMA(NULL,PERIOD_H4,12,0,MODE_EMA,PRICE_
CLOSE,i); return(MA1);}
//Fifth neuron of the second network
int Neiron_2_5SAR(inti){intNeir=0; double SAR=iSAR(NULL,PERIOD_H4,0.02,0.2,i);
if(iClose(Symbol(), PERIOD_H4, i)> SAR) Neir= 1; if(iClose(Symbol(), PERIOD_
H4, i)< SAR) Neir=-1; return(Neir);}
double pNeiron_2_5SAR(inti){ double SAR=iSAR(NULL,PERIOD_H4,0.02,0.2,i);
return(SAR);}
//The sixth neuron of the second network
int Neiron_2_6Ishim(inti){intNeir=0; double SpanA=iIchimoku(NULL,PERIOD_
H4,9,26,52,MODE_SENKOUSPANA,i); double SpanB=iIchimoku(NULL,PERIOD_
H4,9,26,52,MODE_SENKOUSPANB,i); double Supremum=MathMax(SpanA,SpanB); double
Infinum =MathMin(SpanA,SpanB); if(iClose(Symbol(), PERIOD_H4, i)>Supremum)
Neir= 1; if(iClose(Symbol(), PERIOD_H4, i)<Infinum)  Neir=-1; return(Neir);}
//The first neuron of the third network
```

```
int Neiron_2_7Heiken_Ashi(inti){intNeir=0; double OpenHA=iCustom(NULL,PERIOD_
H4,"Heiken Ashi",2,i); double CloseHA=iCustom(NULL,PERIOD_H4,"Heiken
Ashi",3,i); if(OpenHA<CloseHA)  Neir= 1; if(OpenHA>CloseHA)  Neir=-1;
return(Neir);}
double pNeiron_2_7Heiken_Ashi(inti){ double OpenHA=iCustom(NULL,PERIOD_
H4,"Heiken Ashi",2,i); return(OpenHA);}
double p2Neiron_2_7Heiken_Ashi(inti){ double CloseHA=iCustom(NULL,PERIOD_
H4,"Heiken Ashi",3,i); return(CloseHA);}
//The second neuron of the third network
int Neiron_2_8Bollinger(inti){intNeir=0; double H=iBands(NULL,PERIOD_
H4,20,1,0,PRICE_CLOSE,MODE_UPPER,i); double L=iBands(NULL,PERIOD_
H4,20,1,0,PRICE_CLOSE,MODE_LOWER,i); if(iHigh(Symbol(), PERIOD_H4, i)>H)  Neir=
1; if(iLow(Symbol(), PERIOD_H4, i) <L)  Neir=  -1; return(Neir);}
double pNeiron_2_8Bollinger(inti){ double H=iBands(NULL,PERIOD_H4,20,1,0,PRICE_
CLOSE,MODE_UPPER,i); return(H);}
double p2Neiron_2_8Bollinger(inti){ double L=iBands(NULL,PERIOD_
H4,20,1,0,PRICE_CLOSE,MODE_LOWER,i); return(L);}
//The third neuron of the third network
int Neiron_2_9_38parrots(inti){ intNeir=0; double P38=iCustom(NULL, PERIOD_H4,
"38 parrots",7,100,70,0,i); double H  =iCustom(NULL, PERIOD_H4, "38
parrots",7,100,70,1,i); double L  =iCustom(NULL, PERIOD_H4, "38
parrots",7,100,70,2,i);  if(P38>H)  Neir= -1; if(P38<L)  Neir= 1;
return(Neir);}
//The fourth neuron of the third network
int Neiron_2_10CCI(inti){ intNeir=0; double CCI=iCCI(Symbol(),PERIOD_
H4,14,PRICE_CLOSE,i); if(CCI>100)  Neir=   1; if(CCI<-100)  Neir=  -1;
return(Neir);}
double pNeiron_2_10CCI(inti){ double CCI=iCCI(Symbol(),PERIOD_H4,14,PRICE_
CLOSE,i); return(CCI);}
//The fifth neuron of the third network
int Neiron_2_11RSI(inti){intNeir=0; double RSI=iRSI(NULL,PERIOD_H4,9,PRICE_
CLOSE,i); if(RSI>50)  Neir=   1; if(RSI<50)   Neir=  -1; return(Neir);}
double pNeiron_2_11RSI(inti){double RSI=iRSI(NULL,PERIOD_H4,9,PRICE_CLOSE,i);
return(RSI);}
//The sixth neuron of the third network
int Neiron_2_12Stoch(inti){intNeir=0; double Stoch1=iStochastic(NULL,PERIOD_
H4,5,3,3,MODE_SMA,0,MODE_MAIN,i); double Stoch2=iStochastic(NULL,PERIOD_
H4,5,3,3,MODE_SMA,0,MODE_SIGNAL,i); if(Stoch1>Stoch2)   Neir=   1;
if(Stoch1<Stoch2)   Neir=  -1; return(Neir);}
double pNeiron_2_12Stoch(inti){double Stoch1=iStochastic(NULL,PERIOD_
H4,5,3,3,MODE_SMA,0,MODE_MAIN,i); return(Stoch1);}
double p2Neiron_2_12Stoch(inti){double Stoch2=iStochastic(NULL,PERIOD_
H4,5,3,3,MODE_SMA,0,MODE_SIGNAL,i); return(Stoch2);}
```

APPENDIX 2

Creation and Testing of a Neural Network in the MatLab Package

Code Block Number 1

```
TrainDataInput = xlsread('D:\Data\Data.xlsx','TrainData','B4:DW786'); Train-
DataTarget = xlsread('D:\Data\Data.xlsx','TrainData','DX4:DY786'); TestDataIn-
put = xlsread('D:\Data\Data.xlsx','TestData','B4:DW33');
TestDataTarget = xlsread('D:\Data\Data.xlsx','TestData','DX4:DY33');
```

Code Block Number 2

```
TrainDataInputForNet1 = TrainDataInput(1:end,1:42); TrainDataIn-
putForNet2 = TrainDataInput(1:end,43:84); TrainDataInputFor-
Net3 = TrainDataInput(1:end,85:126); TrainDataTargetForNet1 =
TrainDataTarget(1:end,1);
TrainDataTargetForNet2 = TrainDataTarget(1:end,2); TrainDataTargetForNet3 =
TrainDataTarget(1:end,2);
```

Code Block Number 3

```
TestDataInputForNet1 = TestDataInput(1:end,1:42); TestDataInputForNet2 =
TestDataInput(1:end,43:84); TestDataInputForNet3 = TestDataInput(1:end,85:126);
```

Code Block Number 4

```
TrainDataInputForNet1 = TrainDataInputForNet1'; TrainDataInputForNet2 = Train-
DataInputForNet2'; TrainDataInputForNet3 = TrainDataInputForNet3'; Train-
DataTargetForNet1 = TrainDataTargetForNet1'; TrainDataTargetForNet2 = Train-
DataTargetForNet2'; TrainDataTargetForNet3 = TrainDataTargetForNet3';
TestDataInputForNet1 = TestDataInputForNet1'; TestDataInputForNet2 = TestDa-
taInputForNet2'; TestDataInputForNet3 = TestDataInputForNet3';
```

Code Block Number 5

```
OutputTrainNet1 = []; OutputTrainNet2 = []; OutputTrainNet3 = []; OutputTrain-
Net4 = []; OutputTestNet1 = []; OutputTestNet2 = []; OutputTestNet3 = []; Out-
putTestNet4 = [];
```

Code Block Number 6

```
Net1 = newelm(TrainDataInputForNet1,TrainDataTargetForNet1,[200,200,200],{'logs
ig','logsig','logsig'},'traingd'); Net1.trainParam.epochs = 10000; Net1.train-
Param.goal=0; Net1.performFcn='mae'; Net1.divideParam.trainRatio = 100/100;
Net1.divideParam.valRatio = 0/100; Net1.divideParam.testRatio = 0/100;
```

Code Block Number 7

```
Net1 = train(Net1,TrainDataInputForNet1,TrainDataTargetForNet1); Out-
putTrainNet1 = sim(Net1,TrainDataInputForNet1); OutputTestNet1 =
sim(Net1,TestDataInputForNet1);
```

Code Block Number 8

```
Net2 = newelm(TrainDataInputForNet2,TrainDataTargetForNet2,[200,200,200],{'logs
ig','logsig','logsig'},'traingd'); Net2.trainParam.epochs = 10000; Net2.train-
Param.goal=0; Net2.performFcn='mae'; Net2.divideParam.trainRatio = 100/100;
Net2.divideParam.valRatio = 0/100; Net2.divideParam.testRatio = 0/100;
```

Code Block Number 9

```
Net2 = train(Net2,TrainDataInputForNet2,TrainDataTargetForNet2); Out-
putTrainNet2 = sim(Net2,TrainDataInputForNet2); OutputTestNet2 =
sim(Net2,TestDataInputForNet2);
```

Code Block Number 10

```
Net3 = newelm(TrainDataInputForNet3,TrainDataTargetForNet3,[200,200,200],{'logs
ig','logsig','logsig'},'traingd'); Net3.trainParam.epochs = 10000; Net3.train-
Param.goal=0; Net3.performFcn='mae'; Net3.divideParam.trainRatio = 100/100;
Net3.divideParam.valRatio = 0/100; Net3.divideParam.testRatio = 0/100;
```

Code Block Number 11

```
Net3 = train(Net3,TrainDataInputForNet3,TrainDataTargetForNet3); Out-
putTrainNet3 = sim(Net3,TrainDataInputForNet3); OutputTestNet3 =
sim(Net3,TestDataInputForNet3);
```

Code Block Number 12

```
TrainDataInputForNet4 = TrainDataInput(1:end,1:126); TrainDataTargetForNet4 =
TrainDataTarget(1:end,2);
```

Code Block Number 13

```
TrainDataInputForNet4 = TrainDataInputForNet4'; TrainDataTargetForNet4 =
TrainDataTargetForNet4';
```

Code Block Number 14

```
TrainDataInputForNet4 = [TrainDataInputForNet4; OutputTrainNet1; OutputTrain-
Net2; OutputTrainNet3];
```

Code Block Number 15

```
TestDataInputForNet4 = TestDataInput(1:end,1:126); TestDataInputForNet4 = Tes-
tDataInputForNet4'; TestDataTargetForNet4 = TestDataTarget(1:end,2);
```

Code Block Number 16

```
TestDataInputForNet4 = [TestDataInputForNet4; OutputTestNet1; OutputTestNet2;
OutputTestNet3];
```

Code Block Number 17

```
Net4 = newelm(TrainDataInputForNet4,TrainDataTargetForNet4,[200,200,200],{'logs
ig','logsig','logsig'},'traingd'); Net4.trainParam.epochs = 10000; Net4.train-
Param.goal=0; Net4.performFcn='mae'; Net4.divideParam.trainRatio = 100/100;
Net4.divideParam.valRatio = 0/100; Net4.divideParam.testRatio = 0/100;
```

Code Block Number 18

```
Net4 = train(Net4,TrainDataInputForNet4,TrainDataTargetForNet4); Out-
putTrainNet4 = sim(Net4,TrainDataInputForNet4); OutputTestNet4 =
sim(Net4,TestDataInputForNet4);
```

Code Block Number 19

```
OutputTestNet4 = OutputTestNet4';
```

Code Block Number 20

```
Result = [OutputTestNet4 TestDataTargetForNet4];
```

Code Block Number 21

```
fori = 1:30 OutputTestNet4v2 = Result(i,1); if OutputTestNet4v2 > 0.5
Result(i,3) = 1; else Result(i,3) = 0; end cend
```

Code Block Number 22

```
fori = 1:30 OutputTestNet4Value2 = Result(i,3); TestDataTargetForNet4I =
Result(i,2); if TestDataTargetForNet4I == OutputTestNet4Value2 Result(i,4) =
1; else Result(i,4) = 0; end end
```

Code Block Number 23

```
NumberOfCorrectAnswers = sum(Result(1:end,4))
```

Code Block Number 24

```
ShareOfCorrectAnswers = NumberOfCorrectAnswers / 30
```

Chapter 11
Neuro-Linguistic-Programming-Based Modeling of Stock Markets

Mamed Babaev
Southern Federal University, Russia

Oxana Savenko
Southern Federal University, Russia

ABSTRACT

Speaking of the preliminary findings of this chapter and considering empirical data observed in the course of stock exchange trading, it is possible to forge a connection between the actual behavior of the investors and the NLP meta-program classes. In particular, there is a palpable correspondence between the 13 distinct meta-program categories affecting workplace motivation and performance, commonly known as the language and behaviour profile or LAB Profile, of Shelle Rose Charvet, and the behavior patterns of investors. This may represent a development of the idea of "the conventional wisdom" formulated in his time by John Kenneth Galbraith. However, in its new incarnation of collective conscious and collective subconscious, it may affect decision making processes around the globe.

INTRODUCTION

The world at this stage is barely starting to sluggishly recover from the raging tempest of the financial crisis of 2007-2008. One is witnessing a unique process: the world financial system is being slowly dismantled, and in its stead something new is being created. But what is it? For the time being, none can tell. The only thing that one knows for sure is that the rules of the game are drastically changing. This is particularly true for the stock market. Sooner or later, any and every most reliable financial system would malfunction. When this is the case, the majority of its participants falls into panic and is very likely to make multiple mistakes. Only a select few, having the expertise, know-how and innovative analytical implements manage to stay ahead of time and to make profit while the rest fall prey to the depredations of crisis.

DOI: 10.4018/978-1-5225-3767-0.ch011

The purpose of this work is to outline a path to a method, based on institutional and in particular behavioral economic approach that would allow one to effectively predict the conduct of investors in a given stock market, be it a local entity or a transboundary unit with a focus on the erratic behavior and how it can be used in day-to-day investment practice.

BACKGROUND

For those economists and financiers who may not be familiar with neuro-linguistic programming, it would be opportune to give a brief description of this branch of knowledge.

Let us begin with programming. Each person, by the instrumentality of different means and factors has managed to program themselves to excel at a certain number of tasks, to be mediocre at various tasks and awful in other areas. If one observes and listen carefully to how a person behaves and communicates by means of their language (both verbally and nonverbally), one may glean an understanding of how, from the neurological prospective, said person puts his or her experience together to be excellent, mediocre or awful at the tasks that he or she performs. Hence the name of this field of study - neuro-linguistic programming (NLP).

The applications are many, various and profound. It means if someone is highly proficient/skilled at something and you are trained in certain NLP protocols, you can perform something that is called modeling. In this context, modeling means determining how it is possible for said person to do what they do. NLP searches for the answers to questions such as: "What are the absolute essentials" or "What is that person paying attention to or ignoring, sequentially and/or simultaneously, to be able to do it?" If one could discover the answers to these and some other questions, then one could teach that skill-set or behavior to other people and even learn it yourself, but what is more important - one could predict behavior responses of different individuals.

In more scientific terms, NLP is an approach to communication, personal development, and psychotherapy created by Richard Bandler and John Grinder in California, United States in the 1970s. NLP's creators claim there is a connection between neurological processes (neuro-), language (linguistic) and behavioral patterns learned through experience (programming), and that these can be changed to achieve specific goals in life (Tosey & Mathison, 2006) (Dilts, Grinder, Delozier, & Bandler, (980).

The Language and Behaviour Profile (LAB Profile) is a model developed by Rodger Bailey who was an avid researcher and developer in the NLP field. He created the LAB Profile in early 1980's. It is based on a set of NLP patterns called Meta Programs, developed at the time by Leslie Cameron-Bandler and others (Bandler & Grinder, 1979). These represent the filters that we use to make up our model of the world.

Every person has a certain number of filters by which they let in certain parts of the tangible world. In his book called "The Logical Structure of Linguistic Theory" (Chomsky, 1975), Noam Chomsky states that there are three processes by which people establish the filters of their individual Model of the World, in other words, their personal reality:

The first process is called deletion. One deletes much information from the surrounding environment and internally. In his 1956 work entitled "The Magical Number Seven, Plus or Minus Two: Some Limits on Our Capacity for Processing Information" (Miller, 1956), George Miller, an American Psychologist, predicated that our conscious minds are only capable of handling seen plus or minus two bits of information at any one time and that we delete the rest. Which may be translated in simple terms that on a good day a person may handle nine and on a bad day, probably, only five.

This explains why most phone numbers are around seven digits. Nevertheless, in the home town of the author the administration once decided to add one additional number to the beginning of the phone numbers - chaos ensued. Everyone had to decide whether to remember phone numbers by groups of two, four or simply to add the new city code (incidentally, it was 2) onto the front of their old number. Virtually no one found a simple method of keeping the new digit set in their mind at once. Each person had to come up with their one way of breaking it down.

Therefore, seven plus or minus two bits of information would be what one can comfortably hold in one's aware mind at one time. By means of deletion, one filters out a multitude of things without even being aware of this fact or consciously electing to do so.

The second process is called distortion. Our mind, having an access but to the reflection of the objective reality within the limitations of its sensory and logical perception (Luriya, 1998), tends to distort said reflection. Distortion twists the truth in that the speaker draws conclusions or makes assumptions. By his statement he appears to know the truth about what another person or other people think, feel or believe. It is easy for someone to believe that his or her distorted belief is the truth rather than one's own perception because they have compared it with their own map of the world and decided it is so. An example would be more illuminating. The reader may have moved to a new place and gone into the living room prior to moving the furniture in and pictured what it would look like furnished. The furniture was not actually there. Therefore, reality was being distorted by the conscious effort of the mind.

Two broader cases of distortion are hallucination and creativity. They are both similar in terms of the subject's modifying the external information to something else. This, in essence, represents the process of distortion.

The last of the three aforementioned processes that one uses to make up one's model of the surrounding world is generalization. It is the opposite of Cartesian Logic (where one can go from a general rule to specific examples but not the other way around). Generalization is when one takes a few examples and then establishes a general rule. This is the way learning occurs. A would-be-driver, having learnt to start from one to three passenger cars, will know how to start cars in general. Until, that is, he or she is asked to drive a truck or a more complex vehicle and that person will have to relearn to tackle those exceptions.

Generalisation is about how people make rules for themselves, outside of their awareness. People have a certain number of experiences and they make a rule that from now on constitutes a part of how they create their own model of the world.

It would now be logical to ask what the link between the three abovementioned processes, NLP and LAB profile is. Richard Bandler and John Grinder used deletion, distortion and generalization to create a map (Meta Model) aimed at discovering and influencing a person's perceptions and interpretations of their experiences (Bandler & Grinder, 1975). Leslie Cameron-Bandler used them to build the Meta Programs, which she developed from her work in therapeutic settings. Rodger Bailey, in his turn, a student of Leslie Cameron-Bandler, adapted and applied her work in a business framework. He created a tool which makes it possible to understand what people are communicating about their reality.

Initially, there were approximately 60 Meta Program patterns and students of NLP had to spend a considerable amount of time trying to learn to determine the particular programs used by each person. To make detection and use of said patterns easier, Rodger Bradley managed to reduce the number of patterns from 60 to 14 and to develop a small set of specific questions. Regardless of what people answer to these well-made questions, their unconscious personality patterns are revealed in the structure of the utterances they use (Charvet, 1995). Therefore, one pays attention to the way the person answers, instead of what they actually say. Thus, after asking a few simple questions, one can determine what will activate and maintain someone's motivation.

Secondly, Rodger Bradley developed the concept of Influencing Language (IL). Once one knows a person's patterns, you can then modify one's own way of speaking so that it has a maximum impact on that person. Using the right IL is exceedingly powerful since one is speaking on someone's own personal style.

According to the NLP researchers, most prominently Shelle Rose Charvet, as one becomes skilled at using the LAB profile, one will not always need to ask questions. One will be able to *simply listen* in a focused way and pick the necessary patterns. The key lies in the way a person communicates.

When someone talks about their experience, they only communicate a minuscule part of the experience itself. They have to edit out the majority of the information related to the event in question in order to be able to communicate it in an acceptable time limit. What this means is that for one to be able to tell somebody about doing something (for example, reading a book on financial markets), the speaker will have to eliminate most of what he or she experienced. The speaker may say that the book was interesting, leaving out all the things the book in question made him or her think about without speaking about whether or not said person was comfortable at the time of reading the book. It is worthwhile recalling those times when you did not fully understand what your interlocutor was taking about due to the fact that they had left out the elements necessary for full understanding.

People achieve this editing in several ways, primarily through deletion, distortion and generalization. Being aware of these processes may make it possible to communicate with and predict the behavior of other humans, including investors in financial markets, much easier. People tend to transform their actual experiences, their opinions and many other things so that it corresponds to their own specific patterns of deletion, distortion and generalization. LAB profile makes it possible to know what it means when they edit their experiences in certain ways. Rodger Bailey showed that people who use the same linguistic patterns when speaking have similar if not the same behaviors (Charvet, (1995). From a simple conversation one can determine how a person communicates and detect what will activate and keep his or her motivation or what will make him change his mind.

Application of LAB profiles enables one to predict and influence behavior of other individuals. The LAB profiles represent a set of tools that can be learned as a skill and used many different ways including automation of the profiling process. However, since the LAB profile is a set of skills, it has to be used with a certain measure of circumspection, paying attention to the shifts that people male as they move from situation to situation. That is why it is also important to consider the notion of Context.

In the field of Neuro-Linguistics, one especially highlights the role of context. The Context may be defined as a frame of reference that a person puts around a certain situation. Owing to the fact that humans are flexible by their nature, they may display different behavior at different times.

Yeager (1985) gives a more appropriate definition of the Context:

Thinking occurs within a context, purpose or frame of reference that is unique to the individual. If you don't know the context of another's thing, many things may seem illogical… When you think of what you want for dinner you think in terms of the context of where (location), with whom or when (time) or even in terms of good nutrition (biochemistry). These are all contextual factors. Yet the definition of a context is typically subjective…

Some people think of time mostly in the past tense. Others in the present tense... This characteristic is a learned performance and it "frames" the range of behaviours ... possible within that subjective context. In this sense, a context is a set of limits that defines what is, reciprocally, defines what is not an issue.

Context is a stabilizing reference point that locates where you are or are not in your subjective worls. If an individual habitually thinks in terms of precedent (the past tense), it will be difficult for the person to imagine "possibilities" (future tense) if history isn't imagined into the changed future.

Yeager's description leads one to realize that one's Meta-Programs function as one's thinking contexts. It is necessary to consider the significance of this. When one speaks about these "programs" ("thinking" sorting styles, one's operating system for how one uses one's nervous system) that lie meta to one's primary level of thinking - one essentially identifies the contexts for thinking, one's "thinking contexts".

According to Hall M. and Bodenhamer B. (1997), this leads to a set of most important questions to ask oneself or any person with whom one communicates:

- Within what context do you do your thinking?
- Within what frame-of-reference does your thinking occur?
- As you think about things right now - do you use a global or specific frame?
- Do you use a match or a mismatch frame?
- Do you use a past, present or future frame? etc.

Hall M. and Bodenhamer B. state that these questions are most important because if we do not know a person's frame-of reference for their "thinking" - we will not understand their meanings, emotions and responses. We will not know or understand the person's stabilizing reference point. Human thinking always inevitably and inescapably occurs within some frame. And, in that frame, the "thinking" (emoting, perceiving, behaving) makes perfect sense. It functions "logically" to that internal frame.

It only seems illogical to other frames. If you think and feel that someone's way of thinking seems really illogical, it is necessary to endeavor to do so from a different frame of reference, from a different model of the world.

The value of the context becomes particularly important for the financial markets, where one may take various macro, meso- and mega-economic conditions and trends that would define the broader frame for investor's decision making. Be it a period of leverage or deleverage, short-term or long-term debt cycle (Dalio, 2015), positive or negative growth, technological cycle, balance of payments of a particular state or set of states etc.

MAIN FOCUS OF THE CHAPTER

Some people may reasonably ask whether they always behave that way. The answer to that question is negative. People change and are subject to significant events that may affect the way that they function. Owing to the fact that our behavior may vary in different situations, one has to ensure while using the LAB profile that one is bale to clearly and unequivocally identify the Context.

The first six categories in the LAB profile (broadly assembled under the name of Motivation traits) demonstrate how different people activate their motivation and what linguistic patterns one needs to use to capture their interest or inversely to identify their appurtenance to a specific LAB profile.

However, as it is pointed out by Rose Shelle Charvet (1995), behavior predictions are only valid in the same Context in which the subject was profiled. Further description will borrow extensively from her works.

The first motivation category is the LEVEL of activity. It is about what will get you going and what will make you think. Here one distinguishes two patterns:

- **Proactive:** Proactive people initiate. They tend to act with little to no consideration; to jump into situations without thinking or carrying out an analysis of the way the land lies, so to speak. They may discomfiture some people since they charge on with what they want to do. Such people are good at getting the job done. They do not wait for others to initiate.
- **Reactive:** Reactive people wait for others to initiate or until the situation is right before they act. They may consider and analyse without acting. They want to fully understand and assess the situation before they act. They believe in chance and luck. They will spend a lot of time waiting. Some people may get upset with them because they do not get started. They will wait for others to initiate and then respond. In the extreme, they operate with extra caution and study situations endlessly. They make good analysts.

As to the patterns recognition, since there is no specific question to ask for this particular category, one may pay attention to the target's sentence structure and in case of a tête-à-tête contact, to their body language, as they will be demonstrating their pattern throughout the conversation. However, since the face to face contacts between financial market investors are rare these days, we shall concentrate on linguistic patterns.

- **Proactive Sentence Structure:**
    - Short sentences: noun, active verb, tangible object
    - Speaks as if they in control of the world
    - Crisp and clear sentence structure
    - Direct
    - At the extreme, they "bulldoze"
- **Reactive Sentence Structure:**
    - Incomplete sentences, subject or verb missing
    - Passive verbs or verbs transformed into nouns
    - Lots of infinitives
    - Speaks as if the world controls them, things happen to them, believe in change or luck
    - Long and convoluted sentences
    - Talks about thinking about, analyzing, understanding or waiting, or the principle of the thing
    - Conditionals, would, could, might, may
    - Overly cautious, need to understand and analyse

The second motivation category is the CRITERIA of activity. Criteria is the term that one uses to describe a person's way of making decisions about what is good, bad, awful, wrong, right, etc. They are personal labels. Many of these words are labels for our values.

A person's criteria, as Rose Shelle puts it, are those words which incite a physical and emotional reaction. They are the so called "Hot Buttons". The words themselves are associated, or struck in one's memory, to a series of emotionally similar events that one has experienced through one's life. So when a person hears one of their criteria, the word itself will trigger the emotional response attached to it.

Each of us has their own definition for each Criterion. A single criterion consists of many elements, both conscious and unconscious. One may never need to know a person's definition of their Criteria in those situations where one simply wants to find out how they describe something they are excited about in a given Context.

It is important to note the specific questions that may help elicit Criteria, here there are some of them:

- What do you want in (a job, financial transaction, interaction with other traders)?
- What is important to you?
- What counts?
- What has to be there?
- What would you like to have, be or do?

However, knowing the Criteria is insufficient for a give Context. It is equally important to be able to find out their hierarchy - discover what holds more value for the subject and what is less consequential. One of the ways of achieving this is to use the technique called Hierarchy of Criteria which essentially allows one to confront the subject with a choice between two Criteria and to find out which one of them is more important. Through several iterations of this process a complete hierarchy may be established.

The third motivation category is the CRITERIA of activity. There are two patterns in this category which describe the Direction a person is moving in a given Context. They are either moving toward an objective, or away from problems.

Toward persons stay focused on their objective. They think in terms of goals that have to be achieved. They have the motivation to have, get, achieve, attain, etc. Because of their concentration on goals to be accomplished, they tend to be good at managing priorities. Moreover, they are energized by their goals.

People with a strong Toward pattern often have trouble either recognizing what should be avoided or in identifying problems. At the extreme, they are perceived as naïve by others because they do not take potential obstacles into account.

Away From individuals notice what should be avoided, got rid of and otherwise not to happen. Their motivation is triggered when there is a problem to be solved or when there is something to move away from. They are energized by threats. For example, for a stock trader the relevant triggers may be the deadlines for completing a transaction, stock's reaching a certain buying or selling threshold etc. People with an Away From Pattern in a given Context are good at trouble-shooting, solving problems, and at pinpointing possible obstacles during planning because they automatically pick up on what is wrong.

They may have trouble maintaining focus on their goals because they are easily distracted and are compelled to respond to negative situations. This is the kind of person who will drop everything to fix something. At the extreme they forget what the priorities are and only concentrate on treating crises. If this person is at the top of a department or an organization, the entire organization will be run by crisis management. Away From people have some difficulty managing priorities because whatever is wrong will attract most of their attention. People who have a strong Away From orientation in a given Context are often perceived as jaded or cynical, particularly by Toward people.

The two categories in question have the following patterns of recognition

- **Questions:**
  - Why is having that (their Criteria) important?
    Or
  - What will having that (their Criteria) do for you?
- **Toward—Sentence Structure:**
  - Talks about what they gain, achieve, get, have etc.
  - Inclusion
  - What they want, goals
- **Away From—Sentence Structure:**
  - Will mention situations to be avoided, gotten rid of
  - Exclusion of unwanted situations, things
  - Problems

Direction of Motivation

When you master this category, you will be able to prevent and avoid many problems as well as know how to reach your goals more effectively.

There are two patterns in this category which describe the Direction a person is moving, in a given Context. They are either moving toward a goal, or away from problems. Each pattern is first described in its pure form.

Toward

Toward people stay focused on their goal. They think in terms of goals to be achieved. They are motivated to have, get, achieve, attain, etc. Because of their concentration on goals to be accomplished, they tend to be good at managing priorities. Moreover, they are energized by their goals.

People with a strong Toward pattern often have trouble either recognizing what should be avoided, or in identifying problems. At the extreme, they are perceived as naive by others because they do not take potential obstacles into account.

Away From

Away From people notice what should be avoided, gotten rid of and otherwise not happen. Their motivation is triggered when there is a problem to be solved or when there is something to move away from. They are energized by threats. A salesperson told me: "If I don't get out there and sell, I won't be able to pay my bills at the end of the month." Deadlines, for example, get this people into action. People with an Away From pattern in a given Context are good at trouble-shooting, solving problems, and at pinpointing possible obstacles during planning because they automatically pick up on what is wrong.

They may have trouble maintaining focus on their goals because they are easily distracted and are compelled to respond to negative situations. This is the kind of person who will drop everything to fix something. At the extreme they forget what the priorities are and only concentrate on treating crises. If this person is at the top of a department or an organization, the entire organization will be run by crisis management. Away From people have some difficulty managing priorities because whatever is wrong will attract most of their attention. People who have a strong Away From orientation in a given Context are often perceived as jaded or cynical, particularly by Toward people.

Pattern Recognition

- **Questions:**
    - Why is having that (their Criteria) important?
      Or
    - What will having that (Their Criteria) do for you?
- **Toward—Sentence Structure:**
    - Talks about what they gain, achieve, get, have etc.
    - Inclusion
    - What they want, goals
- **Away From—Sentence Structure:**
    - Will mention situations to be avoided, gotten rid of
    - Exclusion of unwanted situations, things
    - Problems

Source of Motivation

This category deals with the source of motivation, or in other words, its location. Where are judgements made; inside a person's body or from outside? These patterns affect how you make judgements and decisions. As you try them out, you will be able to decide how best to use them and others will notice your increased effectiveness. Here are the two patterns:

Internal

People with an Internal pattern in a given context provide their own motivation from within themselves. They decide about the quality of their work. They have difficulty accepting other people's opinions and outside direction. When they get negative feedback on work they feel has been well done, they will question the opinion or judge the person giving the feedback.

They are motivated to gather information from outside sources and then they decide about it, based on internal standards. Because they take orders as information, they can be hard to supervise. "My boss wants this out by Tuesday? That's interesting."

Since they do not need external praise, they tend not to give much feedback as managers.

Internal people hold standards somewhere within themselves, for the things that are important to them. Their motivation is triggered when they get to gather information from the outside, process it against their own standards, and make judgments about it.

External

External people need other people's opinions, outside direction and feedback from external sources to stay motivated. In the Context of work, if they do not get that feedback, they will not know how well they are doing. They take information as orders. "He said the green paper matches the decor. I'd better go get some."

They are motivated when someone else decides. They have trouble starting or continuing an activity without outside feedback of some kind.

External people do not hold standards within themselves. They gather them from the outside. In the absence of external feedback, they will experience something akin to sensory deprivation.

Pattern Recognition

- **Questions:**
    - How do you know that you have done a good job? (at work, at choosing a car, etc.)
- **Internal—Sentence Structure:**
    - They decide or know themselves, "I know"
    - They evaluate their own performance based on their own standards and Criteria
    - They resist when someone tells them what to do, or decides for them
    - Outside instructions are taken as information
- **External—Sentence Structure:**
    - Other people or external sources of information decide or judge for them
    - Need to compare their work to an external norm or standard i.e., a checklist or a quota
    - Outside information is taken as a decision or order

Reason of Motivation

This category will lead you to unlimited possibilities and show you the right way to get there. There are two patterns:

Options

Options people are motivated by opportunities and possibilities to do something in a different way. There is always another better way to do things. They love to create procedures and systems but have great difficulty following them. If you give an Options person a guaranteed way to make a million dollars, they will try to improve it.

They are thrilled by unlimited possibilities and ideas. The thing that is totally irresistible to Options people is breaking or bending the rules.

Options people like to start a new idea or a new project. However, they do not necessarily feel compelled to finish it. They much prefer to do development and set up rather than maintenance activities. Sometimes they will have difficulty committing themselves because they believe this will reduce their options. At the extreme, they might avoid deciding anything (particularly if they are also Reactive). Alternatively, they can be totally committed to an idea or project, until the next new idea comes along.

Procedures

Procedures people like to follow set ways. They believe there is a "right" way to do things. Once they have a procedure they can follow it over and over again. These are people who are interested in how to do things, not in why things are the way they are.

A procedure has a beginning and an end. It may have choice points in which you gather more information and make a decision. Without one, Procedure people feel lost or get stuck. When they commence a procedure, the most important thing for them is to get to the end of the procedure. They are the ones who will complete and finish what they start.

Procedures people feel personally violated when it is suggested to them to break, or go around the rules. Once they know the procedure to follow, they are happy doing that.

Pattern Recognition

- **Question:**
  - Why did you choose your present job? (or house, vacation, car, etc.)

You may recall that, in the introduction of this book, we discussed Reality. In order for people to create their own models of reality, they use three processes called Deletion, Distortion and Generalization. This category deals with Distortion. The question we ask for this pattern is: "Why did you choose ..." and the rest refers to a particular Context.

For Options people, when they hear the question Why did you choose?, they hear the question why and give you a list of Criteria for their answer.

When Procedures people hear the question, they delete the word why and substitute how did it come to be? They answer the question: "How?" Sometimes, the first thing they will say is, "Well I didn't choose." They will tell you a story or series of events that led them to getting the job. "I was working at my brother-in-law's and there was an opening in this other company at the time I finished a contract, so I took it."

- **Options:**
  - List of Criteria
  - Opportunities, possibilities
  - Expanding options and choice
- **Procedures:**
  - Did not choose
  - Answers the question "why" by telling "how" it came to pass
  - The facts, events leading to, a story

How you recognize someone who is equally Options and Procedures is interesting. They might tell you a story with Criteria imbedded in it. I happen to be quite in the middle on this pattern, and if someone asked me why I chose to set up my own company after I came back to Canada, I would say: "I was working in another company for a while when I came back and I wasn't happy. And being organized is important to me. I realized that 1 could probably make more money on my own, and decrease expenses and be independent. So I set up my company, Success Strategies." I have used both patterns throughout the answer.

Motivation Decision Factors

The Decision Factors category is about your internal time clock and how often the bell rings for change. Are you motivated by evolution, revolution, both or stability? There are four patterns:

Sameness

Sameness people want their situation in a given Context to stay the same. They do not like change and may refuse to adapt. They may accept a major change once every ten years, but they will provoke change only once every 15 to 25 years.

Sameness With Exception

Sameness with Exception people like a given Context to stay mainly the same but will accept change once a year, if the change is not too drastic. They prefer their situations to evolve slowly over time. They tend to resist major changes except when they are perceived to be progressive or gradual. They need major change once every five to seven years. This is by far the largest category in the work Context and probably in many other Contexts.

Difference

People with a Difference pattern love change; they thrive on it and want it to be constant and major. They will resist static or stable situations. They need drastic change about every one to two years, and if they do not get it, they may leave. They like change to be revolutionary, dramatically different.

Sameness With the Exception and Difference (the Double Pattern)

People with this double pattern like change and revolutionary shifts but are also comfortable where things are evolving. They are happy with both revolution and evolution. They need major change every three to four years, on average.

Pattern Recognition

- **Question:**
    - What is the relationship between your work this year and last year? (vacation, this home and the last one, etc.)
      or
    - What is the relationship between this job and your last one?

The question asks: "What is the relationship between . . . ?" In this framework, the word relationship has the connotation of similarity. People will either naturally understand the word and tell you how it is the same or similar, or alternatively they will not know what you mean or reinterpret it to mean how is it different.

- **Sameness:**
  - How they are the same, identical
  - What they have in common
  - How it has not changed
- **Sameness with Exception:**
  - How is has evolved over time
  - It is the same except more; less; better; worse; improving; etc. (comparisons on a sliding sealer)
  - Focus on the trip more than arriving at the destination
- **Difference:**
  - Doesn't understand the word relationship
  - Will describe how it is completely different
  - New, different, changed, transformed, revolutionary
  - Language points to an immediate switch
  - Focus on the destination, ignore the trip
- **Difference and Sameness with Exception:**
  - Use both Difference and Sameness with Exception language

NOW WE SHALL PASS TO THE SECOND SUBSET OF TRAITS: WORKING TRAITS

The next eight categories of the LAB Profile will tell you how people deal with information; what type of tasks and environment they need to be most productive in a given Context, and how they get convinced about something. These categories will demonstrate how to maintain someone's motivation.

Working Scope

Using the Scope category, you can determine whether someone can handle overviews and grand designs, or whether the details make more sense. There are two patterns in this category:

Specific

Specific people handle small pieces of information well. At the extreme, they cannot perceive or create an overview. They treat information in linear sequences, step by step, in all its detail. A Specific person perceives the trees, branches and twigs, rather than the forest. They may have difficulty prioritizing as a result. If they are interrupted in the middle of a sequence, they tend to either start over at the beginning, or resume the telling from the point at which they were interrupted. Specific people work well where details must be attended to, in tasks such as organizing events or handling logistics.

Note: There is a difference between a sequence, which Specific people use and procedure, used by Procedures people. While both have a beginning and an end, there maybe choice points, branches and several end points in a procedure. A sequence is chronological, linear, narrow and uni-directional. It is possible for a person to have both Specific and Options patterns or a Procedures pattern in combination with a General pattern.

General

People with a General pattern in a given Context prefer to work on the overview, or at the conceptual level, though they can concentrate on details for finite periods of time. Because they see the big picture all at once, they may present ideas in a random order without stating the link between one thought and another. They concentrate on the forest, and having to deal with the trees for long periods of time irritates them.

Pattern Recognition

While there are no specific questions for this category, you will be able to recognize the patterns within virtually each sentence spoken.

Hint: One simple way you can know for sure is to time your

LAB Profile interview. On average, a full interview takes about 20 minutes to complete, not including giving the person your feedback. With a Specific person, the interview will be at least 40 minutes.

Here is how to recognize these patterns in conversation:

- **Specific:**
    - Speak in sequences, step-by-step
    - Many modifiers, adverbs, adjectives
    - Proper nouns for people, places and things
    - If they lose the sequence, they will start over again or continue from where they left off
    - Only seem to be aware of the step before and after the one they are on, not much perception of the overview
- **General:**
    - May present things in random order
    - Overview, summaries
    - Concepts, abstracts
    - Simple sentences, few modifiers or details

Figure 1. Sequence: A — B — C — D etc.

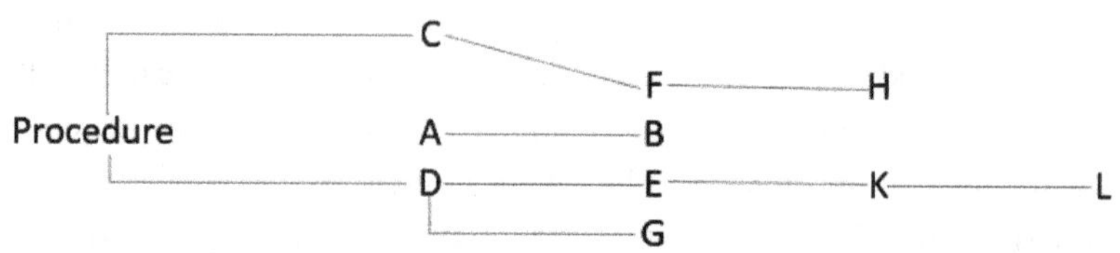

Working Attention Direction

The Attention Direction category reveals whether a person can perceive, and respond automatically to the body language and voice tone of other people or not. There are two patterns:

Self

Self people do not show many emotions although they do have feelings. There is sometimes a time gap between when they receive a stimulus and when they respond to it. They respond based on what they consider to be appropriate. These people are convinced only by the content of what people say, rather than the accompanying tone, body language or level of rapport. They have difficulty establishing rapport because they do not notice other people's body language, and therefore they miss many clues. People with this pattern simply do not pick up hints.

They know how well the communication is going based on their own feelings. As a result, they tend not to be adept at interpersonal communication. At work, many Self people become technical experts in fields where communication skills are not essential.

Other

Other people have automatic, reflex reactions to people's behavior. They are animated (for their culture) and respond to others with facial expressions, body movements and shifts in voice tone. They know how the communication is going based on the responses they consciously or unconsciously observe from the other person. These people are good at creating and maintaining rapport, provided they also have the other appropriate patterns.

Pattern Recognition

There is no verbal test for this pattern as it shows up in body language or the absence thereof. To test for Attention Direction, I usually drop a pencil accidently on purpose. People with an Other pattern will spontaneous) bend down and pick it up, provided they have seen it or heard it drop. Self people will not pick it up. When speaking to someone on the phone, I will sneeze or have a coughing fit in the middle of a sentence. Does the person say their cultural equivalent of Gezundheit or continue on as if nothing had happened?

The following clues will also be apparent throughout the conversation:

- **Self:**
    - Absence of culturally appropriate behavioral responses such as head nodding, saying "uh-huh", etc.
    - Reacts only to the content of what you say
    - Doesn't "pick up the pencil"
    - Doesn't notice or respond to your voice tone
    - Little or no facial expression or voice variation

- **Other:**
    - Responds to both content and nonverbal aspects of the communication
    - Will nod head, move body, say "uh-huh", etc., as a response
    - Animated (for their culture)

Let us say the authors were speaking to someone who had a Self pattern and the authors, with our shoulders drooped, our bottom lip sticking out, and a whiny tone of voice, said: "I'm really happy to be here." She or he would think that I was really happy to be here. For this kind of person, unless you actually say: "I am annoyed and irritated," explicitly, they do not pick it up. Hinting will not work, nor will sarcasm as a method of communication.

While only 7% of the population at work are Self, probably many more are borderline; that is, having some of the Self pattern. You can recognize those people when you drop the pencil. They look at the pencil, they look at you, they look at the pencil and then they may eventually decide to pick it up. It is not spontaneous; not a reflex. A reflex action is something outside of voluntary control. Sometimes you can also recognize these borderline people because, although they may exhibit little nonverbal behavior themselves (i.e.: facial expressions, gestures and voice variation), they may be able to notice and respond to the nonverbal behavior of others.

Working Stress Response

Wouldn't you like to have a way to find out in five minutes or less, if someone can handle a high stress job?

The Stress Response category examines how you respond to the pressures, at work or elsewhere, that are typical for the Context you are in? This is not about how someone would respond to major life dramas, since almost everyone would have an emotional response in those situations. People respond to these "normal" pressures in the following three ways:

Feeling

People with a Feeling pattern have emotional responses to the normal levels of stress at work. They go into their emotions and stay there. High stress jobs can therefore be difficult for them to handle over the long term. To many other people, they seem to overreact to situations or be hyper-sensitive. They are well suited for artistic or creative work where emotion provides the juice. As salespeople, they find it difficult to handle rejection and may not, as a result, prospect for new customers as often as they should.

Choice

Choice people first have an emotional response to the normal stresses at work and then either return to an unemotional state or not as they desire, in a given situation. They have choice. Because they feel emotions themselves, they can empathize with others, or choose not to. They tend to perform well as people managers, as they can combine the personal side of the job and distance themselves when necessary.

Thinking

People with a Thinking pattern do not have emotional responses to the normal stressful situations for a given Context. They have trouble empathizing with others as they themselves do not go into emotional states. They will not panic in most emergencies, but keep a cool head. They are reliable performers in high stress jobs.

Pattern Recognition

- **Question:**
    - Tell me about a work situation that caused you trouble.

For Contexts other than work, simply substitute the Context for "work situation", i.e. Tell me about a buying decision that caused you trouble.

When you are asking this question, avoid having the person tell you about all the times they had a particular kind of problem. "Whenever a customer is unhappy with our service, I get nervous." Make sure the person picks one specific troublesome (not catastrophic) situation that he remembers. What you are to determine, as they review that situation, is if they go into an emotional state and get stuck there, have an emotional response and come out of it, or not have an emotional reaction at all.

Working Style

The Working Style category will allow you to discover, or confirm, how you can be at your best; whether you want to share your work with other people, do it yourself while involving others, or work all on your own.

In this category many people have more than one pattern; a dominant and a secondary Style in a given Context. There are three patterns:

Independent

People who have an Independent pattern in the work Context want to work alone and have sole responsibility. Their productivity suffers if others are around or if they have to share responsibility. When interrupted, they may lose their train of thought. They prefer to work in an office with the door closed. At the extreme, they may forget to consult with others (especially if they are also Internal). At work, they can go for long periods of time without craving contact with others.

The expression: "A camel is a horse designed by a committee," was probably authored by someone with this pattern. As a manager, an Independent person will do most of the work by himself and probably not establish rapport easily.

Proximity

Proximity people want a dear territory of responsibility but need to have others involved at around, in proximity. They need well-defined responsibilities, and to be productive and stay motivated, their tasks must involve other people. Their productivity will fall if others share responsibility and authority, or if they have to work totally alone.

Of the three patterns, this one is most suited for people and project management positions. They will make sure that everyone knows what they are responsible for. Proximity people do well as the boss, or when they have a boss, as long as territories are well-established.

Co-Operative

Co-operative people want to work and share responsibility with others. They believe in the 2+2=5, the whole is greater than the sum of its parts, Synergy Principle. They have trouble with deadlines and finishing tasks if they have to work on their own. They do not need a territory to be in charge of, and as managers, will want to do everything with their employees.

The Californian light bulb joke describes these people. How many Californians does it take to change a light bulb? It takes six. One to change the light bulb, and five to share the experience.

Co-operative does not necessarily mean that the person co-operates, in the usual sense of the term, just that she needs to do an activity with someone else. When my oldest son, Jason, was about five years old, he spent an hour building a Lego boat. My youngest son, Sammy, then about two and a half, came down from his nap and kicked the boat to pieces. Jason was very upset, but from Sammy's point of view, Jason had the nerve to do it all by himself. He did not wait for Sammy to do it with him. This was intolerable for Sammy, who usually did not go off by himself and play. He may sometimes be disruptive when he plays, but he needs the company.

Pattern Recognition

- **Questions:**
  - Tell me about a work experience that was (their criteria)?
    - Listen to the answer.
  - What did you like about it?

For Contexts other than work, you can simply insert the Context into the question, i.e. "Tell me about an experience in a relationship that was ..."

Make sure the person you are profiling picks a specific example of a situation that meets their Criteria. If they have numerous Criteria for the Context you are discussing, simply use one Criterion. Some people may never have had an experience that meets all of their Criteria.

For this category, you will need to listen carefully to what the person talks about. Listen to the answer for the first question and then ask the second question. Does the person talk about doing something totally alone, in charge, with others around or together with others?

- **Independent:**
    - Says I, I did it, myself, my responsibility
    - Won't talk about or mention other people
    - The activity presupposes that they did it on their own
- **Proximity:**
    - Other people are present but "I did it"
    - May or may not mention others, but the nature of the activity necessitates the presence of others (i.e. Sales or teaching)
- **Co-operative:**
    - Will say: we, us, our job, together, etc.
    - Includes other people and shares responsibility

Working Organization

Working Organization is about how people work, either by getting the job done or focusing on feelings. There are two patterns in this category:

Individuals with a Person pattern pay attention to the feelings and thoughts of either themselves or others. Feelings take on such an importance that they become the task itself. They will organize their work so that they can focus on people and their feelings. They are good at establishing rapport.

Thing people concentrate on products, ideas, tools, tasks and systems (things). They treat people and ideas as objects, and believe that emotions have no place in the world of work. They want to get things done, and have a task orientation.

Pattern Recognition

Since 55% of the population are mainly Thing-oriented at work, you will hear this pattern more commonly, although specific professions will have their own cultural pattern (i.e. many social workers have a Person pattern). The questions are the same as for the Style category. In other words, when you ask the questions below, you will receive both the Style and Organization categories of information simultaneously.

- **Questions:**
    - Tell me about a work experience that was (their criteria)?
        - Listen to the answer.
    - What did you like about it?
- **Person:**
    - Speak about people, emotions, feelings
    - Will name people, use personal pronouns
    - People are the object of their sentences
- **Thing:**
    - Talk about processes, systems, tools, ideas, tasks, goals
    - Will not mention people often except as impersonal pronouns i.e. "they", or "you"
    - People become objects, parts of a process

Working Rule Structure

Rule Structure will give you some information regarding the ability or willingness to manage oneself and others. There are four patterns in this category:

My/My: My Rules for Me / My Rules For You

My/My people have rules for themselves and for others. They are willing to communicate their rules to others. Because they believe that people are similar, they think that what is good for themselves will also suit other people. They will say things such as: "If I were you, I would ..." A large majority of people at work have this pattern and probably in other Contexts as well.

My/.: My Rules for Me / I Don't Care

The My/(period) people have rules for themselves and do not care about others. They do not necessarily harbour malicious intent towards others; it is simply not their problem or concern. These people often get on with what they need to do without thinking about others. Sometimes they are called selfish by others because they simply did not consider anyone else.

I was often awakened in the middle of the night in my sixth floor Paris apartment, by people whom I believe had this pattern. They would honk loudly at 2 o'clock in the morning, in a residential district, surrounded by hundreds of people who had, presumably, been sleeping.

No/My: No Rules or Don't Know Rules for Me / My Rules for You

No/My people do not know or do not have guidelines for themselves, but once given the rules, are quite willing to pass them on to others. As a result, they may have difficulty providing direction for themselves or making decisions. Instead, they may get stuck and not know what to do.

My/Your: My Rules for Me / Your Rules for You

People with a My/Your pattern know the rules and policies to follow at work but are reluctant or unable to communicate them to others. They operate from a "Different strokes for different folks" perspective. Because they believe everyone is different, they would consider it arrogant to tell others what to do. As a result, other people are often unclear as to their expectations.

These are the people who can understand both sides of an argument, as annoying as this may seem to those of us who take strong positions.

Pattern Recognition

You will need to ask these two questions and notice whether the person answers both or one only, as follows:

- **Questions:**
    - What is a good way for you to increase the chances for your success (at work)?
        - Listen to the answer.
    - What is a good way for someone else to increase the chances for their success (at work)?

To identify these patterns, you will need to compare the answers to these questions. When someone does not know, they will hesitate a long time and respond with a questioning tone.

- **My/My:**
    - Same kind of answer to both questions or answers both questions easily.
- **My/.:**
    - Clear response to question one. Indicates lack of interest for question two.
- **No/My:**
    - Doesn't know the answer to question one. Has rules for question two.
- **My/Your:**
    - Has rules for question one. Doesn't know or wouldn't presume to say for question two.

Having outlined the agent groups and subgroups as well as the methods for their procedural linguistic analysis based determination; it would be interest to know the distribution (percentage) of agents in each of the groups in question.

SOLUTIONS AND RECOMMENDATIONS

In their works Roger Bailey and Shelle Rose Charvet provide the following findings (Table 1).

However, it is of paramount importance to consider the importance of the context since the aforementioned figures are, first of all dated, and secondly are too broad to be deemed adequate for financial market evaluation, since it is not the general public that has to be examined, but a very specific portion of it, the target market actors. The situation would appear to also be complicated by the following circumstances:

- The LAB profile might need minor adjustments to reflect the cultural and communication differences between different English speaking countries/markets, die to the fact that works of Roger Bailey and Shelle Rose Charvet were oriented mainly towards the American society in terms of the object of study.
- The LAB profile will need further research and adjustment when one will pass from the English speaking community to another one (German, French, Italian, Japanese, Korean Chinese), necessitating new procedures for placing the market actors within a certain LAB profile category.

These are formidable challenges in and of themselves, but they are not insurmountable. The solution would comprise two facets, namely: use of chat bots within the framework of social media (Facebook, Twitter, Mobage, Cyworld, Weibo etc.) and applicative social networking.

A chatbot (also known as an Artificial Conversational Entity) is a computer program which conducts a conversation via auditory and/or textual methods. Such programs are often designed to convincingly

Table 1. Percentage distribution of LAB profiles of Roger Bailey and Shelle Rose Charvet

<table>
<tr><th colspan="4">Level</th></tr>
<tr><td colspan="2">Proactive
15-20%</td><td>Equally Procative/Reactive
60-65%</td><td>Reactive
15-20%</td></tr>
<tr><th colspan="4">Direction</th></tr>
<tr><td colspan="2">Toward
40%</td><td>Equally Toward/Away From
20%</td><td>Away From
40%</td></tr>
<tr><th colspan="4">Source</th></tr>
<tr><td colspan="2">Internal
40%</td><td>Equally Internal/External
20%</td><td>External
40%</td></tr>
<tr><th colspan="4">Reason</th></tr>
<tr><td colspan="2">Options
40%</td><td>Equally Options/Procedures
20%</td><td>Procedures
40%</td></tr>
<tr><th colspan="4">Decision Factors</th></tr>
<tr><td>Sameness
5%</td><td>Sameness with exception
65%</td><td>Difference
20%</td><td>Sameness with Exception and Difference
10%</td></tr>
<tr><th colspan="4">Scope</th></tr>
<tr><td colspan="2">Specific
15%</td><td>Equally Specific/General
25%</td><td>General
60%</td></tr>
<tr><th colspan="4">Attention</th></tr>
<tr><td colspan="2">Self
7%</td><td>Other
93%</td><td></td></tr>
<tr><th colspan="4">Stress Response</th></tr>
<tr><td colspan="2">Feeling
15%</td><td>Choice
70%</td><td>Thinking
15%</td></tr>
<tr><th colspan="4">Style</th></tr>
<tr><td colspan="2">Independent
20%</td><td>Proximity
60%</td><td>Co-operative
20%</td></tr>
<tr><th colspan="4">Organisation</th></tr>
<tr><td colspan="2">Person
15%</td><td>Equally Person/Thing
30%</td><td>Thing
55%</td></tr>
<tr><th colspan="4">Rule Structure</th></tr>
<tr><td>My/My
75%</td><td>My/.
30%</td><td>No/My
7%</td><td>My/Your
15%</td></tr>
<tr><th colspan="4">Convincer Channel</th></tr>
<tr><td>See
55%</td><td>Hear
30%</td><td>Read
30%</td><td>Do
12%</td></tr>
<tr><th colspan="4">Convincer Mode</th></tr>
<tr><td>Number of Examples
52%</td><td>Automatic
8%</td><td>Consistent
15%</td><td>Period of Time
25%</td></tr>
</table>

simulate how a human would behave as a conversational partner, thereby passing the Turing test. Chatbots are typically used in dialog systems for various practical purposes including customer service or information acquisition. Some chatbots use sophisticated natural language processing systems, but many simpler systems scan for keywords within the input, then pull a reply with the most matching keywords, or the most similar wording pattern, from a database.

Chatbots are part of virtual assistants such as Google Assistant, and are accessed via many organizations' apps, websites, and on instant messaging platforms such as Facebook Messenger. Non-assistant applications include chatbots used for entertainment purposes, for research, and social bots which promote a particular product, candidate, or issue.

Other companies explore ways how they can use chatbots internally, for example for Customer Support, Human Resources, or even in Internet-of-Things (IoT) projects. Overstock, for one, has reportedly launched a chatbot named Mila to automate certain simple yet time-consuming processes when requesting for a sick leave (Greenfield, 2017). Other large companies such as Lloyds Banking Group, Royal Bank of Scotland, Renault and Citroën are now using automated online assistants instead of call centres with humans to provide a first point of contact.

The chatbot that would be particularly useful for the aforementioned purposes is "socbot". A socbot (also: socialbot or social bot) is a particular type of chatbot that is employed in social media networks to automatically generate messages (e.g. tweets) or in general advocate certain ideas, support campaigns, and public relations either by acting as a "follower" or even as a fake account that gathers followers itself ("What is socialbot?", 2017).

FUTURE RESEARCH DIRECTIONS

However, in order to efficiently apply the LAB profiling procedure for large groups of users belonging to a given set of target financial market social media related communities, it will be necessary to repurpose the socbot for leading a conversation oriented towards asking the questions associated with each profile, collecting the necessary statistical data and compiling a database judging by the results of each conversation. One might argue, that one could use a survey for LAB profiling the market agents, however, this would be associated with, firstly, limited number of respondents and therefore smaller statistical pool, and secondly, intentional consideration of the questions asked and potentially distorting the true nature of their attitude.

In order to take this data-base analysis one step further and to expose the ties on the micro and meso social group level, one could complement the results of the socbot LAB profile with social networking techniqies, in particular, sociograms, whose relations would be those of the LAB profile relations.

The ties through which any given social unit connects represent the convergence of the various social contacts of that unit. This theoretical approach is, necessarily, relational. An axiom of the social network approach to understanding social interaction is that social phenomena should be primarily conceived and investigated through the properties of relations between and within units, instead of the properties of these units themselves. Thus, one common criticism of social network theory is that individual agency is often ignored although this may not be the case in practice (see agent-based modeling), but owing to

LAB profiling procedures this particular weak point of the social network can be eliminated, resulting in a sociogram that reflects both the collective trends and individual relations and qualities.

For example, in the situation of an impending lift tactics to occur in a given stock market, LAB profiling will make it possible to determine the percentage of the "Reactive" and "Toward" people who would be more likely to fall prey to the lift, but it could do more. The sociogram would allow to pinpoint specific actors and, running a cross-reference with the stock exchange data base regarding funds available to the actors in question, one could achieve an estimation of amount that will be involved in the series of lift transactions.

An additional benefit of the applicative LAB profiling is partial solution of deviations from the Nash equilibrium within the framework of the Game Theory. When game theorists refer to a game's 'solution', or a 'solution concept', they have in mind how *RATIONAL PEOPLE* might play the game. A 'rational solution' corresponds to a set of strategies (one for each player) that the players of the game have no (rational) reason to regret choosing. It is in this sense that a game's 'solution' is often used interchangeably with the term 'equilibrium'. The most important solution concept in game theory is the Nash equilibrium. Nash's most influential 'solution' is based on the idea that, in equilibrium, players' strategies must be best replies to one another (Hargreaves & Varoufakis, 2003).

A strategy for player R, R_i, is best reply (or best response) to C's strategy C_j if it gives R the largest pay-off given that C has played C_j. (Similarly for player C.)

CONCLUSION

In conclusion it is necessary to note that, in real life, many decisions do not entirely correspond to the rational part of the human intelligence and therefore LAB profiling may play the role of a rectifier that makes it possible to "weaponise" the concept of the Nash equilibrium within the framework of stock trading.

Nonetheless, it is important to note that not all the NLP modeling techniques prove fruitful due to their intrinsic peculiarities and therefore the results of this research are for the time being highly experimental and tentative at best. And yet, something somewhat similar, however far from the proposed analytical paradigm, has already been used with astonishing success by the proponent of the theory of reflexivity George Soros (Soros, 2008).

ACKNOWLEDGMENT

The authors would like to express a special gratitude to Prof. Inna V. Nekrasova (Southern Federal University, Russian Federation) for her invaluable guidance and patience.

REFERENCES

Bandler, R., & Grinder, J. (1975). *The Structure of Magic I: A Book About Language and Therapy*. Palo Alto, CA: Science & Behavior Books.

Bandler, R., & Grinder, J. (1979). *Frogs into Princes: Neuro Linguistic Programming*. Moab, UT: Real People Press.

Charvet S. (1995). *Words That Change Minds – Mastering the Language of Influence.* Kendall/Hunt Publishing Company.

Chomsky, N. (1975). *The Logical Structure of Linguistic Theory*. Springer.

Dalio, R. (2015). How the Economic Machine Works - Leveragings and Deleveragings. *Economic Principles, How the Economic Machine Works*. Retrieved July 7, 2017, from http://www.economicprinciples.org/wp-content/uploads/ray_dalio__how_the_economic_machine_works__leveragings_and_deleveragings.pdf

Dilts, R., Grinder, J., Delozier, J., & Bandler, R. (1980). Neuro-Linguistic Programming: Vol. 1. *The Study of the Structure of Subjective Experience*. Cupertino, CA: Meta Publications.

Greenfield, R. (2016). Chatbots Are Your Newest, Dumbest Co-Workers. *Bloomberg*. Retrieved July 11, 2017, from https://www.bloomberg.com/news/articles/2016-05-05/chatbots-are-your-newest-dumbest-co-workers

Hall, M., & Bodenhamer, B. (1997). Figuring Out People - Design Engineering With Meta-Programs. Comer Press.

Hargreaves, S., & Varoufakis, Y. (2003). *Game Theory – A Critical Introduction.* Taylor & Francis Group.

Luriya, A. (1998). Language and Conscious. Publishing House of the Moscow State University.

Miller, G. (1956). The magical number seven, plus or minus two: Some limits on our capacity for processing information. *Psychological Review*, *63*(2), 81–97. doi:10.1037/h0043158 PMID:13310704

Soros, S. (2008). The New Paradigm for Financial Markets: The Credit Crisis of 2008 and What It Means. *Public Affairs*, 208.

Tosey, P., & Mathison, J. (2006). *Introducing Neuro-Linguistic Programming*. Centre for Management Learning & Development, School of Management, University of Surrey.

What is socialbot? (n.d.). In *WhatIs.com*. Retrieved from http://whatis.techtarget.com/definition/socialbot

Yeager, J. (1985). *Thinking About Thinking with NLP*. Cupertino, CA: Meta Publications.

Chapter 12
Volatility Spillover Between Developed and Developing Markets During Crisis Period

Rıfat Karakuş
Çankırı Karatekin University, Turkey

Şeyma Yılmaz Küçük
Çankırı Karatekin University, Turkey

İbrahim Bozkurt
Çankırı Karatekin University, Turkey

ABSTRACT

The purpose of this chapter is to determine how the volatility spillover between developed and developing markets behaved during times of crisis. For this purpose, daily returns of indices from the Group of Seven countries and the fragile five countries between 2004 and 2016 are used. The volatility spillover between the markets is examined by the Lagrange multiplier-based causality-in-variance test. As a result of the study, it is determined that the volatility of emerging markets is less influenced by the developed markets in the crisis period than before the crisis and after the crisis. Furthermore, in the post-crisis period, an increase in the volatility spillover to the developed markets from the developing markets is detected.

INTRODUCTION

When investors turn their savings into an investment, they prefer the highest return at a given risk level or the lowest risk at a given return level. Markowitz (1952, 1959) discloses this preference with efficient frontier. Investors prefer their portfolios on the point upon the efficient frontier. So the level of risk is also determinative as well as the expected return in making a new investment decision. Risk is generally defined as the probability of encountering unexpected outcomes in the future, in other words risk is a deviation from expected situations. That is, risk shows the probability of deviation from the expected outcome within the uncertainty state. Volatility is used as a measure of risk. Volatility measures the

DOI: 10.4018/978-1-5225-3767-0.ch012

extent to which changes in prices occur, and the magnitude of the difference between price movements of financial products. The high volatility of a financial asset return indicates that the asset is risky, while the low volatility indicates existence of low risk.

It is possible to classify volatility in two ways as historical and implied. If volatility is calculated using a historical dataset, this is called historical volatility. Historical volatility does not give information on how an asset will behave in the future while giving information about past price behavior. Option pricing models include the expected volatility of the underlying assets until the due date. Since the other parameters in the option pricing models are known, the volatility value that gives the market price of the option can be determined and is called implied volatility. In other words, implied volatility refers to the volatility of the underlying asset for an option which investors expect to realize to the maturity.

It is known that as a risk measure, financial asset volatility has increased considerably during times of crisis when uncertainty has increased. The mortgage crisis in the US in 2008 caused significant increase in volatility for US stock market as well as other markets because US market is the largest economy in the World and has the potential to affect other economies. The fact that rising of volatility, or riskiness, causes investors to increase their expectations of returns, leading to significant decreases in the value of financial assets during times when volatility is increasing. Value of the S&P 500 index which are the primary indicators for the US stock market, the historical volatility of index returns and Chicago Board Options Exchange Volatility Index (VIX), which represents the implied volatility of the S&P 500 index are shown in Figure 1. In Figure 1, it is seen that between 2008 and 2009, the S&P 500 index values showed a serious decline and in the same period both the historical volatility and the implied volatility have reached very high values.

The effects of the 2008 US mortgage crisis were not limited to just the US. Since the American economy is of a size that guides the world economy, this crisis affects the world economy negatively. Developing countries, along with developed countries, were significantly affected by this crisis. Many variables, such as growth, imports, exports, unemployment, stock market indexes, have shown that this

Figure 1. S&P 500 index, its historical volatility and VIX index: 2004-2016

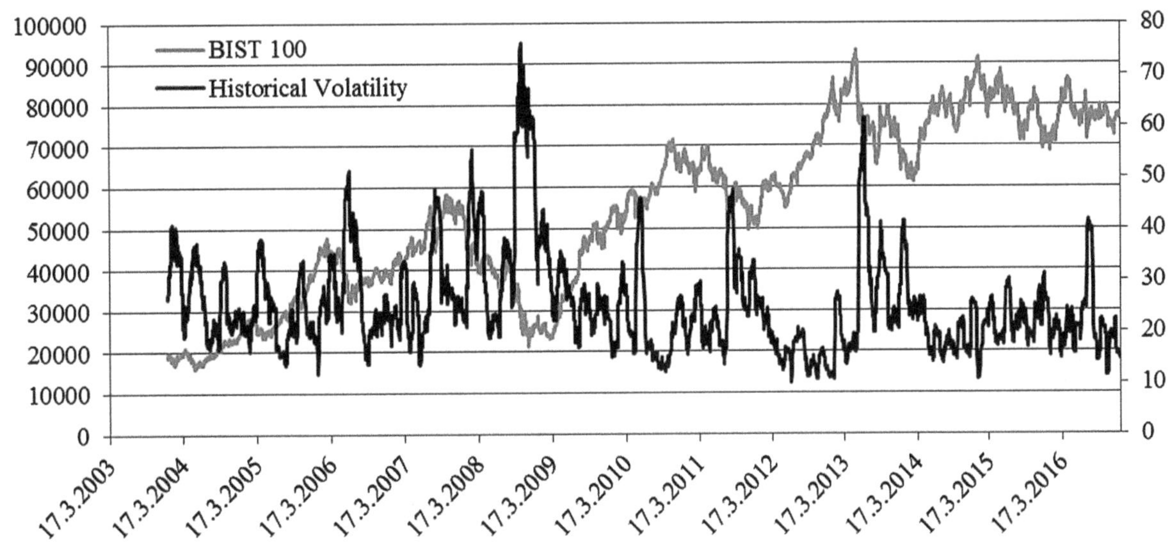

**For a more accurate representation see the electronic version.*

crisis has a global impact (Yıldırım, 2010). As seen in the Turkish case, which is one of the developing countries, it is seen in Figure 2 that BIST-100 index, which is the basic indicator of the stock market of Turkish showed a significant decrease economy during 2008-2009 and the volatility of the index returns peaked in the same period. The situation is similar not only for Turkey, but for many emerging economies.

As seen from the example of the 2008 crisis, the extent of financial liberalization, international capital movements and information technologies has resulted in an increase in the interaction of financial markets on a global basis. In the literature, the effect of volatility of one country's stock market on any other stock market volatility is called volatility spillover (Korkmaz ve Çevik, 2009). The assumption that information transfer from one market to another is realized by volatility channels (Ross, 1989) has made the issue of volatility transmission important in the literature and the volatility spillover which provided important information, such as whether circumstances in a market affect other markets and the degree of the effect (Büberkökü, 2013) has been the objective of many studies.

As is known, diligent investors who know the effect of other markets volatility on their own market volatility should also examine the situation of these markets in their portfolio management processes. For this reason, it is important for investors to disclose the volatility spillover between markets. It is also important for market makers to establish the volatility spillover between the markets. Market makers must take protective measures to protect their own markets against the volatility that will affect the investment behavior of individual and institutional investors and therefore the markets, especially during the periods of increasing uncertainty. At this point, it is thought that this work will provide important information to market makers, revealing the foreign volatility that local volatility is reacting to. The volatility spillover effect will also affect trade volume between countries, even indirectly. The increase in volatility in the stock market makes the stock market risky. The riskier stock market, the leading indicator of the economy, will also point to the increased risk in economic activity and increase the risk of asset

Figure 2. BIST-100 index and its historical volatility: 2004-2016

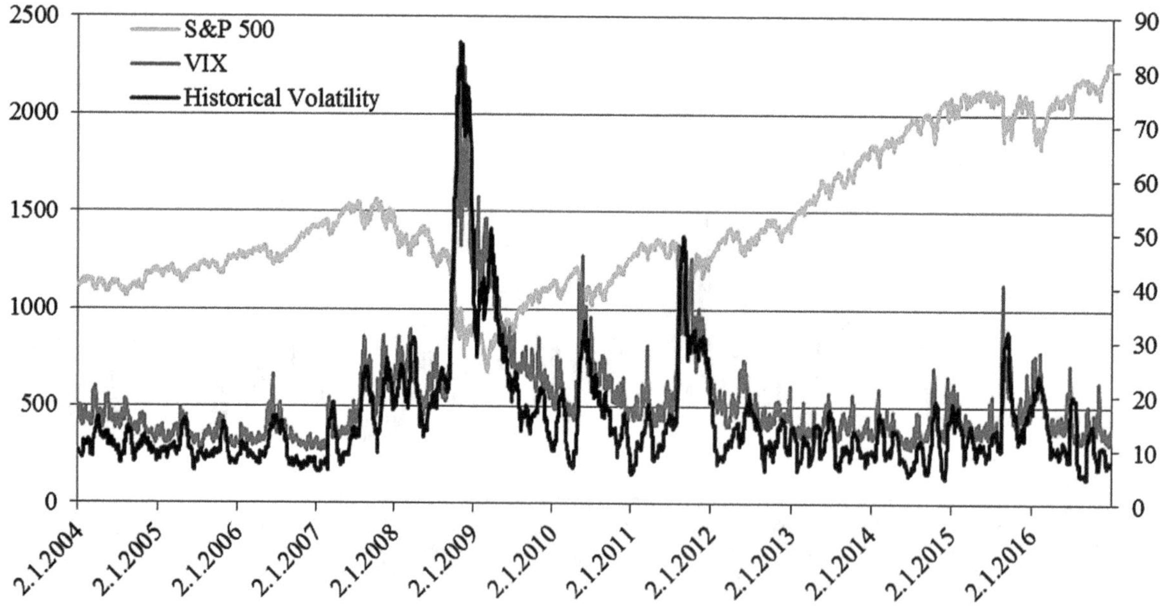

**For a more accurate representation see the electronic version.*

prices. This situation will also affect the foreign exchange prices and will affect the activities of the firms whose foreign exchange position is short. For this reason, companies operating in international trade will also want to know the extent of the volatility spillover between the markets in order to be able to take an appropriate position. In another respect, volatility affects all kinds of financial decisions, especially investment expenditures of firms, by affecting capital costs. For this reason it is important to know the effect of volatility spillover and to take necessary precautions when it is necessary for companies. In brief, volatility spillover between markets is important for individual and institutional investors' portfolio management strategies and affects growth of economies by influencing economic indicators such as countries' balance of payments, interest and inflation rates, and also affects firms' investment decisions and activities. Due to these and other reasons, the determination of volatility spillover becomes important. In this study, whether there has been a volatility spillover from the developed economies stock markets to developing country markets during the crisis of 2008 is examined by comparing with pre-crisis and post-crisis period. Thus, it is aimed to determine the effects that may be experienced in financial asset prices during crisis periods and to be a guide for investors.

BACKGROUND

Whether there is volatility spread between markets have importance for monetary policy decisions taken by countries, hedging of international investors and portfolio diversification decisions. International investors who wish to reduce their systematic risks through portfolio diversification will review their diversification decisions according to the volatility spillover among markets. Also, the flow of funds provided by global funds will be determinant for the monetary policy of countries. For these reasons, the volatility spillover between the markets is extremely important. In recent years, studies in the literature have focused on this subject, which will show the importance of this topic. Especially, the level of influence of the markets on the other markets during the crisis period is important for decision makers. The 2008 crisis also generated important data for research on the global scale. In literature, There are important studies that examine the volatility spillover between markets and the effect of the crisis period on this phenomenon.

Miyakoshi (2003) studied the return and volatility spillovers from Japan and the US to seven Asian equity markets. A bivariate EGARCH model is used by stock market indexes from Korea, Taiwan, Singapore, Thailand, Indonesia, Malaysia, Hong Kong, Japan, and US for the period from 1 January 1998 to 30 April 2000. According to results of the study, returns of the selected Asian equity markets are influenced by US markets. There is no influence of Japanese market return on other Asian markets returns. However, the volatility of the Asian markets is affected more by the Japanese market volatility than US market volatility. Also Japanese market volatility is influenced by other Asian markets volatility.

Al-Deehani (2006) investigated the volatility spillovers among the stock markets of Bahrain, Kuwait, and Saudi Arabia by using the concept of stochastic volatility and structural time-series modeling. With the sample between January 1, 2000 and April 15, 2003, the study revealed that volatility of the Kuwait market affects the volatility of other two markets. In addition, the effect of the Saudi market's volatility on volatility of Kuwait market and the Bahrain market's volatility on volatility of the Kuwait market is significant. However, it is seen that volatility in each of these markets cannot be explained totally by volatility in the other two markets.

Diebold and Yılmaz (2008) used daily nominal local-currency stock market indexes data from January 1992 to November 2007 for seven developed stock markets (for the U.S., U.K., France, Germany, Hong Kong, Japan and Australia) and twelve emerging markets (Indonesia, South Korea, Malaysia, Philippines, Singapore, Taiwan, Thailand, Argentina, Brazil, Chile, Mexico, and Turkey) and investigated whether return and volatility spillovers occur between markets. By using vector autoregressive (VAR) models and distilling all of the various cross-country spillovers into a single Spillover Index for full 1992-2007 data sample, they find that almost forty percent of forecast error variance comes from spillovers, both for returns (36 percent) and volatilities (40 percent). They also revealed the return and volatility spillovers for each separate market. Return spillover contribution from others is highest in Germany market and lowest in Us market. Volatility spillover from other contributors is highest also in Germany and lowest in Hong Kong market. In addition the study express that return spillovers display a gently increasing trend with the gradually increasing financial market integration of the last fifteen years, but no bursts. Volatility spillovers, in contrast, display no trend but clear bursts associated with readily-identified "crisis" events like the East Asian currency crisis in late 1997, the June-August 1998 Russian crisis etc.

Kutlar and Torun (2014) investigated the volatility spillover between developed and developing markets by using closing value of indexes for NSYE Composite Index, FTSE 100, DAX 30, CAC 40, Nikkei 225, BOVESPA, SSE Composite Index, HSI Index, RTSI Index and BIST 100 between 01.2000 and 01.2014. The results of CCC GARCH and BEKK GARCH models revealed that while there is a strong volatility spillover between the developed country markets, there is a weak volatility spillover from the developed countries towards the developing countries. In addition, domestic shocks in the previous period and the volatility of the previous period affect the current period volatility.

Natarajan (2014) used GARCH-M model to analyze the mean and volatility spillovers among Australia, Germany and USA with data between January 2002 and December 2011. They found that mean spillover from USA market to Germany and Australia markets exists. In addition there are volatility spillovers among all markets. Also significant negative own-spillovers are present in all stock markets. Past-market volatility shocks in the USA influence current volatility in both the Australian and German market with varying degrees of intensity.

Li and Giles (2015) examined the volatility spillover between USA, Japan and six Asian developing countries (China, India, Indonesia, Malaysia, the Philippines and Thailand) for the period from 1993 to 2012 by using an asymmetric multivariate GARCH model. The analysis revealed that there are significant unidirectional shock and volatility spillovers from the US market to both the Japanese and the Asian emerging markets. Besides the volatility spillovers between the US market and the Asian markets are stronger and bidirectional during the Asian financial crisis. The study also emphasized the increasing linkages between the Japanese market and the Asian emerging markets in recent years.

Güloğlu et al. (2016) investigated the volatility spillovers among five major Latin American stock markets by using the index values of MERVAL (Argentina), BOVESPA (Brazil), COLCAP (Colombia), IPC (Chile) and IPSA (Mexico). They used DCC GARCH model with taking into the breaks in variance for the period from 1/16/2008 to 5/20/2015. The results of the study revealed that there is one way causality in mean from BOVESPA to all markets, whereas causality in variance occurs only from BOVESPA to IPSA and IPSA to MERVAL. According to researchers this implies that there exists interdependence rather than contagion effects among the Latin American stock markets. In another study related to Latin American markets, Gamba-Santamaria et al. (2017) and researched the stock market volatility between United States and four Latin American countries as Brazil, Chile, Colombia and

Mexico by creating spillover indexes with DCC GARCH model the period January 2nd, 2003 to January 27th, 2016. Analysis results showed that the total system's spillover is higher between third quarter of 2008 and second quarter of 2012 which is the period United States subprime crisis and the European sovereign bonds crisis. So volatility spillover from US to these four Latin American countries increased in the crisis period. In addition for the whole sample Brazil is a net transmitter while Chile, Colombia and Mexico are net receivers of volatility spillovers.

VOLATILITY SPILLOVER FROM DEVELOPED TO DEVELOPING MARKETS

Data and Methodology

In this study which is aimed to examine the volatility spillover to developing markets from developed markets and whether this spillover has changed in the crisis period, Group of Seven (G7) countries as developed countries has been included to data set. The G7 is a group consisting of Canada, France, Germany, Italy, Japan, the United Kingdom, and the United States. These countries are the seven major advanced economies of the world. S&P TSX COMPOSITE index for Canada, CAC 40 index for France, DAX index for Germany, FTSE MIB index for Italy, NIKKEI 225 index for Japan, FTSE 100 index for the UK and the S&P 500 index for the US are selected to represent these countries capital markets. Fragile five countries are also included to sample to represent the developing countries. The fragile five was used in Morgan Stanley's August 2013 economic report to refer to the countries that are most affected by US monetary policy, and includes India, Indonesia, Brazil, South Africa and Turkey. NIFTY 50, JAKARTA COMPOSITE, BOVESPA, FTSE SOUTH AFRICA and BIST 100 indices are included in the data set for representing the capital markets of these countries respectively.

Daily returns of indices which represent the developed and developing markets between January 1, 2004 and December 31, 2016 have been calculated. The closing values of the indices were obtained from investing.com and the closing values were used to obtain logarithmic returns. Related data range is divided into three considering the 2008 US mortgage crisis. The period from January 1, 2004 to December 31, 2007 was used to express the pre-crisis period, the period from January 1, 2008 to December 31, 2009 is to express the crisis period, and the period January 01, 2010 to December 31 is to express the post-crisis period.

Table 1 presents the descriptive statistics of indices returns. When the mean of the index returns is examined, it is seen that the means differ significantly in the crisis period. While the average return of all indices was positive in the pre-crisis period, it is seen that all the index returns except for BOVESPA returned to negative during the crisis period. In the post-crisis period, all index returns other than the BOVESPA and FTSE MIB indexes have returned to positive again. This indicates that index returns are adversely affected by the crisis as expected. For BOVESPA, the effect of the crisis has been priced with time-lagged. For the FTSE MIB, it is observed that the negativity experienced during the crisis period has been also reflected in the post-crisis period. When the mean data is examined, it is seen that the positive returns experienced in the post-crisis period are still low compared to the pre-crisis period. It can be said that the stock markets did not reach the performance of the pre-crisis period except for the S&P 500 index.

When the standard deviation data, which is an indicator of risk, was examined, the standard deviation of each index increased in the crisis period. It is observed that the standard deviation of both developed and developing market indices shows a significant increase during the crisis period. In the post-crisis period, standard deviations declined with respect to the crisis period, but for some markets they have not yet returned to the pre-crisis period levels.

Skewness is another indicator of risk. Investors try to choose indices which their return series have a negative skewness coefficient to their portfolios because negative skewness is to increase the likelihood of selection of positive return investments (Altaylıgil, 2008). When the skewness values are examined, it appears that there is a similar situation to the interpretations made on the standard deviation data. For all indexes, it was found that the skewness, which was negative in the pre-crisis period, returned to positive or negativity has decreased in the crisis period. In the post-crisis period, skewness became negative again and risk is decreased in all the markets (except S&P TSX COMPOSITE).

The kurtosis coefficient is also considered as a measure of risk. In the periods when severe price movements are experienced, the kurtosis coefficient is observed to be high (Evci and Kandır, 2015). In Table 1, it is seen that the kurtosis values which were low in the pre-crisis period increased during the crisis period (except FTSE SOUTH AFRICA, JAKARTA COMPOSITE, NIFTY 50 indices). In the post-crisis period, the kurtosis values of some markets decreased compared to the crisis period and increased in some markets.

Descriptive statistics are shown that in general, for most of the markets; the risk has risen in the crisis period compared to the pre-crisis period and in the post-crisis period it showed a decrease compared to the crisis period. It is even more important for investors to know the level of increasing risk interaction between the markets during the crisis. Because investors seeking to mitigate risk through diversification can increase the efficiency of diversification by taking into account the transmission of risk between markets. Otherwise, diversification with investing in two markets with risk transitivity will have no advantage in periods when risk increases.

Table 2 reveals the correlation coefficients between the returns indices. Correlation coefficients between returns provide an indication of whether there is a risk transfer between the markets. When Table 2 is examined, in the pre-crisis period, the highest correlation was found between the CAC 40, DAX and FTSE 100 indices. The lowest correlations are found in the S&P 500 index between JAKARTA COMPOSITE, NIFTY 50 and NIKKEI 225 indices. In the crisis period, all the correlation coefficients increased. This suggests that the index returns during the crisis period moved in the same direction. In the post-crisis period, the correlation coefficients between the index returns fall for many markets. A striking feature of the correlation coefficients for post-crisis period is that the correlation coefficient between the S&P 500 index and other developed market indices increased unlike other observations. It is understood that in the post-crisis period, the interaction between the US market and other developed markets continued above the crisis period.

Before testing ARCH and GARCH effects, the stationarity of the series has been tested by the unit root test. The augmented Dickey-Fuller (ADF) statistic was used to test for a unit root. The test results are shown in the Table 3. The results of the ADF unit root test analysis reveal that all of the series are stationary at 1% statistical significance level for all of the periods.

The univariate GARCH (1,1) models were estimated before the volatility spillover in the study, so whether the volatility effects of the shocks experienced in the markets were short or long term is deter-

Table 1. Descriptive statistics for return series

	BIST 100	BOVESPA	CAC 40	DAX	FTSE 100	FTSE MIB	FTSE SOUTH AFRICA	JAKARTA COMPOSITE	NIFTY 50	NIKKEI 225	S&P 500	S&P TSX COMPOSITE
Panel A: Pre-Crisis 06 January 2004-31 December 2007												
Mean	0.000998	0.000961	0.000425	0.000667	0.000345	0.000324	0.000928	0.001281	0.001101	0.000333	0.000259	0.000482
Median	0.000828	0.000787	0.000640	0.001154	0.000508	0.000887	0.001092	0.001252	0.001346	0.000000	0.000539	0.000596
Maximum	0.064844	0.051643	0.032178	0.026051	0.034441	0.022919	0.055441	0.067338	0.079691	0.036031	0.028790	0.023725
Minimum	-0.086708	-0.068565	-0.033109	-0.035162	-0.041850	-0.038360	-0.064580	-0.078005	-0.130539	-0.055695	-0.035343	-0.035803
Std. Dev.	0.017039	0.016103	0.008914	0.009206	0.007888	0.007779	0.010640	0.012961	0.015241	0.010826	0.007480	0.007634
Skewness	-0.282479	-0.314683	-0.397768	-0.396365	-0.428988	-0.667428	-0.542602	-0.713239	-0.970349	-0.364601	-0.312351	-0.561115
Kurtosis	4.281108	3.930430	4.130728	3.836725	5.943222	4.914178	6.640031	8.047071	10.94345	4.650496	4.972672	4.133025
Jarque-Bera	84.86963	54.62548	82.74877	57.51421	406.8843	235.7630	624.5903	1190.857	2894.683	140.9522	185.3614	110.0970
Probability	0.000000	0.000000	0.000000	0.000000	0.000000	0.000000	0.000000	0.000000	0.000000	0.000000	0.000000	0.000000
Observations	1039	1039	1039	1039	1039	1039	1039	1039	1039	1039	1039	1039
Panel B: Crisis Period 02 January 2008-31 December 2009												
Mean	-9.61E-05	0.000136	-0.000681	-0.000339	-0.000582	-0.000971	-1.69E-05	-0.000154	-0.000318	-0.000715	-0.000528	-0.000314
Median	0.000000	0.000415	-0.000101	0.000000	0.000000	0.000000	0.000000	0.000000	0.000000	0.000000	0.000343	0.001039
Maximum	0.121274	0.136766	0.105946	0.093842	0.107975	0.108742	0.056705	0.076231	0.163343	0.132346	0.109572	0.093703
Minimum	-0.090138	-0.120961	-0.094715	-0.092646	-0.074335	-0.085991	-0.077297	-0.109540	-0.130142	-0.121110	-0.094695	-0.097880
Std. Dev.	0.022839	0.026586	0.021341	0.019437	0.020784	0.021476	0.017156	0.020085	0.024336	0.023414	0.021630	0.020719
Skewness	0.044800	0.106149	0.194958	-0.009377	0.291941	0.114907	-0.048626	-0.518910	0.185697	-0.320038	-0.118493	-0.487492
Kurtosis	5.753555	7.272525	7.370181	7.413206	7.685016	6.674931	4.536704	7.850483	8.507261	8.301392	7.533445	6.694911
Jarque-Bera	164.7683	397.2525	417.8967	422.8075	483.8850	294.3201	51.46864	534.1173	661.4064	619.0013	447.3716	317.0059
Probability	0.000000	0.000000	0.000000	0.000000	0.000000	0.000000	0.000000	0.000000	0.000000	0.000000	0.000000	0.000000
Observations	521	521	521	521	521	521	521	521	521	521	521	521
Panel C: Post-Crisis 04 January 2010-30 December 2016												
Mean	0.000215	-7.14E-05	0.000116	0.000360	0.000152	-0.000104	0.000351	0.000405	0.000249	0.000327	0.000383	0.000145
Median	0.000158	0.000000	0.000264	0.000679	0.000170	0.000000	0.000363	0.000581	0.000000	0.000000	0.000294	0.000413
Maximum	0.068951	0.063887	0.092208	0.052104	0.050323	0.106840	0.055367	0.070136	0.037380	0.106654	0.046317	0.039410
Minimum	-0.110637	-0.084307	-0.083844	-0.070673	-0.061994	-0.133314	-0.041504	-0.092997	-0.060973	-0.111534	-0.068958	-0.041227
Std. Dev.	0.014435	0.014311	0.013467	0.012905	0.010015	0.016819	0.010343	0.011030	0.010194	0.014266	0.009647	0.008154
Skewness	-0.564454	-0.021285	-0.156305	-0.269834	-0.255649	-0.342483	-0.136598	-0.596478	-0.170220	-0.457284	-0.440916	-0.344153
Kurtosis	7.180212	4.514349	6.566336	5.362995	5.890385	6.943044	4.611777	9.314911	4.627818	10.43524	7.424186	5.261630
Jarque-Bera	1421.769	174.0424	971.9160	445.5198	653.3611	1214.606	202.6617	3132.011	209.7317	4255.711	1543.288	423.8126
Probability	0.000000	0.000000	0.000000	0.000000	0.000000	0.000000	0.000000	0.000000	0.000000	0.000000	0.000000	0.000000
Observations	1820	1820	1820	1820	1820	1820	1820	1820	1820	1820	1820	1820

mined. Then, the Lagrange multiplier-based causality-in-variance test, which was developed by Hafner and Herwartz (2006) is used to determine volatility spillover between markets. Hafner and Herwartz (2006) proposed the following LM (Langrange Multiplier) test to reveal the volatility spillover between two series as x and z (Kamışlı, Kamışlı and Sevil, 2016).

$$\lambda_{LM} = \frac{1}{4T}\left(\sum\nolimits_{t=1}^{T}\left(\xi_{it}^{2}-1\right)z_{jt}'\right)V(\theta_{i})^{-1}\left(\sum\nolimits_{t=1}^{T}\left(\xi_{it}^{2}-1\right)z_{jt}\right) \tag{1}$$

Table 2. Correlation matrix for return series

	BIST 100	BOVESPA	CAC 40	DAX	FTSE 100	FTSE MIB	FTSE SOUTH AFRICA	JAKARTA COMPOSITE	NIFTY 50	NIKKEI 225	S&P 500	S&P TSX COMPOSITE
Panel A: Pre-Crisis 06 January 2004-31 December 2007												
BIST 100	1.000000											
BOVESPA	0.306196	1.000000										
CAC 40	0.440955	0.408126	1.000000									
DAX	0.393087	0.390252	0.924321	1.000000								
FTSE 100	0.448072	0.415077	0.881458	0.820488	1.000000							
FTSE MIB	0.406757	0.405178	0.892410	0.863607	0.833073	1.000000						
FTSE SOUTH AFRICA	0.358556	0.294996	0.506392	0.456932	0.520714	0.467987	1.000000					
JAKARTA COMPOSITE	0.335926	0.198706	0.304315	0.271823	0.297261	0.287032	0.373385	1.000000				
NIFTY 50	0.334326	0.218117	0.349470	0.317172	0.329860	0.326696	0.360516	0.456287	1.000000			
NIKKEI 225	0.304598	0.203162	0.345996	0.326621	0.302921	0.277549	0.353493	0.419262	0.371306	1.000000		
S&P 500	0.211597	0.628272	0.480496	0.470121	0.476188	0.462111	0.212718	0.054092	0.116352	0.090557	1.000000	
S&P TSX COMPOSITE	0.284214	0.579480	0.457807	0.422901	0.487807	0.438435	0.327137	0.153820	0.224562	0.196832	0.665154	1.000000
Panel B: Crisis Period 02 January 2008-31 December 2009												
BIST 100	1.000000											
BOVESPA	0.443647	1.000000										
CAC 40	0.618868	0.652528	1.000000									
DAX	0.622803	0.624128	0.943262	1.000000								
FTSE 100	0.617890	0.656826	0.920669	0.876511	1.000000							
FTSE MIB	0.586877	0.596231	0.923468	0.871262	0.859023	1.000000						
FTSE SOUTH AFRICA	0.561499	0.486713	0.618857	0.626129	0.589727	0.590265	1.000000					
JAKARTA COMPOSITE	0.365846	0.270348	0.335251	0.344431	0.302471	0.350563	0.428687	1.000000				

continued on following page

Table 2. Continued

	BIST 100	BOVESPA	CAC 40	DAX	FTSE 100	FTSE MIB	FTSE SOUTH AFRICA	JAKARTA COMPOSITE	NIFTY 50	NIKKEI 225	S&P 500	S&P TSX COMPOSITE
NIFTY 50	0.438627	0.364648	0.441258	0.443690	0.462627	0.454066	0.459312	0.484456	1.000000			
NIKKEI 225	0.356652	0.212663	0.380148	0.392002	0.354597	0.386690	0.457583	0.498084	0.376960	1.000000		
S&P 500	0.388778	0.743184	0.585244	0.559892	0.639172	0.541730	0.354613	0.150049	0.333025	0.115446	1.000000	
S&P TSX COMPOSITE	0.406623	0.708114	0.585127	0.590410	0.568152	0.560322	0.461059	0.223206	0.297215	0.282533	0.746000	1.000000
Panel C: Post-Crisis 04 January 2010-30 December 2016												
BIST 100	1.000000											
BOVESPA	0.300687	1.000000										
CAC 40	0.444118	0.451048	1.000000									
DAX	0.437253	0.421960	0.931832	1.000000								
FTSE 100	0.439237	0.460548	0.866377	0.840024	1.000000							
FTSE MIB	0.393925	0.424125	0.888252	0.836834	0.762810	1.000000						
FTSE SOUTH AFRICA	0.422733	0.368859	0.565792	0.559447	0.611883	0.492220	1.000000					
JAKARTA COMPOSITE	0.277038	0.202338	0.291278	0.279052	0.321050	0.221797	0.359171	1.000000				
NIFTY 50	0.298776	0.273884	0.407701	0.405241	0.424301	0.350772	0.411789	0.417979	1.000000			
NIKKEI 225	0.153937	0.118270	0.256572	0.243026	0.263539	0.206870	0.244940	0.335641	0.295288	1.000000		
S&P 500	0.334637	0.556173	0.654320	0.638053	0.619843	0.592270	0.378750	0.171870	0.277469	0.141725	1.000000	
S&P TSX COMPOSITE	0.304385	0.542558	0.579334	0.563640	0.589997	0.518182	0.423072	0.243193	0.308287	0.166462	0.758431	1.000000

Table 3. Unit root test results

	Pre-Crisis Period		Crisis Period		Post-Crisis Period	
	t-Statistic	Prob.	t-Statistic	Prob.	t-Statistic	Prob.
S&P 500	-35.00781	0.0000	-19.46720	0.0000	-44.77385	0.0001
DAX	-33.29229	0.0000	-23.87264	0.0000	-40.83429	0.0000
CAC 40	-34.84334	0.0000	-25.11533	0.0000	-42.41091	0.0000
FTSE 100	-35.81825	0.0000	-11.14471	0.0000	-41.17544	0.0000
FTSE MIB	-34.94299	0.0000	-10.60053	0.0000	-43.80830	0.0001
NIKKEI 225	-32.48848	0.0000	-23.06009	0.0000	-46.22364	0.0001
S&P TSX COMPOSITE	-32.00927	0.0000	-24.72549	0.0000	-40.02724	0.0000
BIST 100	-31.94386	0.0000	-20.72712	0.0000	-43.39657	0.0001
BOVESPA	-31.86670	0.0000	-23.74411	0.0000	-42.98259	0.0000
NIFTY 50	-24.17772	0.0000	-22.08641	0.0000	-40.53981	0.0000
FTSE SOUTH AFRICA	-31.00110	0.0000	-21.56797	0.0000	-32.89681	0.0000
JAKARTA COMPOSITE	-29.24520	0.0000	-19.04465	0.0000	-24.76545	0.0000

Here, ξ_{it} = standardized errors of GARCH model and;

$$V(\theta_i) = \frac{K}{4T}\left(\sum_{t=1}^{T} z_{jt} z'_{jt}\right) - \left(\sum_{t=1}^{T} z_{jt} x'_{it}\left(\sum_{t=1}^{T} x_{it} x'_{it}\right)^{-1} \sum_{t=1}^{T} x_{it} z'_{it}\right) \tag{2}$$

$$K = \frac{1}{T}\sum_{t=1}^{T}\left(\xi_{it}^2 - 1\right)^2 \tag{3}$$

Asymptotic distribution of λ_{LM} depends on the number of misspecification indicators in z_{jt}. If there are two misspecification indicators therefore test will have asymptotic chi-square distribution

Empirical Results

Before testing the volatility spillover between markets, the response of markets to shocks was investigated using the univariate GARCH (1,1) model and the results are presented in Table 4. In Table 4, α variable represents the ARCH parameter and β variable represents the GARCH parameter. A high ARCH parameter implies that the effects of a shock are more pronounced in the subsequent period. In contrast, a high GARCH parameter implies that the effects of a shock are more persistent (Enders, 2004). When the GARCH model results are examined, it is shown that the GARCH effect for all market indices is applicable, that is, the effect of shocks on volatility is long-lasting and permanent.

Table 4. Results for variance equations

	Pre-Crisis		Crisis		Post-Crisis	
	Coefficient	Probability	Coefficient	Probability	Coefficient	Probability
			S&P 500			
ω	1.74E-06	0.0005	2.36E-06	0.1991	4.31E-06	0.0000
α	0.051756	0.0000	0.094246	0.0000	0.140235	0.0000
β	0.916596	0.0000	0.901355	0.0000	0.811347	0.0000
			DAX			
ω	4.49E-06	0.0075	9.50E-06	0.0003	3.08E-06	0.0000
α	0.075823	0.0001	0.102537	0.0000	0.081497	0.0000
β	0.871220	0.0000	0.875811	0.0000	0.900510	0.0000
			CAC 40			
ω	4.42E-06	0.0032	7.91E-06	0.0107	5.28E-06	0.0000
α	0.084298	0.0000	0.104321	0.0000	0.105720	0.0000
β	0.858514	0.0000	0.879351	0.0000	0.866981	0.0000
			FTSE 100			
ω	2.54E-06	0.0011	5.53E-06	0.0291	4.52E-06	0.0000
α	0.112440	0.0000	0.095451	0.0000	0.139267	0.0000
β	0.843376	0.0000	0.889392	0.0000	0.816642	0.0000
			FTSE MIB			
ω	3.47E-06	0.0006	7.88E-06	0.0165	1.06E-05	0.0000
α	0.086404	0.0000	0.115856	0.0000	0.093276	0.0000
β	0.854903	0.0000	0.868531	0.0000	0.869610	0.0000
			NIKKEI 225			
ω	2.94E-06	0.0033	1.06E-05	0.0275	1.21E-05	0.0000
α	0.078804	0.0000	0.116117	0.0000	0.134583	0.0000
β	0.898657	0.0000	0.860035	0.0000	0.813043	0.0000
			S&P TSX COMPOSITE			
ω	2.33E-06	0.0119	4.65E-06	0.0821	1.52E-06	0.0000
α	0.055853	0.0001	0.109723	0.0000	0.096128	0.0000
β	0.905220	0.0000	0.879099	0.0000	0.881003	0.0000
			BIST 100			
ω	1.40E-05	0.0016	1.94E-05	0.0118	1.92E-05	0.0000
α	0.061484	0.0000	0.105631	0.0000	0.104297	0.0000
β	0.888610	0.0000	0.857993	0.0000	0.803963	0.0000
			BOVESPA			
ω	1.33E-05	0.0078	9.52E-06	0.0286	6.19E-06	0.0001
α	0.058017	0.0000	0.093639	0.0000	0.061854	0.0000
β	0.890881	0.0000	0.890756	0.0000	0.907864	0.0000

continued on following page

Table 4. Continued

	Pre-Crisis		Crisis		Post-Crisis	
	Coefficient	Probability	Coefficient	Probability	Coefficient	Probability
NIFTY 50						
ω	9.14E-06	0.0000	1.48E-05	0.0207	3.49E-06	0.0001
α	0.127001	0.0000	0.152344	0.0000	0.054928	0.0000
β	0.831841	0.0000	0.841764	0.0000	0.912033	0.0000
FTSE SOUTH AFRICA						
ω	4.14E-06	0.0006	1.60E-06	0.3571	2.46E-06	0.0002
α	0.139521	0.0000	0.080115	0.0001	0.080312	0.0000
β	0.828131	0.0000	0.915281	0.0000	0.898067	0.0000
JAKARTA COMPOSITE						
ω	1.70E-05	0.0000	8.75E-06	0.0005	5.70E-06	0.0000
α	0.174604	0.0000	0.088961	0.0000	0.134697	0.0000
β	0.726004	0.0000	0.890844	0.0000	0.822779	0.0000

While Table 2 shows that there is information that allows for a prediction of risk pass-through based on the correlation between the returns of the indices, Table 5 provides statistical information on risk transfer between market indices. Table 5 shows the volatility spillover between indices.

The volatility spillover between the BIST 100 index and the developed market indices showed a significant differentiation during the crisis period. It is seen that there is volatility spillover from all developed market indices to BIST 100 index in the pre-crisis period. In analyzes made during the crisis period, none of the spillovers to BIST 100 index were statistically significant. In the post-crisis period, BIST 100 volatility affected again by the volatility of the whole developed market indices. In other words, it is seen that the volatility spillovers, which is from the developed markets to BIST 100 in pre-crisis and post-crisis period, is not valid during the crisis period. When the volatility spillover for developed markets from BIST-100 is examined, it is seen that only 3 markets (FTSE 100, NIKKEI 225 and S&P TSX COMPOSITE) were affected before the crisis and 3 markets (DAX, NIKKEI 225, FTSE MIB) were affected by the volatility of BIST 100 during the crisis period. In the post-crisis period, volatility spillover from BIST 100 to all developed market indices except NIKKEI 225 is statistically significant.

It is seen that the volatility that occurred in the indices of developed markets affects the volatility of the BOVESPA index in the pre-crisis period. The volatility spillover towards the BOVESPA from the whole of the seven developed market indices included in the sample was found significant for the pre-crisis period. In the crisis period, the effects of the European market indices DAX, CAC 40 and FTSE MIB on volatility of BOVESPA were not statistically explained. In other words, it was seen that there was a decrease in the developed markets where the index was affected. In the post-crisis period, unlike BIST 100, it is seen that the volatility spillovers that existed in the pre-crisis period did not occur again. In the post-crisis period, the volatility of BOVESPA was influenced only by the volatility of the S&P 500, FTSE 100 and S&P TSX COMPOSITE indices. The influence of volatility of BOVESPA on volatility of developed markets seems to be intense in the post-crisis period. In the pre-crisis period, the volatility of five of the seven developed market indices is affected by the volatility of the BOVESPA index. In the

Table 5. Result for the volatility spillover test

	Pre-Crisis				Crisis				Post-Crisis			
	LM-Stat	**p-Value**	**LM-Stat**	**p-Value**	**LM-Stat**	**p-Value**	**LM-Stat**	**p-Value**	**LM-Stat**	**P-Value**	**LM-Stat**	**p-Value**
	Spillover to BIST 100		**Spillover From BIST 100**		**Spillover to BIST 100**		**Spillover From BIST 100**		**Spillover to BIST 100**		**Spillover From BIST 100**	
S&P 500	17.12933	0.000191	4.340374	0.114156	2.386449	0.303242	4.456553	0.107714	13.85233	0.000982	6.809267	0.033219
DAX	11.96277	0.002525	3.542535	0.170117	3.563463	0.168346	5.298547	0.070703	10.69131	0.004769	14.33278	0.000772
CAC 40	13.22344	0.001345	1.962574	0.374828	2.015212	0.365092	4.115320	0.127753	10.63780	0.004898	10.88504	0.004329
FTSE 100	11.68785	0.002897	7.604431	0.022321	2.832813	0.242584	2.959179	0.227731	14.64939	0.000659	22.09590	1.59E-05
FTSE MIB	11.87331	0.002641	1.873121	0.391974	1.851941	0.396147	7.083016	0.028970	10.73972	0.004655	6.920792	0.031417
NIKKEI 225	8.740164	0.012650	60.53194	7.17E-14	2.407094	0.300128	9.058895	0.010787	12.66880	0.001774	1.994450	0.368902
S&P TSX C.	10.46300	0.005346	6.363571	0.041511	3.166215	0.205336	1.710417	0.425194	12.12254	0.002331	8.060922	0.017766
	Spillover to BOVESPA		**Spillover From BOVESPA**		**Spillover to BOVESPA**		**Spillover From BOVESPA**		**Spillover to BOVESPA**		**Spillover From BOVESPA**	
S&P 500	13.17913	0.001375	1.034887	0.596042	6.826786	0.032929	2.621374	0.269635	4.744540	0.093269	7.586935	0.022517
DAX	12.27608	0.002159	18.62281	9.04E-05	4.155394	0.125218	6.491260	0.038944	2.992574	0.223960	14.95563	0.000565
CAC 40	13.28692	0.001303	17.25078	0.000179	4.306133	0.116128	4.572376	0.101653	4.062485	0.131172	11.96830	0.002518
FTSE 100	15.93777	0.000346	17.29151	0.000176	5.181971	0.074946	3.222632	0.199625	17.53167	0.000156	23.96222	6.26E-06
FTSE MIB	15.81451	0.000368	17.47128	0.000161	4.108612	0.128182	7.454093	0.024064	4.448749	0.108135	9.356113	0.009297
NIKKEI 225	11.73458	0.002831	68.47829	1.33E-15	6.415618	0.040445	41.31830	1.07E-09	2.363957	0.306671	8.355103	0.015336
S&P TSX C.	12.20253	0.002240	2.903495	0.234161	4.983992	0.082745	1.714381	0.424353	7.732717	0.020934	7.953208	0.018749
	Spillover to NIFTY 50		**Spillover From NIFTY 50**		**Spillover to NIFTY 50**		**Spillover From NIFTY 50**		**Spillover to NIFTY 50**		**Spillover From NIFTY 50**	
S&P 500	47.01876	6.17E-11	1.786854	0.409251	1.200081	0.548789	0.764955	0.682169	28.69557	5.87E-07	6.210721	0.044808
DAX	28.66316	5.97E-07	3.267627	0.195184	1.618580	0.445174	4.515660	0.104577	11.05845	0.003969	15.68948	0.000392
CAC 40	28.62437	6.09E-07	1.322832	0.516120	2.156561	0.340180	3.993073	0.135805	11.21683	0.003667	12.97347	0.001524

continued on following page

Table 5. Continued

	Pre-Crisis				Crisis				Post-Crisis			
	LM-Stat	p-Value	LM-Stat	p-Value	LM-Stat	p-Value	LM-Stat	p-Value	LM-Stat	P-Value	LM-Stat	p-Value
FTSE 100	26.82762	1.49E-06	6.451072	0.039734	2.145880	0.342002	2.705595	0.258516	13.23284	0.001338	21.95889	1.70E-05
FTSE MIB	22.98004	1.02E-05	2.314219	0.314394	3.375226	0.184960	6.035066	0.048922	8.883704	0.011774	7.024807	0.029825
NIKKEI 225	15.15275	0.000512	18.06525	0.000119	0.437786	0.803407	3.202648	0.201629	6.712412	0.034867	2.149529	0.341378
S&P TSX C.	20.37106	3.77E-05	3.485538	0.175035	1.272503	0.529273	6.626742	0.036393	17.54969	0.000155	7.724150	0.021024
	Spillover to FTSE SOUTH AFRICA		Spillover From FTSE SOUTH AFRICA		Spillover to FTSE SOUTH AFRICA		Spillover From FTSE SOUTH AFRICA		Spillover to FTSE SOUTH AFRICA		Spillover From FTSE SOUTH AFRICA	
S&P 500	22.14455	1.55E-05	2.610099	0.271159	2.062800	0.356508	1.481172	0.476834	28.65448	5.99E-07	6.434304	0.040069
DAX	10.21271	0.006058	4.533871	0.103629	0.504950	0.776875	10.05331	0.006561	27.68527	9.73E-07	15.10837	0.000524
CAC 40	9.487962	0.008704	4.702848	0.095233	0.386259	0.824375	5.197932	0.074350	30.58833	2.28E-07	11.23827	0.003628
FTSE 100	10.94825	0.004194	7.968353	0.018608	0.770608	0.680244	3.581883	0.166803	28.57170	6.25E-07	23.16623	9.32E-06
FTSE MIB	9.803909	0.007432	5.990057	0.050035	0.307342	0.857554	9.048478	0.010843	39.25947	2.98E-09	6.723231	0.034679
NIKKEI 225	7.904413	0.019212	20.98904	2.77E-05	0.223728	0.894166	8.366636	0.015248	30.29010	2.65E-07	5.131478	0.076862
S&P TSX C.	34.11617	3.91E-08	7.323353	0.025689	0.884908	0.642458	1.447596	0.484907	32.67136	8.04E-08	9.183395	0.010136
	Spillover to JAKARTA COMPOSITE		Spillover From JAKARTA COMPOSITE		Spillover to JAKARTA COMPOSITE		Spillover From JAKARTA COMPOSITE		Spillover to JAKARTA COMPOSITE		Spillover From JAKARTA COMPOSITE	
S&P 500	69.09669	9.99E-16	2.073503	0.354605	8.670352	0.013100	7.323543	0.025687	88.84403	0.000000	7.023889	0.029839
DAX	29.79306	3.39E-07	3.612460	0.164272	7.012539	0.030009	5.626699	0.060004	28.66712	5.96E-07	14.38889	0.000751
CAC 40	29.98253	3.09E-07	2.546471	0.279925	7.916199	0.019099	5.308429	0.070354	22.93291	1.05E-05	10.97691	0.004134
FTSE 100	28.97260	5.11E-07	6.826819	0.032929	7.840987	0.019831	4.415557	0.109945	29.29883	4.34E-07	22.17363	1.53E-05
FTSE MIB	22.89971	1.07E-05	3.640789	0.161962	7.540037	0.023052	6.925294	0.031347	21.06422	2.67E-05	6.921386	0.031408
NIKKEI 225	16.54616	0.000255	21.86380	1.79E-05	6.980215	0.030498	7.183772	0.027546	12.66916	0.001774	1.921451	0.382615
S&P TSX C.	52.47949	4.02E-12	5.117085	0.077417	10.51660	0.005204	1.155271	0.561224	48.30909	3.23E-11	9.373201	0.009218

crisis period, only the volatility of the DAX, FTSE MIB and NIKKEI 225 indices is influenced by the volatility of the BOVESPA index. In the post-crisis period, the volatility of the BOVESPA index has an impact on the volatility of all developed market indices.

It is seen that the volatility of the NIFTY 50 index, which is included in the study to represent the Indian market, is influenced by the volatility of the seven developed markets in the pre-crisis period, in other words there is the volatility spillover to NIFTY 50 from these developed market indices. A similar situation exists in the post-crisis period. Also in this period, the volatility spillover from all of the developed market indices to NIFTY 50 is valid. However, this situation seems to be completely invalid during the crisis period. No significant volatility spillover to NIFTY 50 was found from any of the developed market indices during the crisis period. When effect of the volatility of the NIFTY 50 index on the volatility of developed markets is examined, an important difference in the post-crisis period is seen obviously. In the pre-crisis period, there is volatility spillover only to the FTSE 100 and NIKKEI 225 indices and in the crisis period, only to the FTSE MIB and the S&P TSX COMPOSITE index from the NIFTY 50 index. However, in the post-crisis period, the volatility of all developed market indices excluding the NIKKEI 225 index is affected by the volatility of the NIFTY 50 index.

The FTSE SOUTH AFRICA index is similar to the NIFTY 50 index in terms of volatility spillover from developed market indices. It is seen that the volatility of all developed market indices in the pre-crisis and post-crisis period has an effect on the FTSE SOUTH AFRICA index volatility. In the crisis period, there is no volatility spillover to the FTSE SOUTH AFRICA index from any developed market index. It is seen that the volatility of all developed market indices other than the S&P 500 and DAX indices in the pre-crisis period is influenced by the volatility of the FTSE SOUTH AFRICA index. While the impact on the volatility of the FTSE 100 and S&P TSX COMPOSITE indices has disappeared during the crisis, the volatility of the DAX index has been affected by the volatility of the FTSE SOUTH AFRICA index. In the post-crisis period, it is statistically explained that FTSE SOUTH AFRICA index has volatility spillover to all developed market indices.

The effect of the volatility of the developed market indices on the volatility of the JAKARTA COMPOSITE index do not differ between periods. The impact of the volatility of all developed market indices on the volatility of the JAKARTA COMPOSITE index is statistically significant for all three periods. The volatility spillover to developed market indices from the JAKARTA COMPOSITE index varies between periods. In the pre-crisis period, only the volatility of the FTSE 100, NIKKEI 225 and S&P TSX COMPOSITE indexes is affected by the volatility of the JAKARTA COMPOSITE index. During the crisis period, the volatility of all developed market indices excluding the FTSE 100 and S&P TSX COMPOSITE index is influenced by the volatility of the JAKARTA COMPOSITE index. In the post-crisis period, there is the volatility spillover from the JAKARTA COMPOSITE index to all developed market indices excluding the NIKKEI 225 index.

SOLUTIONS AND RECOMMENDATIONS

In this study, volatility spillover between developed and emerging markets is examined. In addition, by investigating whether this volatility spillover is differentiated during the crisis period creates a guide for investors and market authorities in terms of management strategies during crisis periods. Firstly, when descriptive statistics related to developed and emerging market indexes were evaluated, it was found that risk for all markets increased during the crisis period. Standard deviation, skewness and kurtosis values reveals that in the crisis period the risk for all markets increases compared to the previous period and decreases again in the post-crisis period.

Next, estimates made with the GARCH (1,1) model shows that shocks at all markets causes long-term volatility. It is observed that the ARCH effect is limited for all periods, although it varies slightly between periods, and the GARCH effect is high for all of the market. In other words, the shocks that are happening in the markets cause the volatilities to be affected from long periods.

Volatility spillover between markets is the main focus of study. According to the results of the analysis, the volatility spillover to developing markets from the developed markets in the crisis period differs significantly from the pre-crisis period. The effect of the volatility of developed market indices on the volatility of emerging market indices in the pre-crisis period is obvious for all samples. However, in the crisis period, the volatility spillover from developed to developing markets lose validity except one market. In the post-crisis period, the volatility spillover from developed to developing markets reoccur. This suggests that developing markets are an important investment alternative for investors, especially during periods of crisis. Investors who want to minimize their risks through portfolio diversification can reach these objectives by investing in developing markets during times of crisis when volatility, or risk transfer, is not taking place. Again, since investors' behavior will provide a flow of investment in developing markets during crisis periods, relevant market authorities can determine appropriate actions in these periods.

Another outcome of the analysis results is related to the impact of developing market volatility on developed market volatility. While the volatility spillover from the developing markets to developed markets in the pre-crisis period and the crisis period was limited, it is seen that this spread has increased after the crisis. The volatility of all developed market indices except NIKKEI 225 is influenced by the volatility of developing markets in the post-crisis period. In contrast to other developed market indices, in post-crisis period the volatility of the NIKKEI 225 index, which represents the Japanese market, is not influenced by the developing market indices which affect volatility of NIKKEI 225 index in the pre-crisis and crisis period.

FUTURE RESEARCH DIRECTIONS

This study will provide significant benefits to the literature and its users by demonstrating how the volatility spillover between developed and developing markets differs during the crisis period. However, usage of the limited number of developed and developing markets data is a significant limitation of the study. Again, focusing on one crisis period constitutes another limitation. The further studies will be able to make a significant contribution to this area with wider country sets and different crisis periods. In addition, the method used in this study only determines the existence and absence of volatility spillover and the use of methods that can detect the magnitude of this effect in further studies will make them valuable.

CONCLUSION

This study examines the volatility spillover between developed and developing markets. It is focused on whether this volatility spillover is changed during periods of crisis. According to the results of the analysis, it is seen that volatility spillover from the developed markets to the developing markets is not valid in crisis period. In the post-crisis period, it is determined that these spillovers commence again. In addition, it was observed that the volatility spillover to developed markets from the developing markets which were limited in the crisis period and in the crisis period increases in the post-crisis period. These results suggest that developing markets constitute an important alternative for investors during the crisis.

REFERENCES

Al-Deehani, T., & Moosa, I. A. (2006). Volatility Spillover in Regional Emerging Stock Markets: A Structural Time-Series Approach. *Emerging Markets Finance & Trade*, *42*(4), 78–89. doi:10.2753/REE1540-496X420404

Altaylıgil, B. (2008). Portföy seçimi için ortalama-varyans-çarpıklık modeli. *İstanbul Üniversitesi İşletme Fakültesi Dergisi, 37*(2), 65-78.

Büberkökü, Ö. (2013). Kriz Döneminde Yükselen Piyasa Ekonomileri, Euro Bölgesi ve ABD piyasaları Arasındaki Volatilite Yayılmasının İncelenmesi: Varyansta-Granger-Nedensellik Testinden Kanıtlar. *EY International Congress on Economics I "Europe and Global Economic Rebalancing"*, Ankara, Turkey.

Diebold, F. X., & Yilmaz, K. (2008). *Measuring financial asset return and volatilty spillovers, with application to global equity markets.* CFS Working Paper, No.2008/26.

Enders, W. (2004). *Applied Econometric Time Series.* John Wiley & Sons Inc.

Evci, S., & Kandır, S. Y. (2015). Altın Piyasasında Piyasa Riskinin Ölçülmesi: Riske Maruz Değer (VAR) Yöntemi İle Bir Uygulama. *Bankacılar Dergisi*, *92*, 53–70.

Gamba-Santamaria, S., Gomez-Gonzalez, J. E., Hurtado-Guarin, J. L., & Melo-Velandia, L. F. (2017). Stock market volatility spillovers: Evidence for Latin America. *Finance Research Letters*, *20*, 207–216. doi:10.1016/j.frl.2016.10.001

Güloğlu, B., Kaya, P., & Aydemir, R. (2016). Volatility transmission among Latin American stock markets under structural breaks. *Physica A*, *462*, 330–340. doi:10.1016/j.physa.2016.06.093

Hafner, C. M., & Herwartz, H. (2006). A Lagrange multiplier test for causality in variance. *Economics Letters*, *93*(1), 137–141. doi:10.1016/j.econlet.2006.04.008

Kamışlı, M., Kamışlı, S., & Sevil, G. (2016). The Effects of Crises on Volatility Spillovers between Borsa Istanbul Sector Indexes. *Advances in Economics and Business*, *4*(7), 339–344. doi:10.13189/aeb.2016.040703

Korkmaz, T., & Çevik, E. İ. (2009). Zımni Volatilite Endeksinden Gelişmekte Olan Piyasalara Yönelik Volatilite Yayılma Etkisi. *BDDK Bankacılık ve Finansal Piyasalar Dergisi*, *3*(2), 87–105.

Kutlar, A., & Torun, P. (2014). *Gelişmiş Piyasa Ekonomileri İle Yükselen Piyasa Ekonomileri Arasindaki Volatilite Dinamiklerinin Ekonometrik Analizi.* The 24th International Conference of The International Trade and Finance Association, Kayseri, Turkey.

Li, Y., & Giles, D. E. (2015). Modelling Volatility Spillover Effects between Developed Stock Markets and Asian Emerging Stock Markets. *International Journal of Finance & Economics*, *20*(2), 155–177. doi:10.1002/ijfe.1506

Markowitz, H. (1952). Portfolio Selection. *The Journal of Finance*, *7*, 77–91.

Markowitz, H. (1959). *Portfolio Selection: Efficient Diversification of Investment*. New York: Wiley.

Miyakoshi, T. (2003). Spillovers of stock return volatility to Asian equity markets from Japan and the US. *Journal of International Financial Markets, Institutions and Money*, *13*(4), 383–399. doi:10.1016/S1042-4431(03)00015-5

Natarajan, V. K., Azariah, R. R. S., & Nagarajan, C. P. (2014). Examining mean-volatility spillovers across national stock markets. *Journal of Economics, Finance and Administrative Science*, *19*(36), 55–62. doi:10.1016/j.jefas.2014.01.001

Ross, S. A. (1989). Information and volatility: The no-arbitrage martingale approach to timing and resolution irrelevancy. *The Journal of Finance*, *44*(1), 1–17. doi:10.1111/j.1540-6261.1989.tb02401.x

Yıldırım, S. (2010). 2008 Yılı Küresel Ekonomi Krizinin Dünya ve Türkiye Ekonomisine Etkileri. *KMÜ Sosyal ve Ekonomik Araştırmalar Dergisi*, *12*(18), 47–55.

KEY TERMS AND DEFINITIONS

Fragile Five: A group consisting of Turkey, Brazil, India, South Africa, and Indonesia.

Group of Seven: A group consisting of Canada, France, Germany, Italy, Japan, the United Kingdom, and the United States.

Volatility: The magnitude of the difference between price movements of financial products.

Volatility Spillover: The effect of one market's volatility on other markets' volatility.

Volatility Spillover Test of Hafner and Herwartz: The Lagrange multiplier-based causality-in-variance test which is used to determine volatility spillovers.

Chapter 13
Riesz Potential in Generalized Hölder Spaces

Boris Grigorievich Vakulov
Southern Federal University, Russia

Galina Sergeevna Kostetskaya
Southern Federal University, Russia

Yuri Evgenievich Drobotov
Southern Federal University, Russia & Azov Sea Research Fisheries Institute, Russia & SPE Vibrobit LLC, Russia

ABSTRACT

The chapter provides an overview of the advanced researches on the multidimensional Riesz potential operator in the generalized Hölder spaces. While being of interest within mathematical modeling in economics, theoretical physics, and other areas of knowledge, the Riesz potential plays a significant role for analysis on fractal sets, and this aspect is briefly outlined. The generalized Hölder spaces provide convenient terminology for formalizing the smoothness concept, which is described here. There are constant and variable order potential type operators considered, including a two-pole spherical one. As a sphere is, in some sense, a convenient set for analysis, there are two results, proved in detail: the conditions for the spherical fractional integral of variable order to be bounded in the generalized Hölder spaces, whose local continuity modulus has a dominant, which may vary from point to point, and the ones for the constant-order two-pole spherical potential type operator.

INTRODUCTION

The chapter called "Potentials and Their Basic Properties" of a monograph by N.S. Landkof (1973) begins with basic definition of a potential as "some integral operator, acting on a space of signed measures" (p. 43). Such an operator is defined by a kernel $K(x,y)$ which is usually dependent only on the distance. It is emphasized, that M. Riesz kernels include the ones of the classical theory (Newtonian and logarithmic kernels) and Green kernels associated with a region as special and limiting cases. This is

DOI: 10.4018/978-1-5225-3767-0.ch013

what makes the Riesz potential to be of a great interest for mathematical physics, but while fractal models are being considered, the new reasons for the operator to be important are being discovered.

Fractal sets significantly widen the scope of functional analysis and its applications as they generalize the classical vision and approaches, which are based on the concept of a smooth manifold. According to this, fractional calculus appears to provide a convenient theory of integration over fractal sets and non-integer-dimensional spaces, as well as to define key methods for solving specific problems. Let the definitions of functions and integrals on fractals be recalled here following (Tarasov, 2010) in order to outline briefly the motivation for fractional calculus operators to be considered while studying fractal models in any field of science.

Let Ω be a non-empty subset of the n-dimensional Euclidian space $\mathbf{R}^n$, and let $\{E_i\}$ be a countable family of disjoint subsets of diameter at most ε, covering Ω:

$$\Omega \subset \bigcup_{i=1}^{\infty} E_i, \ \operatorname{diam}(E_i) \leq \varepsilon \text{ for all } i,$$

where

$$\operatorname{diam}(E_i) = \sup\{r(x,y) : x, y \in E_i\}, \ r(\cdot)$$

denotes a metric. A characteristic function of a subset E_i is defined as follows:

$$\chi_{E_i}(x) = \begin{cases} 1, & x \in E_i; \\ 0, & x \notin E_i. \end{cases}$$

A function $f(x)$, defined on Ω, is called a simple one, if

$$f(x) = \begin{cases} f_k, & x \in E_k; \\ 0, & x \notin \bigcup_{k=1}^{\infty} E_k, \end{cases} \quad f_i \in \mathbf{R}, \text{ and } f(x) = \sum_{i=1}^{\infty} f_i \chi_{E_i}(x).$$

Let the simple function $f(x)$ be integrable on Ω, i.e. $\forall i : f_i \neq 0 \ \mu_H(E_i) < \infty$, where

$$\mu_H(\Omega, D) = \lim_{\varepsilon \to 0} \mu_H^{(\varepsilon)}(\Omega, D),$$

$$\mu_H^{(\varepsilon)}(\Omega, D) = \inf\left\{\sum_{i=1}^{\infty} \omega(D)\left[\operatorname{diam}(E_i)\right]^D : \Omega \subset \bigcup_{i=1}^{\infty} E_i, \operatorname{diam}(E_i) \leq \varepsilon\right\},$$

$$\omega(D) = \frac{\pi^{D/2}}{2^n \Gamma(D/2+1)}.$$

Then the Lebesue-Stieltjes integral can be defined by

$$\int_{\Omega} f(x)\, d\mu_H(x) = \sum_{i=1}^{\infty} f_i \mu_H(E_i, D) = \omega(D) \lim_{\operatorname{diam}(E_i)\to 0} \sum_{E_i} f(x_i)\left[\operatorname{diam}(E_i)\right]^D.$$

As it is always possible to divide the subset $\Omega \subset \mathbf{R}^n$ into the collection of ε-cubes, defined as

$$E_{i_1 \dots i_n} = \left\{(x_1, \dots, x_n) \in \Omega : i_k \varepsilon \le x_k \le (i_k + 1)\varepsilon,\ i_k \in \mathbf{N}\right\},$$

then

$$d\mu_B(x) = \lim_{\operatorname{diam}(E_{i_1 \dots i_n})\to 0} \omega(D)\left[\operatorname{diam}(E_i)\right]^D,$$

where

$$\mu_B(\Omega, D) = \omega(D)\zeta^D(\Omega),\ \zeta^D(\Omega) = \lim_{\varepsilon\to 0} \zeta_\varepsilon^D(\Omega),$$

$$\zeta_\varepsilon^D(\Omega) = \inf\left\{\sum_{i=1}^{\infty}\left[\operatorname{diam}(E_{i_1 \dots i_n})\right]^D : \Omega \subset \bigcup_{i_1 \dots i_n}^{\infty} E_{i_1 \dots i_n}, \operatorname{diam}(E_{i_1 \dots i_n}) \le \varepsilon\sqrt{n}\right\}.$$

Considering the Euclidian metric $|x-y| = \left(\sum_{i=1}^{n} |x_i - y_i|^2\right)^{\frac{1}{2}}$, $x, y \in \mathbf{R}^n$, Ω can be parameterized using polar coordinates: $\rho = r(x,0) = |x|$ and the angle φ. Within this parametrization, a spherically symmetric covering by the $E_{\rho,\Omega}$ sets around the centre at the origin can be considered, thus

$$d_{\mu_B}(\rho, \varphi) = \lim_{\operatorname{diam}(E_{r,\varphi})\to 0} \omega(D)\left[\operatorname{diam}(E_{\rho,\varphi})\right]^D = d\varphi^{D-1}\rho^{D-1}dr.$$

If $f(x)$ is assumed to be symmetric with respect to some centre $x_0 \in \Omega$, i.e. $f(x) = \text{const}$ for $\forall x \in \Omega : |x - x_0| = \rho$ and arbitrary ρ, then the centre of symmetry can be shifted with the transformation

$$\Omega \to \Omega' : x \to x' = x - x_0,$$

and the integral over Ω can be represented as

$$\int_{\Omega} f\, d\mu_B = \frac{2\pi^{D/2}}{\Gamma(D/2)} \int_0^{\infty} f(\rho)\rho^{D-1} dr = \frac{2\pi^{D/2}\Gamma(D)}{\Gamma(D/2)} (I_-^D f)(0),$$

where

$$(I_-^\alpha f)(x) := \frac{1}{\Gamma(D)} \int_x^{\infty} \frac{f(t)}{(t-x)^{1-\alpha}} dt, \; -\infty < x < \infty,$$

is the right-sided Riemann-Liouville integral, studied in fractional calculus (Samko, Kilbas, & Marichev, 1993, p.).

Along with I_-^α, the left-sided fractional integral is considered as (Samko et al., 1993, p. 33)

$$(I_+^\alpha f)(x) := \frac{1}{\Gamma(D)} \int_{-\infty}^{x} \frac{f(t)}{(x-t)^{1-\alpha}} dt, \; -\infty < x < \infty.$$

Fractional integration operators with constant limits occur in many fields of mathematics, and due to analogies in mathematical physics, such operators came to be called potential type operators. Thus, the Riesz potential over the real axis is considered as the following operator (Samko et al., 1993, p. 214):

$$(I^\alpha f)(x) = \frac{1}{2\Gamma(\alpha)\cos(\alpha\pi/2)} \int_{-\infty}^{\infty} \frac{f(t)}{|t-x|^{1-\alpha}} dt, \; \mathrm{Re}\alpha > 0, \; \alpha \neq 1,3,5,\ldots,$$

and its useful modification is

$$(H^\alpha f)(x) = \frac{1}{2\Gamma(\alpha)\sin(\alpha\pi/2)} \int_{-\infty}^{\infty} \frac{\mathrm{sgn}(x-t)}{|x-t|^{1-\alpha}} f(t)\,dt, \; \mathrm{Re}\alpha > 0, \; \alpha \neq 2,4,6,\ldots,$$

with I^α and H^α being defined on functions $f \in L_p(\mathbf{R})$, $1 \le p < 1/\mathrm{Re}\alpha$, while $0 < \mathrm{Re}\alpha < 1$.

The connection between these operators and the fractional integrals, defined above, is evident:

$$I^\alpha = \left[2\cos\left(\frac{\alpha\pi}{2}\right)\right]^{-1} (I_+^\alpha + I_-^\alpha), \; H^\alpha = \left[2\sin\left(\frac{\alpha\pi}{2}\right)\right]^{-1} (I_+^\alpha - I_-^\alpha).$$

In the multidimensional case, the Riesz potentials appear to be negative fractional powers of the Laplace operator (Samko et al., 1993, p. 483):

$$\left(-\Delta\right)^{\alpha/2} f = \mathsf{F}^{-1}\left|x\right|^{-\alpha} \mathsf{F} f = \begin{cases} I^{\alpha} f, & \operatorname{Re}\alpha > 0; \\ D^{-\alpha} f, & \operatorname{Re}\alpha < 0, \end{cases}$$

where F denotes the multidimensional Fourier transform:

$$\mathsf{F} f = \hat{f}\left(x\right) = \int_{\mathbf{R}^n} e^{ix\cdot t} f\left(t\right) dt \,,\; x \in \mathbf{R}^n \,,$$

and D^{α} is the hypersingular integral:

$$\left(D^{\alpha} f\right)\left(x\right) = \frac{1}{d_{n,l}\left(\alpha\right)} \int_{\mathbf{R}^n} \frac{\left(\Delta_t^l f\right)\left(x\right)}{\left|t\right|^{n+\alpha}} dt \,,$$

where $d_{n,l}\left(\alpha\right)$ is a normalizing constant. It is remarkable, that the Laplace operator on a fractal set is usually a starting point for developing the analysis on fractals.

In the present chapter, the Riesz potential type operator is considered in a general form:

$$\left(I_{\Omega}^{\alpha(\cdot)} c f\right)\left(x\right) = \int_{\Omega} \frac{c\left(x,t\right) f\left(t\right)}{r^{n-\alpha(x)}\left(x,t\right)} dt \,,\; \alpha\left(x\right) \neq n, n+2, n+4, \ldots,$$

where $n \geq 2$, $r\left(\cdot\right)$ is a metric and $A\left(\cdot\right)$ is called a characteristic of the potential.

The constant order two-pole potential type operator is also considered:

$$\left(K_{\Omega}^{\alpha} cf\right)\left(\xi\right) = \int_{\Omega} \frac{c\left(\xi,\sigma\right) f\left(\sigma\right)}{\left|\xi-\sigma\right|^{n-\alpha} \left|\xi+\sigma\right|^{n-\alpha}} d\sigma \,,\; \operatorname{Re}\alpha > 0 \,.$$

The set Ω is assumed afterwards to be a unit sphere $\mathrm{S}^n \subset \mathbf{R}^{n+1}$ or the space $\dot{\mathbf{R}}^n$. In this chapter, standard terms and notions of multidimensional geometry and harmonic analysis are used, such that an n-dimensional unit sphere S^n, embedded to the $\left(n+1\right)$-dimensional Euclidian space $\mathbf{R}^{n+1}$, represents hereinafter the set of points which are equidistant from the origin and the radius of the sphere is equal to 1: $\mathrm{S}^n = \left\{x \in \mathbf{R}^{n+1} : \left|x\right| = 1\right\}$, while $\dot{\mathbf{R}}^n$ stands for one point compactified $\mathbf{R}^n$: $\dot{\mathbf{R}}^n = \mathbf{R}^n \cup \left\{\infty\right\}$. It is worth noticing, that the constant order operator $I_{\mathrm{S}^n}^{\alpha}$, both with the two-pole one $K_{\mathrm{S}^n}^{\alpha}$ are of convolution type. In studying such operators, their Fourier-Laplace multipliers play a great role, as there is an opportunity to calculate them explicitly. Since $\alpha(\cdot)$ is admitted being variable, the correspondent operators are not the spherical convolutions. Then, some other method of studying needs to be applied, i.e. the Zygmund type estimates, considered below in this chapter.

The theory of the generalized Hölder spaces is currently relevant as, inter alia, it provides sufficiently convenient terms for defining and studying the concept of smoothness. Thus, even if the "usual" smoothness (i.e. the integer order differentiability) on a sphere is considered, it is essential to clarify this concept. It can be defined through two, fundamentally different, groups of terms (Vakulov & Samko, 1987): the ones, connected with the Laplace-Beltrami operator, and the terms, which relate to differentiation in Cartesian coordinates. Both ways lead to, in general, equivalent Hölderian classes $H^{\lambda}\left(\mathrm{S}^{n}\right)$ with non-integer λ and to certainly different ones with integer $\lambda \geq 1$.

The generalized Hölder spaces are based on the concept of the local continuity modulus. Both the modulus and the spaces are defined in the section "Key Terms and Definitions" of the present work.

BACKGROUND

The fractional calculus operators in functional spaces with variable (dependent on coordinates) defining characteristics have been being studied intensively in recent years. Thus, for one-dimensional case, such problems have been considered in (Ginsburg & Karapetyants, 1994; Karapetyants & Ginsburg, 1995; Samko, 1998). The variable order fractional calculus operators have been considered in papers (Odzijewicz, Malinowska, & Torres, 2013; Ross & Samko, 1995; Samko & Vakulov, 2011; Samko & Ross, 1993; Samko, 1995; Vakulov & Kochurov, 2010; Vakulov & Kochurov, 2011; Vakulov, Kochurov, & Samko, 2011).

The works by A. Chegolin and V. Nogin (2001; 2002; 2013) are also to be mentioned in connection with the subject, as well as the paper (Chegolin, 2011).

Specific Hölderian spaces on a sphere were considered, in particular, in (Werens, 1980, 1981) as special cases of more generalized functional spaces. They were proposed also in (Samko, 2008; Nikolsky, & Lizorkin, 1984). Equivalent normalizations in the Hölderian spaces on a sphere were studied in (Vakulov & Samko, 1987; Samko & Vakulov, 2000).

The generalized Hölder spaces are defined in the terms of an arbitrary continuity modulus. Such spaces on a sphere are considered in (Dzhafarov, 1966, 1985).

The spherical potential type operator of a constant real order in the generalized Hölder spaces was considered in (Vakulov, 1986). In this paper, the Zygmund type estimates were studied and theorems on homeomorphism and boundedness were proven. The result was widened to the Riesz potential over $\dot{\mathbf{R}}^{n}$ with specific exponential weights.

The spherical Riesz potential of a constant complex order was studied in works by Shankishvili and the Zygmund type estimates were carried out. In the paper (Vakulov, 2001) the result was extended to the case of the spatial Riesz potential.

In the paper (Samko & Vakulov, 2008), the spherical convolution operator with specified asymptotics of its multiplicator was considered, and the conditions for it to be bounded in the generalized Hölder spaces were studied.

The Zygmund type estimates and smoothness properties of the spherical convolution operator of the potential type with an exponentially-logarithmic kernel were studied in (Vakulov, Karapetjanc, & Shankishvili, 2003), and the theorems on boundedness were proven.

The spherical potential type operator in the weighted generalized variable Hölder spaces was studied in paper (Vakulov, 2005), in which the results on the isomorphism were carried out.

The spherical potential type operator of a constant order with weak singularities on the poles was considered in (Vakulov & Karapetjanc, 2004), and the theorems on the isomorphism were carried out.

In the paper (Samko & Vakulov, 2011), the spherical variable order Riesz potential was studied in the weighted and non-weighted generalized variable Hölder spaces, and the theorems on boundedness were proven.

CONSTANT ORDER POTENTIAL TYPE OPERATORS IN GENERALIZED VARIABLE HÖLDER SPACES

Constant Order Potential Type Operator With a Constant Characteristic

There are two principal problems to be considered in studying the smoothness property of the reflection by an operator: in what terms the concept can be defined and what methods are to be used for investigating the existence of the property. For the first problem, the terms of the Hölder spaces are proposed, while the second one can be solved by applying so-called Zygmund type estimates, in which the local continuity modulus of an image is to be majorized by some integral operator of the modulus of its preimage. Vakulov carried out the following Zygmund type estimates for the constant order potential type operator (1986):

Theorem 1: *For* $I^{\alpha}_{\mathrm{S}^n} f$, $f \in C\left(\mathrm{S}^n\right)$, *the following estimate holds:*

$$M_r\left(I^{\alpha}_{\mathrm{S}^n} f, x, h\right) \le ch \int_h^2 \frac{M_r\left(f, x, t\right)}{t^{2-\alpha}} dt \, , \; c > 0 \, .$$

Theorem 2: *Let* $w\left(x\right) = \left|x - e_{n+1}\right|^{\mu}$, $0 < \mu < n+1$, and f satisfies the following conditions:

$$wf \in C\left(\mathrm{S}^n\right), \; \left(wf\right)\left(e_{n+1}\right) = 0 \, .$$

Then, the following estimate holds:

$$M_r\left(I^{\alpha}_{\mathrm{S}^n} f, x, h\right) \le ch^{\gamma} \int_h^2 \frac{M_r\left(wf, x, t\right)}{t^{1-\alpha+\gamma}} dt \, , \; \gamma = \min\left\{1, \mu\right\}.$$

Basing on Theorems 1 and 2, the following results are revealed (Vakulov, 1986):

Theorem 3: *Let* $\varphi_\alpha(t) = t^{\alpha-[\alpha]}\varphi(t)$. $\varphi \in \Phi^0_{1-\alpha+[\alpha]}$, *then* $I^\alpha_{S^n}$, $\alpha > 0$, *is isometric between* $H^\varphi_k(S^n)$ and $H^{\varphi_\alpha}_{k+[\alpha]}(S^n)$:

$$I^\alpha_{S^n}\left(H^\varphi_k(S^n)\right) = H^{\varphi_\alpha}_{k+[\alpha]}(S^n),\ \alpha > 0.$$

Theorem 4: *Let* $\varphi_\alpha(t) = t^{\alpha-[\alpha]}\varphi(t)$, $\varphi \in \Phi^{\max\{0,\mu-n\}}_{\min\{1-\alpha+[\alpha],\mu-\alpha+[\alpha]\}}$, $w(x) = |x-\sigma|^\mu$, $x,\sigma \in S^n$, $\alpha - [\alpha] < \mu < n - \alpha + [\alpha] + 1$, then

$$I^\alpha_{S^n}\left(H^\varphi_k(S^n, w)\right) = H^{\varphi_\alpha}_{k+[\alpha]}(S^n, w),\ \alpha > 0,\ \alpha \neq 1,2,\ldots.$$

For $\alpha = 1,2,\ldots$, $\varphi \in \Phi^{\max\{0,\mu-n\}}_1$, $0 < \mu < n+1$, $\mu \neq n$, the following holds:

$$I^\alpha_{S^n}\left(\overset{\circ}{H}{}^\varphi_k(S^n, w)\right) = \overset{\circ}{H}{}^\varphi_{k+\alpha}(S^n, w),$$

where

$$\overset{\circ}{H}{}^\varphi_k(S^n, w) = \left\{ f \in C(S^n) : wD^k f \in H^\varphi(S^n), \lim_{x\to a_m} wD^k f = 0 \right\},$$

$w(x) = \prod_{m=0}^{l} |x - a_m|^{\beta_m}$, $x, a_m \in S^n$, D^k is a spherical convolution operator with the Fourier-Laplace multiplier $\{d_m\} \in W_{k,N}$, $N > n+2$, $d_m \neq 0$, $m = 0,1,2,\ldots$.

The Zygmund-Bary-Stechkin class Φ^δ_β, $W_{k,N}$ and $H^\varphi_k(S^n)$ are defined in the section "Key Terms and Definitions".

Vakulov (1986) used the stereographic projecting, which is also described there, to extend the result for the spherical potential to the case of the spatial one. They relate to the following modification of the integral operator:

$$\left(I^\alpha_{\dot{\mathbf{R}}^n} f\right)(x) = \frac{1}{\gamma_n(\alpha)} \int_{\dot{\mathbf{R}}^n} \frac{f(t)\,dt}{|x-t|^{n-\alpha}},\ 0 < \alpha < n,\ x \in \dot{\mathbf{R}}^n,$$

and the following realization of the functional space:

$$H^{\varphi}\left(\dot{\mathbf{R}}^n\right)=\left\{f\in C\left(\dot{\mathbf{R}}^n\right):\ \sup_{x,y\in\mathbf{R}^n}\left[\left|f(x)-f(y)\right|\cdot\varphi^{-1}\left(\frac{|x-y|}{\sqrt{\left(1+|x|^2\right)\left(1+|y|^2\right)}}\right)\right]<\infty\right\}.$$

Theorem 5: If $\varphi\in\Phi^0_{1-\alpha}$, $\varphi_\alpha(\tau)=\tau^\alpha\varphi(\tau)$, then $I^\alpha_{\dot{\mathbf{R}}^n}$, $0<\alpha<1$, implements a homeomorphic mapping between $H^{\varphi}\left(\dot{\mathbf{R}}^n,\left(|t|^2+1\right)^{\frac{n+\alpha}{2}}\right)$ to $H^{\varphi_\alpha}\left(\dot{\mathbf{R}}^n,\left(|t|^2+1\right)^{\frac{n+\alpha}{2}}\right)$:

$$I^\alpha_{\dot{\mathbf{R}}^n}\left(H^{\varphi}\left(\dot{\mathbf{R}}^n,\left(|t|^2+1\right)^{\frac{n+\alpha}{2}}\right)\right)=H^{\varphi_\alpha}\left(\dot{\mathbf{R}}^n,\left(|t|^2+1\right)^{\frac{n-\alpha}{2}}\right).$$

Consequently, the following result holds:

Theorem 6: *If* $\varphi(\tau)=\tau^\lambda$, $0<\lambda+\alpha<1$, *then*

$$I^\alpha_{\dot{\mathbf{R}}^n}\left(H^{\lambda}\left(\dot{\mathbf{R}}^n,\left(|t|^2+1\right)^{\frac{n+\alpha}{2}}\right)\right)=H^{\alpha+\lambda}\left(\dot{\mathbf{R}}^n,\left(|t|^2+1\right)^{\frac{n-\alpha}{2}}\right).$$

Potential Type Operator With a Differential Characteristic

Consider a constant order potential type operator with differential characteristic $c(x,t)=a(x-t)$, i.e.

$$\left(K^\alpha_a f\right)(x)=\int_{\mathrm{R}^n}\frac{a(x-t)f(t)}{|x-t|^{n-\alpha}}dt,\ 0<\alpha<n,$$

together with a hypersingular integral (HSI)

$$\left(D^\alpha_\Omega f\right)(x)=\frac{1}{d_{n,l}(\alpha)}\int_{\mathrm{R}^n}\frac{\left(\Delta^l_t f\right)(x)}{|t|^{n+\alpha}}\Omega(t)\,dt,\ 0<\alpha<l,$$

where $\left(\Delta^l_t f\right)(x)$ is a finite difference of the function f with the center at the point x of the order l with the step t, $d_{n,l}(\alpha)$ is the normalizing constant such that $D^\alpha_\Omega=\left(-\Delta\right)^{\alpha/2}$, $\Omega\equiv 1$ (46).

The characteristics $a(x)$ and $\Omega(x)$ are assumed to be bounded, sufficiently smooth with respect to the radial variable $r = |x|$ and the angular variable $\sigma = \frac{x}{|x|}$.

The elliptical case is to be considered, which means that the Fourier transform of the kernel of K_c^α multiplied by $|x|^\alpha$ is not equal to zero over R^n:

$$\inf_{x \in R^n} \left(\left| \hat{k}(x) \right| * |x|^\alpha \right) \neq 0, \; k(t) = \frac{a(t)}{|t|^{n-\alpha}}.$$

For D_Ω^α, the ellipticity means that the following condition holds:

$$\inf_{x \in R^n} \left(\left| \mathbf{D}_\Omega^\alpha(x) \right| * |x|^{-\alpha} \right) \neq 0,$$

where $\mathbf{D}_\Omega^\alpha(x)$ is the symbol of D_Ω^α.

Assume the characteristic of K_a^α has discontinuity of a uniform function:

$$a(x) = b(|x|) , \left(\frac{x}{|x|} \right),$$

where $b(r)$ and $,(\sigma)$ are sufficiently smooth, i.e. $b(r) \in C^N(\dot{R}_+^1)$, $,(\sigma) \in C^q(S^{n-1})$, while N and q satisfy specific conditions.

The problem of inverting the potential with the differential characteristic is related to calculating its symbol (the Fourier transform of the kernel of K_a^α). As the function $\theta(\sigma)$ is smooth, by applying the Bochner's formula the symbol of K_a^α can be represented as follows:

$$\mathbf{K}_a^\alpha(x) = \frac{A_\alpha(x)}{|x|^\alpha}, \; A_\alpha(x) = (2\pi)^{\frac{n}{2}} \sum_{m\mu} i^m Y_{m\mu}(x') \theta_{m\mu} \int_0^\infty b\left(\frac{t}{|x|} \right) t^{\alpha - \frac{n}{2}} J_{\frac{n}{2}+m-1}(t)\, dt.$$

The integral converges for $0 < \alpha < \frac{n+1}{2}$, while for $\alpha \geq \frac{n+1}{2}$ the analytical continuation of $A_\alpha(x)$ with respect to α is to be used.

If the ellipticity condition for A_α holds, i.e. $\inf_{x \in R^n} \left| A_\alpha(x) \right| \neq 0$, the inverse operator can be formally considered in the form

$$\left(K_a^{\alpha}\right)^{-1} f = \frac{\left|x\right|^{\alpha}}{\mathrm{A}_{\alpha}\left(x\right)} * f .$$

Specific simple sufficient conditions for the functions $b\left(r\right)$ and $,\left(\sigma\right)$ allow investigating the symbol of the inverse operator while $\left|x\right| \to 0$ as well as exploring its asymptotics at infinity. As a result, the inverse operator can be implemented as the sum of the convolution with the kernel from $L_1\left(\mathrm{R}^n\right)$ and the convolution with a specific generalized function of the form $\frac{\Omega_{\alpha}\left(x\right)}{\left|x\right|^{n+\alpha}}$. For $0<\alpha<2$, $\alpha \neq 1$, the latter can be represented as a sum of a HSI and a bounded characteristic. Thus, the following theorem holds:

Theorem 7: *Let* $b\left(r\right) \in C^{2N+S}\left(\dot{\mathbf{R}}_{+}^{1}\right)$, $N > n + \alpha$,

$$S = \begin{cases} 0, \ \alpha < \dfrac{n-1}{2}; \\ \left[\alpha - \dfrac{n-1}{2}\right] + 1, \ \alpha \geq \dfrac{n-1}{2}, \end{cases} ,\left(\sigma\right) \in C^{2q}\left(\mathrm{S}^{n-1}\right),\ q > \frac{5}{2}n + 2\alpha - \frac{1}{2}.$$

If $\inf\limits_{x \in \mathrm{R}^n} \left|\mathrm{A}_{\alpha}\left(x\right)\right| \neq 0$, then for $f \in I^{\alpha}\left(L_p\right)$, $0<\alpha<1$, $1<\alpha<2$ can be represented as follows:

$$\left(K_a^{\alpha}\right)^{-1} f = D_{\Omega}^{\alpha} f + \mu_{\alpha} * f ,$$

where D_{Ω}^{α} is a HSI with some bounded characteristic $\Omega\left(x\right)$, $\mu_{\alpha}\left(x\right)$ is an integrable function.

Let $\inf\limits_{x \in R^n} \left|A_{\alpha}(x)\right| \neq 0$, then $\left(K_{\alpha}^{\alpha}\right)^{-1}$ defined for $f \in I^{\alpha}\left(L_p\right)$, $0<\alpha<1$, $1<\alpha<2$, can be represented as

$$\left(K_{\alpha}^{\alpha}\right)^{-1} f = D_{\Omega}^{\alpha} f + \mu_{\alpha} * f ,$$

where $D_{\Omega}^{\alpha} f$ is a HIS with a bounded characteristic $\Omega(x)$, $\mu_{\alpha}(x)$ is an integrable function.

For all $0<\alpha<n$, $\alpha \neq 2,4,\ldots$, the operator $\left(K_{\alpha}^{\alpha}\right)^{-1}$ can be represented as a composition with a HIS with a homogeneous characteristic:

Theorem 8: Let $b(r) \in C^{2N+S}\left(\dot{\mathbf{R}}_{+}^{1}\right)$, $N > n+\alpha$, S is from Theorem 7, $\theta(\sigma) \in C^{2q}\left(\mathrm{S}^{n-1}\right)$ and $q > \max\left(2n-2, \frac{3}{2}n+\alpha-1\right)$. Then, in the elliptic case, for $0<\alpha<n$, $\alpha \neq 2,4,6,\ldots$, $f \in K_{\alpha}^{\alpha}\left(L_p\right)$

$$\left(K_{\alpha}^{\alpha}\right)^{-1} f = D_{M}^{\alpha} f + \sum_{k=0}^{[\alpha]} A_k D_{Nk}^{\alpha} f + A_{[\alpha]+1} f,$$

where D_M^{α} and D_{Nk}^{α} are the HSIs with homogeneous characteristics, $A_{k,k=0,\ldots[\alpha]+1}$ are convolutional operators with integrable kernels.

At last, the following representation of $\left(K_{\alpha}^{\alpha}\right)^{-1}$ holds for all $0<\alpha<n$, including $\alpha = 2,4,6,\ldots$:

$$\left(K_{\alpha}^{\alpha}\right)^{-1} f = S_0 D_f^{\alpha} + \sum_{k=1}^{[\alpha]} A_k S_k B^k D_f^{\alpha} + A_{[\alpha]+1} S_{[\alpha]+1} D_f^{\alpha} + A_{[\alpha]+2} f,$$

where D^{α} is a HSI with characteristic $\Omega(t) \equiv 1$, S_k are the CalderonZygmund operators with some homogeneous characteristics, A_k is convolutional operators with integrable kernels ($k = 1,2,\ldots[\alpha]+2$), B^k is the Bessel's potential of order k.

Two-Pole Potential Type Operator of a Constant Order

Assume the spherical harmonics

$$Y_{m\nu}(x),\ m = 0,\ 1,\ 2,\ldots,\ \nu = 1,2,\ldots,(n+m-1)\frac{(n+m-2)!}{m!(n-1)!},$$

which provide an orthonormal basis on a unite sphere S^n, and let a function $f \in C^{\infty}\left(\mathrm{S}^n\right)$ be expanded in the Fourier-Laplace series with coefficients $f_{m\nu}$ through $Y_{m\nu}(x)$. Then the reflection by K is reduced to multiplying of $f_{m\nu}$ by k_m, and, vice versa,

$$k(t) = \frac{\Gamma\left(\frac{n-1}{2}\right)}{4\pi^{\frac{n+1}{2}}} \sum_{m=0}^{\infty} (2m+n-1) k_m C_m^{\left(\frac{n-1}{2}\right)}(t).$$

To reveal the conditions for the spherical potential type operator with weak singularities at the poles

$$\left(K^{\alpha}f\right)(x) = \int_{S^n} \frac{f(\sigma)}{|x-\sigma|^{n-\alpha}|x+\sigma|^{n-\alpha}} d\sigma\,,\ x \in S^n\,,\ \operatorname{Re}\alpha > 0\,,$$

the following lemmas are used:

Lemma 1: If $\left\{k_m\right\}_{m=0}^{\infty} \in W_{\lambda,N}$ and $k_m \neq 0$, $m = 0,1,\ldots$, then $\left\{k_m^{-1}\right\}_{m=0}^{\infty} \in W_{-\lambda,N}$. If $\left\{k_m\right\}_{m=0}^{\infty} \in W_{\lambda,N}$, $\left\{l_m\right\}_{m=0}^{\infty} \in W_{\mu,N}$, then $\left\{k_m l_m\right\}_{m=0}^{\infty} \in W_{\lambda+\mu,N}$.

Lemma 2: If $\left\{k_m\right\}_{m=0}^{\infty} \in W_{-\lambda,N}$, where $\operatorname{Re}\lambda + N > \frac{n}{2}$, $\lambda \neq n+2k$, $k = 0,1,\ldots$, then the operator, corresponding to this multiplicator, is

$$\left(aI + K\right)f = a f(x) + \int_{S^n} k(x\sigma) f(\sigma)\, d\sigma\,,$$

where $k(t) \in L_1\left(\left[-1;1\right],\left(1-t^2\right)^{\frac{n-2}{2}}\right)$, $a = 0$ for $\operatorname{Re}\lambda > 0$ and $a = \lim_{m\to\infty} k_m$ for $\lambda = 0$. If, in addition, $N > n - \operatorname{Re}\lambda$, then

$$k(t) = \begin{cases} (1-t)^{\frac{\lambda-n}{2}}\, l(t), & \operatorname{Re}\lambda > 0; \\ (1-t)^{\frac{1-n}{2}}\, l(t), & \lambda = 0, \end{cases}$$

where

$$l \in C\left(\left[-1;1\right]\right),\ l(1) = \begin{cases} 2^{\frac{\lambda-n}{2}}\gamma_n^{-1}(\lambda) \lim\limits_{m\to\infty} m^{\lambda} k_m, & \text{if } \operatorname{Re}\lambda > 0; \\ 2^{\frac{1-n}{2}}\gamma_n^{-1}(\lambda) \lim\limits_{m\to\infty} m\left(k_m - a\right), & \text{if } \lambda = 0, \end{cases} \quad \gamma_n(\lambda)$$

is a normalizing constant. If $N > n + 2 - \operatorname{Re}\lambda$, then

$$\left(K f\right)(x) = \begin{cases} \int_{S^n} k(x\sigma) f(\sigma)\, d\sigma + \sum\limits_{j=0}^{N} b_j \left(K^{j+\lambda} f\right)(x), & \text{if } \operatorname{Re}\lambda > 0; \\ b_0 f(x) + \int_{S^n} k(x\sigma) f(\sigma)\, d\sigma + \sum\limits_{j=1}^{N} b_j \left(K^{j+\lambda} f\right)(x), & \text{if } \lambda = 0, \end{cases}$$

where $k \in C^1\left[-1;1\right]$, $b_0 = \lim\limits_{m\to\infty} k_m$.

In (Vakulov & Karapetjanc, 2003), it is proposed to consider a generalized two-pole operator

$$\left(K^{\alpha,\beta}f\right)\left(x\right)=\int_{S^n}\frac{f\left(\sigma\right)}{\left|x-\sigma\right|^{n-\alpha}\left|x+\sigma\right|^{n-\beta}}d\sigma,\ x\in S^n,$$

where $0<\mathrm{Re}\,\alpha,\ \mathrm{Re}\,\beta<1$. $K^{\alpha,\beta}$ is the spherical convolution operator and its multiplier $\left\{k_m^{\alpha,\beta}\right\}_{m=0}^{\infty}$ is defined as

$$k_m^{\alpha,\beta}=2^{\frac{\alpha+\beta}{2}-n}\left|S_{n-1}\right|\int_{-1}^{1}(1-t)^{\frac{\alpha-n}{2}}(1+t)^{\frac{\beta-n}{2}}(1-t^2)^{\frac{n-2}{2}}P_m(t)dt\,.$$

The latter can be calculated with using the equality $\left|x-\sigma\right|^2=2\left(1-x\sigma\right)$ and the aforementioned definition of $P_m\left(t\right)$:

$$k_m^{\alpha,\beta}=2^{\alpha+\beta+n+1}\left|S_{n-1}\right|B\left(\frac{\alpha}{2},\frac{\beta}{2}\right){}_3F_2\left(-m,m+n-1,\frac{\alpha}{2},\frac{\alpha+\beta}{2},\frac{n}{2};1\right),$$

where ${}_3F_2\left(\cdot\right)$ is the hypergeometric function.

If $\alpha=\beta$, then $K^{\alpha,\alpha}\equiv K^{\alpha}$, and its multiplier $\left\{k_m\right\}_{m=0}^{\infty}$, $k_m\equiv k_m^{\alpha,\alpha}$, can be calculated as follows:

$$k_m^{\alpha,\beta}=2^{\alpha-n}\left|S_{n-1}\right|(-1)^m\sqrt{\pi}\,\frac{\Gamma\left(\frac{\alpha}{2}\right)}{\Gamma\left(\frac{\alpha+1}{2}\right)}\,{}_3F_2\left(-m,m+n-1,\frac{\alpha}{2},\alpha,\frac{n}{2};1\right)=$$

$$=c_1\left[1+(-1)^m\right]\frac{\Gamma\left(\frac{m+1}{2}\right)\Gamma\left(\frac{m-\alpha+n}{2}\right)}{\Gamma\left(\frac{m+n+2}{2}\right)\Gamma\left(\frac{1+m+\alpha}{2}\right)}\sim am^{-\alpha},\ m\to\infty,$$

where

$$c_1=2^{\alpha-n-1}\left|S_{n-1}\right|\pi^{-1}\Gamma\left(\frac{n}{2}\right)\Gamma\left(\frac{\alpha}{2}\right)\Gamma\left(\frac{\alpha+2-n}{2}\right)\sin\left(\frac{(n-\alpha)\pi}{2}\right).$$

It follows from the Lemmas 1, 2 and the properties of the Gamma function, that the two-pole spherical convolution operator can be represented with using the one-pole potential type operator of a constant order. The theorem on this result, together with the Lemma 2, are used in the paper (Vakulov & Karapetjanc, 2003) to prove the following theorem:

Theorem 9: Let $\mathrm{Re}\,\alpha > 0$, $\alpha \neq n, n+2, \ldots$, $\varphi_\alpha(t) = t^{\mathrm{Re}\,\alpha}\varphi(t)$, $\omega \in \Phi_\delta^\beta$ and $w(x) = |x-a|^\mu$ or $w(x) = |x+a|^\mu$. Then:

$$K^\alpha\left(H_k^\varphi\left(\mathrm{S}^n\right)\right) \to H_{k+[\mathrm{Re}\,\alpha]}^{\varphi_\alpha}\left(\mathrm{S}^n\right),\ \beta = 0,\ \delta = 1 - \mathrm{Re}\,\alpha + [\mathrm{Re}\,\alpha];$$

$$K^\alpha\left(H_k^\varphi\left(\mathrm{S}^n, w\right)\right) \to H_{k+[\mathrm{Re}\,\alpha]}^{\varphi_\alpha}\left(\mathrm{S}^n, w\right),\ \beta = \max\left\{0, \mu - n\right\},$$

$$\delta = \min\left\{1 - \mathrm{Re}\,\alpha + [\mathrm{Re}\,\alpha], \mu - \mathrm{Re}\,\alpha + [\mathrm{Re}\,\alpha]\right\};$$

$$K^\alpha\left(L_p\left(\mathrm{S}^n\right)\right) \to H^{\mathrm{Re}\,\alpha - \frac{n}{p}}\left(\mathrm{S}^n\right),\ \mathrm{Re}\,\alpha > \frac{n}{p},\ 1 < p \leq \infty,\ \mathrm{Re}\,\alpha \neq \frac{n}{p} + m,$$

and

$$K^\alpha\left(L_p\left(\mathrm{S}^n\right)\right) \to \tilde{H}^{\mathrm{Re}\,\alpha - \frac{n}{p}}\left(\mathrm{S}^n\right),\ \mathrm{Re}\,\alpha = \frac{n}{p} + m;$$

$$K^\alpha\left(L_p\left(\mathrm{S}^n, w\right)\right) \to H^{\mathrm{Re}\,\alpha - \frac{n}{p}}\left(\mathrm{S}^n, w\right),\ \mathrm{Re}\,\alpha > \frac{n}{p},\ 1 < p \leq \infty,\ \mathrm{Re}\,\alpha \neq \frac{n}{p} + m,$$

and

$$K^\alpha\left(L_p\left(\mathrm{S}^n, w\right)\right) \to \tilde{H}^{\mathrm{Re}\,\alpha - \frac{n}{p}}\left(\mathrm{S}^n, w\right),\ \mathrm{Re}\,\alpha = \frac{n}{p} + m,$$

where for $0 < \lambda \leq 1$ H^λ stands for the generalized Hölder space on a sphere with $\varphi(t) = t^\lambda$ and $\tilde{H}^\lambda$ denotes $H^\varphi\left(\mathrm{S}^n\right)$ with $\varphi(t) = t^\lambda \ln\frac{3}{t}$.

POTENTIAL TYPE OPERATORS IN GENERALIZED VARIABLE HÖLDER SPACES

Assume $n \geq 2$ and let $\dot{\mathbf{R}}^n$ stand for $\mathbf{R}^n$, compactified following Alexandrov (Kurosh, Markushevich, & Rashevskij), while $\mathrm{S}^n = \left\{x \in \mathbf{R}^{n+1} : |x| = 1\right\}$, $\left|\mathrm{S}^n\right| = 2\pi^{\frac{n}{2}}\Gamma^{-1}\left(\frac{n}{2}\right)$ and $C\left(\Omega, w\right)$ denotes the space of continuous functions with the weight w. Let $\left|\Pi_\alpha\right| = 0$, where $\Pi_\alpha := \left\{x \in \Omega : \mathrm{Re}\left[\alpha\left(x\right)\right] = 0\right\}$.

Variable Order Potential Type Operator

Firstly, let the Zygmund type estimates be considered. Recall, that, in the case of a constant order, the method of Fourier-Laplace multipliers can be applied, but since $\alpha\left(\cdot\right)$ is variable, the correspondent operators are not the spherical convolutions and the Zygmund type estimates provide a technique for studying the operators.

The following theorems have been proposed in (Samko & Vakulov, 2011) and proven with using the results of the spherical geometry and specific inequalities of the number theory:

Theorem 10: Let $f \in C\left(S^n\right)$, let the function $\alpha \in C\left(S^n\right)$ satisfy condition

$$\max_{x \in \mathrm{S}^n} \left|\arg \alpha\left(x\right)\right| < \frac{\pi}{2} - \varepsilon, \ \varepsilon > 0.$$

Then for all points $x, y \in S^{n-1}$ such that $\left|x - y\right| < h$, the following estimate holds:

$$\left|\alpha\left(x\right)\left(I_{\mathrm{S}^n}^{\alpha(\cdot)} f\right)\left(x\right) - \alpha\left(y\right)\left(I_{\mathrm{S}^n}^{\alpha(\cdot)} f\right)\left(y\right)\right| \leq c\left[h^{\mathrm{Re}\left[\alpha(x)\right]} - h^{\mathrm{Re}\left[\alpha(y)\right]}\right] M_r\left(f, x, h\right)$$
$$+ c\, M_r\left(\alpha, x, h\right)\left[\left\|f\right\|_{C(\mathrm{S}^n)} + \int_h^2 \left(\frac{1}{t^{2-\mathrm{Re}\left[\alpha(x)\right]}} + \frac{1}{t^{2-\mathrm{Re}\left[\alpha(y)\right]}}\right) M_r\left(f, x, t\right) dt\right]$$

the constant $c > 0$ does not depend on f, x and h. Under the weak Lipschitz condition ($\alpha \in \mathrm{Lip}\left(\mathrm{S}^n\right)$)

$$\left|\mathrm{Re}\left[\alpha\left(x\right)\right] - \mathrm{Re}\left[\alpha\left(y\right)\right]\right| \leq \frac{A}{\ln \frac{e}{\left|x - y\right|}}, \ x, y \in \mathrm{S}^n, \ A > 0,$$

this estimate is equivalent to

$$M_r\left(\alpha I_{S^n}^{\alpha(\cdot)}, x, h\right) \le ch^{\mathrm{Re}[\alpha(x)]} M_r\left(f, x, h\right) + c\, M_r\left(\alpha, x, h\right)\left[\left\|f\right\|_{C(S^n)} + \int_h^2 \frac{M_r\left(f, x, t\right)}{t^{2-\mathrm{Re}[\alpha(x)]}} dt\right].$$

Theorem 6 provides estimates with the weight $\alpha(\cdot)$ to cover the properties of the potential at the singular points $x \in \Pi_\alpha$, at which the potential does not exist. The following theorem provides another estimate, without weight, but beyond the set Π_α of the degeneracy of $\alpha(\cdot)$. In Theorem 7, in contrast to Theorem 6, purely imaginary values $\alpha(x) = i\theta(x)$ are admitted. The estimate of Theorem 7 is applicable under an assumption on the modulus of continuity of α, which is weaker than $\alpha \in \mathrm{Lip}(S^n)$.

Theorem 11: Let $f \in C(S^n)$ and $\alpha \in C(S^n)$. Then for all points $x, y \in \Pi_\alpha$: $|x - y| < h$, the following estimate holds

$$\left|\left(I_{S^n}^{\alpha(\cdot)} f\right)(x) - \left(I_{S^n}^{\alpha(\cdot)} f\right)(y)\right| \le \frac{c}{\min\left\{\mathrm{Re}\left[\alpha(x)\right], \mathrm{Re}\left[\alpha(y)\right]\right\}}\left[h^{\mathrm{Re}[\alpha(x)]} M_r\left(f, x, h\right) + M_r\left(\alpha, x, h\right)\left\|f\right\|_{C(S^n)}\right].$$

Assume $\omega_\alpha(x,t) := t^{\mathrm{Re}[\alpha(x)]}\omega(x,t)$. In (Samko & Vakulov, 2011, pp. 355-369), Theorems 6 and 7 are used to prove the following results:

Theorem 12: Let $\max\limits_{x\in S^n} \mathrm{Re}\left[\alpha(x)\right] < 1$, $\omega \in \Phi_{1-\mathrm{Re}[\alpha(x)]}$.

$I_{S^n}^{\alpha(\cdot)}$ isboundedfrom $H^{\omega(\cdot)}(S^n)$ to $H^{\omega_\alpha}(S^n, \alpha)$,if $\mathrm{Re}\left[\alpha(x)\right] \ge 0$, $\alpha \in \mathrm{Lip}(S^n)$ and $\max\limits_{x\in S^n}\left|\alpha(x)\right| < \frac{\pi}{2} - \varepsilon$, $\varepsilon > 0$;

$I_{S^n}^{\alpha(\cdot)}$ is bounded from $H^{\omega(\cdot)}(S^n)$ to $H^{\omega_\alpha}(S^n)$, if $\min\limits_{x\in S^n} \mathrm{Re}\left[\alpha(x)\right] > 0$, $M_r(\alpha, x, t) \le Ct^{\mathrm{Re}[\alpha(x)]}\omega(x,t)$.

In the terms of the Matuszewska-Orlicz indices, Theorem 4 can be represented as

Theorem 13: Let $\omega \in W\left(S^n \times [0;2]\right)$ and $m(\omega_\alpha) < 1$.

$I_{S^n}^{\alpha(\cdot)}$ is bounded from $H^{\omega(\cdot)}(S^n)$ to $H^{\omega_\alpha}(S^n, \alpha)$, if $\mathrm{Re}\left[\alpha(x)\right] \ge 0$, $\max\limits_{x\in S^n} \mathrm{Re}\left[\alpha(x)\right] < 1$, $\alpha \in \mathrm{Lip}(S^n)$ and $\max\limits_{x\in S^n}\left|\alpha(x)\right| < \frac{\pi}{2} - \varepsilon$, $\varepsilon > 0$;

$I_{S^n}^{\alpha(\cdot)}$ is bounded from $H^{\omega(\cdot)}(S^n)$ to $H^{\omega_\alpha}(S^n)$, if $\min\limits_{x\in S^n} \mathrm{Re}\left[\alpha(x)\right] > 0$, $\max\limits_{x\in S^n} \mathrm{Re}\left[\alpha(x)\right] < 1$ and $\alpha \in \mathrm{Lip}(S^n)$.

The stereographic mapping is used in (Vakulov & Drobotov, 2017) to extend the represented results to the case of the spatial potential type operator. Let it be recalled, that the stereographic mapping is a transformation of $\dot{\mathbf{R}}^n$ to a unit sphere $S^n \subset \mathbf{R}^{n+1}$: $\xi = v(x)$, where the vector function $v(\cdot)$ is defined by the following formulae:

$$\xi_k = \frac{2x_k}{|x|^2+1},\ k = 1,2,\ldots,n;\ \xi_{n+1} = \frac{|x|^2-1}{|x|^2+1},$$

where $\xi \in S^n$, $x \in \mathbf{R}^n$. The following formulae hold for metrics and volume elements:

$$\rho(\xi,\sigma) := |\xi - \sigma| = \frac{2|x-y|}{\sqrt{|x|^2+1}\sqrt{|y|^2+1}} =: r_{\Pi}(x,y),\ d\sigma = \frac{2^n dy}{\left(|y|^2+1\right)^n},$$

where $x, y \in \dot{\mathbf{R}}^n$, $\xi = v(x)$, $\sigma = v(y)$ and $\xi, \sigma \in S^n$.

Thus, the following results hold for the spatial potential type operator:

Theorem 14: Let $\max_{x\in\dot{\mathbf{R}}^n} \mathrm{Re}\left[\alpha(x)\right] < 1$, $\omega \in \Phi_{1-\mathrm{Re}[\alpha(x)]}$ and

$$c_0(x,y) = \left(1+|y|^2\right)^{-\frac{n+\alpha(x)}{2}},\ w_0(x) = 2^{\alpha(x)}\left(1+|x|^2\right)^{\frac{n-\alpha(x)}{2}}.$$

$I_{\dot{\mathbf{R}}^n}^{\alpha(\cdot)} c_0$ is bounded from $H^{\omega(\cdot)}\left(\dot{\mathbf{R}}^n\right)$ to $H^{\omega_\alpha}\left(\dot{\mathbf{R}}^n, w_0\alpha\right)$, if $\mathrm{Re}\left[\alpha(x)\right] \geq 0$, $\alpha \in \mathrm{Lip}\left(\dot{\mathbf{R}}^n\right)$ and $\max_{x\in\dot{\mathbf{R}}^n}\left|\alpha(x)\right| < \frac{\pi}{2} - \varepsilon$, $\varepsilon > 0$;

$I_{\dot{\mathbf{R}}^n}^{\alpha(\cdot)} c_0$ is bounded from $H^{\omega(\cdot)}\left(\dot{\mathbf{R}}^n\right)$ to $H^{\omega_\alpha}\left(\dot{\mathbf{R}}^n\right)$, if $\min_{x\in\dot{\mathbf{R}}^n} \mathrm{Re}\left[\alpha(x)\right] > 0$, $M_{r_\Pi}(\alpha,x,t) \leq Ct^{\mathrm{Re}[\alpha(x)]}\omega(x,t)$.

Theorem 15: Let $\omega \in W\left(\dot{\mathbf{R}}^n \times [0;2]\right)$ and $m(\omega_\alpha) < 1$.

$I_{\dot{\mathbf{R}}^n}^{\alpha(\cdot)} c_0$ is bounded from $H^{\omega(\cdot)}\left(\dot{\mathbf{R}}^n\right)$ to $H^{\omega_\alpha}\left(\dot{\mathbf{R}}^n, w_0\alpha\right)$, if $\mathrm{Re}\left[\alpha(x)\right] \geq 0$, $\max_{x\in\dot{\mathbf{R}}^n} \mathrm{Re}\left[\alpha(x)\right] < 1$, $\alpha \in \mathrm{Lip}\left(\dot{\mathbf{R}}^n\right)$ and $\max_{x\in\dot{\mathbf{R}}^n}\left|\alpha(x)\right| < \frac{\pi}{2} - \varepsilon$, $\varepsilon > 0$;

$I_{\dot{\mathbf{R}}^n}^{\alpha(\cdot)} c_0$ is bounded from $H^{\omega(\cdot)}\left(\dot{\mathbf{R}}^n\right)$ to $H^{\omega_\alpha}\left(\dot{\mathbf{R}}^n, w_0\alpha\right)$, if $\min_{x\in\dot{\mathbf{R}}^n} \mathrm{Re}\left[\alpha(x)\right] > 0$, $\max_{x\in\dot{\mathbf{R}}^n} \mathrm{Re}\left[\alpha(x)\right] < 1$ and $\alpha \in \mathrm{Lip}\left(\dot{\mathbf{R}}^n\right)$.

REFERENCES

Chegolin, A. P. (2011). - estimates for multideminsional integral operators with the spherical harmonic in oscillating symbol. *Izv. Vuz. Sev.-. Kav. Reg.*, *4*, 3–7.

Chegolin, A. P., & Nogin, V. A. (2001). On the L-characteristic of some multiplier operators with oscillating symbols in the non-elliptic case. *Fractional Calculus & Applied Analysis*, *4*(1), 67–80.

Chegolin, A. P., & Nogin, V. A. (2002). Multidimensional potential integral transforms and description of L-characteristic of multiplier operators. *Integral Transforms and Special Functions*, *13*(3), 193–197. doi:10.1080/10652460213512

Chegolin, A. P., & Nogin, V. A. (2013). The complex powers of some differentional operators in spaces of p-integrable functions. *Differential Equations*, *34*(3), 402–409.

Dzhafarov, A. S. (1966). Nailuchshee priblizhenie konechnymi sfericheskimi summami i nekotorye differencial'nye svojstva garmonicheskih v share funkcij. In L. Kudryavcev (Ed.), *Teoremy vlozhenija i ih prilozhenija. Trudy simpoziuma po teoremam vlozheniya. Baku, 1966 god* (pp. 75–81). Moscow, Russia: Nauka.

Dzhafarov, A. S. (1985). Konstruktivnoe opisanie obobshhennyh klassov Besova na mnogomernoj sfere. *DAN SSSR*, *285*(3), 542–546.

Ginsburg, A. I., & Karapetyants, N. K. (1994). Drobnoe integrodifferencirovanie v gel'derovskih klassah peremennogo porjadka. *Доклады российской академии наук*, *339*(4), 439–441.

Karapetyants, N. K., & Ginsburg, A. I. (1995). Fractional integrodifferentiation in Hölder classes of arbitrary order. *Georgian Mathematical Journal*, *2*(2), 141–150. doi:10.1007/BF02257475

Kurosh, A. G., Markushevich, A. I., & Rashevskij, P. K. (Eds.). (1948). *Matematika v SSSR za 30 let (1917-1947)*. Moscow, Russia: OGIZ.

Landkof, N. S. (1973). *Foundations of Modern Potential Theory*. Providence, RI: Springer.

Nikolsky, S. M., & Lizorkin, P. I. (1984). Priblizhenie spericheskimi polinomami. *Tr. MIAN SSSR*, *166*, 186–200.

Odzijewicz, T., Malinowska, A. B., & Torres, D. F. M. (2013). Fractional variational calculus of variable order. In A. Almeida, L. Castro, & F.-O. Speck (Eds.), *Advances in Harmonic Analysis and Operator Theory: The Stefan Samko Anniversary Volume* (pp. 291–301). Basel, Switzerland: Birkhäuser. doi:10.1007/978-3-0348-0516-2_16

Ross, B., & Samko, S. G. (1995). Fractional integration operator of a variable order in the Holder spaces $H^{\lambda}(S_{n-1})$. *International Journal of Mathematics and Mathematical Sciences*, *18*(4), 777–788. doi:10.1155/S0161171295001001

Samko, N., & Vakulov, B. (2011). Spherical fractional and hypersingular integrals of variable order in generalized Holder spaces with variable characteristic. *Mathematische Nachrichten*, *2-3*(284), 355–369. doi:10.1002/mana.200810113

Samko, S. G. (1995). Fractional integration and differentiation of variable order. *Analysis Mathematica, 21*(3), 213–236. doi:10.1007/BF01911126

Samko, S. G. (1998). Differentiation and integration of variable order and the spaces $L^{p(x)}$. *Contemporary Mathematics, 212*, 203–219. doi:10.1090/conm/212/02884

Samko, S. G., Kilbas, A. A., & Marichev, O. I. (1993). *Fractional integrals and derivatives: theory and applications*. Philadelphia, PA: Gordon and Breach Science Publishers.

Samko, S. G., & Ross, B. (1993). Integration and differentiation to a variable fractional order. *Integral Transforms and Special Functions, 1*(4), 277–300. doi:10.1080/10652469308819027

Samko, S. G., & Vakulov, B. G. (2000). On equivalent norms in fractional order function spaces of continuous functions on the unit sphere. *Fractional Calculus & Applied Analysis, 3*(4), 401–433.

Tarasov, V. E. (2010). *Fractional Dynamics: Applications of Fractional Calculus to Dynamics of Particles, Fields and Media*. Berlin, Germany: Springer. doi:10.1007/978-3-642-14003-7

Vakulov, B. G. (1986). Operator tipa potenciala na sfere v obobshhennyh klassah Gjol'dera. *Izv.vuzov. Matem., 11*, 66–69.

Vakulov, B. G. (2001). O dejstvii operatora rissova potenciala kompleksnogo porjadka po Rn v vesovyh prostranstvah Gjol'dera. *Izvestija vuzov. Severo-Kavkazskij region. Estestvennye nauki, 4*, 47-49.

Vakulov, B. G. (2005). Sfericheskie operatory tipa potenciala v vesovyh prostranstvah Gjol'dera peremennogo porjadka. *Vladikavkazskij matematicheskij zhurnal, 7*(2), 26-40.

Vakulov, B. G., & Drobotov, Yu. E. (2017). Variable order Riesz potential over Ŕn on weighted generalized variable Hölder spaces [Russian, English abstract]. *Siberian Electronic Mathematical Reports, 14*, 647–656.

Vakulov, B. G., & Karapetjanc, N. K. (2003). Operatory tipa potenciala na sfere s osobennostjami na poljusah. *Doklady Akademii Nauk, 392*(2), 151–154.

Vakulov, B. G., Karapetjanc, N. K., & Shankishvili, L. D. (2003). Operatory sfericheskoj svertki so stepenno-logarifmicheskim jadrom v obobshhennyh prostranstvah Gjol'dera. *Izv. vuzov. Matem, 2*, 3–14.

Vakulov, B. G., & Kochurov, E. S. (2010). Operatory drobnogo integrirovanija i differencirovanija peremennogo porjadka v prostranstvah Gjol'dera $H^{\omega(x,t)}$. *Vladikavkazskij matematicheskij zhurnal, 12*(4), 3-11.

Vakulov, B. G., & Kochurov, E. S. (2011). Ocenki tipa Zigmunda dlja operatorov drobnogo integrirovanija i di_erencirovanija peremennogo porjadka. *Izv. vuzov. Sev. Kavk. region. Estestv. nauki. Specvypusk*, 15-17.

Vakulov, B. G., Kochurov, E. S., & Samko, N. G. (2011). Ocenki tipa Zigmunda dlja operatorov drobnogo integrirovanija i differencirovanija peremennogo porjadka. *Izv. vuzov. Matematika, 6*, 25–34.

Vakulov, B. G., & Samko, S. G. (1987). Ob jekvivalentnyh normirovkah v prostranstvah funkcij drobnoj gladkosti na sfere tipa $C^{\lambda}(S_{n-1})$, $H^{\lambda}(S_{n-1})$. *Izv. vuzov. Matem, 12*, 68–71.

Werens, M. (1981). Best approximation on the unit sphere in Rk. In P. Butzer, E. Gärlich, & B. Szökefalvi-Nagy (Eds.), *Proceedings of the Conference held at the Mathematical Research Institute at Oberwolfach Black Forest* (pp. 233-245). Basel, Switzerland: Birkhäuser.

APPENDIX: KEY TERMS AND DEFINITIONS

Gegenbauer Polynomials

The Gegenbauer polynomials are solutions to the Gegenbauer differential equation for integer n, which can be defined in the terms of the Jacobi polynomials:

$$C_n^{(\lambda)}(x) = \frac{\Gamma\left(\lambda + \frac{1}{2}\right)}{\Gamma(2\lambda)} \frac{\Gamma(n + 2\lambda)}{\Gamma\left(n + \lambda + \frac{1}{2}\right)} P_n^{\left(\lambda - \frac{1}{2}, \lambda - \frac{1}{2}\right)}(x),$$

where the Jacobi polynomials can be written in the form:

$$P_n^{(\alpha,\beta)} = \frac{\Gamma(2n + \alpha + \beta + 1)}{n!\Gamma(n + \alpha + \beta + 1)} G_n\left(\alpha + \beta + 1,\ \beta + 1,\ \frac{1}{2}(x + 1)\right),$$

$$G_n(p, q, x) = \frac{n!\Gamma(n + p)}{\Gamma(2n + p)} P_n^{(p-q, q-1)}(2x - 1).$$

Generalized Hölder Space on a Sphere $H_k^{\varphi}(S^n)$

Let D^k be the spherical differentiation operator with a multiplier from $W_{k,N}$, $N \geq \left[\frac{n}{2}\right]$, $D^0 = E$, E is the identity operator, $\omega \in W$. Then:

$$H_k^{\varphi}(\mathrm{S}^n) = \left\{ f \in C(\mathrm{S}^n) : M_r(D^k f, t) \leq c\omega(t) \right\},\ k \in \mathbf{Z}_+,$$

is the generalized Hölder space on a sphere.

Generalized Variable Hölder Space $H^{\omega(\cdot)}(\Omega, w)$ (Weighted)

Let $w(\cdot)$ be a strongly positive function on Ω. The weighted generalized variable Hölder space $H^{\omega(\cdot)}(\Omega, w)$ is a set of functions $f \in C(\Omega, w)$, which satisfy the generalized Hölder condition with a variable characteristics:

$$\forall x \in \Omega,\ t \in \mathbf{R}_+\ M_r\left(w f, x, t\right) \le A \omega\left(x, t\right),\ A = \text{const} > 0 .$$

The norm in $H^{\omega(\cdot)}\left(\Omega, w\right)$ can be defined as:

$$\left\| f \right\|_{H^{\omega(\cdot)}\left(\Omega, w\right)} = \left\| f \right\|_{C\left(\Omega, w\right)} + \sup_{x \in \Omega, t > 0} \frac{M_r\left(w f, x, t\right)}{\omega\left(x, t\right)} .$$

The function $\omega\left(\cdot\right)$ is usually called the characteristics of $H^{\omega(\cdot)}\left(\Omega, w\right)$. Depending on $\omega\left(\cdot\right)$, the generalized Hölder spaces H^{φ}, where $\varphi = \varphi\left(t\right)$, $t \in \mathbf{R}$, and the variable Hölder spaces $H^{\lambda(\cdot)}$, where $\omega\left(x, t\right) = t^{\lambda(x)}$, $t \in \mathbf{R}$, $x \in \Omega$, are defined.

Local Continuity Modulus

Let $r\left(\cdot\right)$ be a metric on Ω, $f\left(\cdot\right)$ be a bounded function on Ω. The local continuity modulus of $f\left(\cdot\right)$ is the functional:

$$M_r\left(f, x, t\right) = \sup_{y \in \Omega : r\left(x, y\right) \le t} \left| f\left(x\right) - f\left(y\right) \right| ,$$

defined for all $t \in \mathrm{T} \subseteq \mathbf{R}_+$ and $x \in \Omega$.

The following notion is used for the Euclidial distance:

$$r\left(x, y\right) = \left| x - y \right| := \sqrt{\sum_{k=1}^{n} \left(x_k - y_k\right)^2} .$$

Matuszewska-Orlicz Indices

The Matuszewska-Orlicz indices of a function $\omega\left(x, t\right)$ with respect to the variable $t \in \left[0; 2\right]$ are:

$$m\left(\omega, x\right) = \sup_{t > 1} \frac{\ln\left[\liminf\limits_{h \to 0} \frac{\omega\left(t h, x\right)}{\omega\left(x, h\right)}\right]}{\ln t} = \lim_{t \to 0} \frac{\ln\left[\limsup\limits_{h \to 0} \frac{\omega\left(t h, x\right)}{\omega\left(x, h\right)}\right]}{\ln t} ,$$

$$M\left(\omega, x\right) = \inf_{t > 1} \frac{\ln\left[\limsup\limits_{h \to 0} \frac{\omega\left(t h, x\right)}{\omega\left(x, h\right)}\right]}{\ln t} = \lim_{t \to \infty} \frac{\ln\left[\limsup\limits_{h \to 0} \frac{\omega\left(t h, x\right)}{\omega\left(x, h\right)}\right]}{\ln t} ,$$

depending on $x \in \Omega$, $m(\omega, x) \leq M(\omega, x)$.

Multiplier

Let K be a spherical convolution operator. The Fourier-Laplace multiplier of K is the sequence $\{k_m\}_{m=0}^{\infty}$, such that:

$$k_m = |\mathrm{S}^{n-1}| \int_{-1}^{1} k(t) P_m(t) (1-t^2)^{\frac{n-2}{2}} dt, \; P_m(t) = \left(C_{m+n-2}^{(m)}\right)^{-1} C_m^{\left(\frac{n-1}{2}\right)}(t).$$

Spherical Convolution Operator

The spherical convolution operator, invariant with respect to all rotations on a sphere, is defined as follows:

$$(Kf)(x) = \int_{\mathrm{S}^n} k(x\sigma) f(\sigma) d\sigma, \; x \in \mathrm{S}^n.$$

$W_{\lambda,N}$ Class

Let $W_{\lambda,N}$ be the class of multipliers $\{k_m\}_{m=0}^{\infty}$ through the spherical harmonics, $|k_m| < \infty$, which allows the following asymptotics:

$$k_m = \sum_{j=0}^{N} c_j m^{\lambda-j} + O\left(m^{\lambda-N-\varepsilon}\right), m \to \infty,$$

where $c_0 \neq 0$, $0 < \varepsilon < 1$, $\lambda \in \mathbf{R}$ and $N = 0,1,\ldots$.

Zygmund-Bary-Stechkin Class

Let $\Omega \subseteq \dot{\mathbf{R}}^n$, $T = \Omega \times [0;l]$, $l \in \mathbf{R}$, and

$$\omega_-(t) = \inf_{x\in\Omega} \omega(x,t), \; \omega_+(t) = \sup_{x\in\Omega} \omega(x,t).$$

A function $\omega : T \to \mathbf{R}_+$ belongs to the class $W = W(T)$, if

- $\omega(x,t)$ is continuous in $t \in [0;l]$ for all $x \in \Omega$;
- $\omega(x,t)$ is almost increasing in $t \in [0;l]$ for all $x \in \Omega$;
- $\omega_-(t) > 0$ for $t > 0$, and $\lim_{t\to+0} \omega(x,t) = 0$ for all $x \in \Omega$.

A function $\omega(\cdot)$ belongs to the generalized Zygmund-Bary-Stechkin class $\Phi_{\beta(\cdot)}^{\delta(\cdot)} = \Phi_{\beta(\cdot)}^{\delta(\cdot)}(T)$, $0 \le \delta(x) < \beta(x)$, $x \in \Omega \subseteq \mathbf{R}^n$, if $\omega \in W$ and

$$\int_0^h \left(\frac{h}{t}\right)^{\delta(x)} \frac{\omega(x,t)}{t} dt \le c\,\omega(x,h);$$

$$\int_h^l \left(\frac{h}{t}\right)^{\beta(x)} \frac{\omega(x,t)}{t} dt \le c\,\omega(x,h)$$

where $0 < h < \frac{l}{2}$ and $c > 0$ is independent on $h \in \left(0; \frac{l}{2}\right]$ and $x \in \Omega \subseteq R^n$.

Compilation of References

Abel-Koch, J., Del Bufalo, G., Fernandez, M., Gerstenberger, J., Lo, V., Navarro, B., & Thornary, B. (2015). SME Investment and Innovation France, Germany, Italy and Spain. Maisons-Alfort Cedex, France: Bpifrance (BPI).

Abrosimova, Dissanaike, & Linowski. (2002). *Testing Weak-Form Efficiency of the Russian Stock Market*. EFA 2002 Berlin Meetings Presented Paper.

Aktan, C., & Huseyin, C. (2001). Ekonomik Kriz; Nedenler ve Çözüm Önerileri. *Yeni Türkiye Dergisi, Ekonomik Kriz Özel Sayısı. Y, 7*, 42.

Al Janabi, M. A. M. (2011a). A generalized theoretical modeling approach for the assessment of economic capital under asset market liquidity risk constraints. *Service Industries Journal, 31*(13), 2193–2221. doi:10.1080/02642069.2010.503878

Al-Deehani, T., & Moosa, I. A. (2006). Volatility Spillover in Regional Emerging Stock Markets: A Structural Time-Series Approach. *Emerging Markets Finance & Trade, 42*(4), 78–89. doi:10.2753/REE1540-496X420404

Alkaraan, F., & Northcott, D. (2006). Strategic capital investment decision making: A role for emergent analysis tool? A study of practice in large UK manufacturing companies. *The British Accounting Review, 38*, 140–173. doi:10.1016/j.bar.2005.10.003

Allen, D. E., Singh, A. K., & Pownell, R. (2011). Extreme market risk-An extreme value theory approach. *Econometrics Society Australasian Meeting*, 1-26.

Altaylıgil, B. (2008). Portföy seçimi için ortalama-varyans-çarpıklık modeli. *İstanbul Üniversitesi İşletme Fakültesi Dergisi, 37*(2), 65-78.

Altman, E. (1968). Financial ratios, discriminant analysis and the prediction of corporate bankruptcy. *The Journal of Finance, 23*(4), 589–609. doi:10.1111/j.1540-6261.1968.tb00843.x

Altman, E. (1984). A further empirical investigation of the bankruptcy cost question. *The Journal of Finance, 39*(4), 1067–1089. doi:10.1111/j.1540-6261.1984.tb03893.x

Andersen, T. G., & Bollerslev, T. (1997). Heterogeneous information arrivals and return volatility dynamics: Uncovering the long-run in high frequency returns. *The Journal of Finance, 52*(3), 975–1005. doi:10.1111/j.1540-6261.1997.tb02722.x

Andersen, T. G., & Bollerslev, T. (1998). Answering the skeptics: Yes, standard volatility models do provide accurate forecasts. *International Economic Review, 39*(4), 885–905. doi:10.2307/2527343

Anderson, R. G., Bordo, M., & Duca, J. V. (2017). Money and velocity during financial crises: From the Great Depression to the Great Recession. *Journal of Economic Dynamics & Control, 81*, 32–49. doi:10.1016/j.jedc.2017.03.014

Andrade, G., & Kaplan, S. (1998). How costly is financial (not economic) distress? Evidence from highly leveraged transactions that become distressed. *The Journal of Finance, 53*(5), 1443–1493. doi:10.1111/0022-1082.00062

Andronache, I. C., Peptenatu, D., Ciobotaru, A.-M., Gruia, A. K., & Gropoşilă, N. M. (2016). Using Fractal Analysis in Modeling Trends in the National Economy. *Procedia Environmental Sciences*, *32*, 344–351. doi:10.1016/j.proenv.2016.03.040

Ang, J. S. (1991). Small business uniqueness and the theory of financial management. *Journal of Small Business Finance*, *1*(1), 11–13.

Ang, J. S. (1992). On the theory of finance for privately held firms. *Journal of Small Business Finance*, *1*(3), 185–203.

Ang, J. S., Chua, J. H., & McConnell, J. J. (1982). The administrative costs of corporate bankruptcy: A note. *The Journal of Finance*, *37*(1), 219–226. doi:10.1111/j.1540-6261.1982.tb01104.x

Asquith, P., Gertner, R., & Scharfstein, D. (1994). Anatomy of financial distress: An examination of junk-bond issuers. *The Quarterly Journal of Economics*, *109*(3), 625–658. doi:10.2307/2118416

Avery, C., & Zemsky, P. (1998). Multidimensional uncertainty and herd behavior in financial markets. *The American Economic Review*, *88*, 724–748.

Bachelier, L. (1900). *Theorie de la Speculation*. Paris: Gauthier-Villars.

Bachelier, L. (1964). *Theory of Speculation. In P. H. Cootner (Ed.), Random Character of Stock Market Prices*. Cambridge, MA: The MIT Press.

Baestaens, D.E., Van Den Berg, W.M., & Wood, D. (1997). Neural Network solutions for trading in financial markets. *Pitman Bublished*, 56-77.

Baillie, R. T., Bollerslev, T., & Mikkelsen, H. O. (1996). Fractionally Integrated Generalized Autoregressive Conditional Heteroskedasticity. *Journal of Econometrics*, *73*(1), 3–20. doi:10.1016/S0304-4076(95)01749-6

Balios, D., Eriotis, N., Fragoudaki, A., & Giokas, D. (2015). Economic efficiency of Greek retail SMEs in a period of high fluctuations in economic activity: A DEA approach. *Applied Economics*, *47*(33), 3577–3593. doi:10.1080/00036846.2015.1019033

Bali, T. G., & Neftçi, S. N. (2003). Disturbing extremal behaviour of spot price dynamics. *Journal of Empirical Finance*, *10*(4), 455–477. doi:10.1016/S0927-5398(02)00070-1

Bandler, R., & Grinder, J. (1975). *The Structure of Magic I: A Book About Language and Therapy*. Palo Alto, CA: Science & Behavior Books.

Bandler, R., & Grinder, J. (1979). *Frogs into Princes: Neuro Linguistic Programming*. Moab, UT: Real People Press.

Barber, B. M., & Odean, T. (2000). Trading is Hazardous to Your Wealth: The Common Stock Investment Performance of Individual Investors. *The Journal of Finance*, *55*(2), 773–806. doi:10.1111/0022-1082.00226

Barberis, N., & Huang, M. (2001). Mental accounting, loss aversion and individual stock returns. *The Journal of Finance*, *56*(4), 1247–1292. doi:10.1111/0022-1082.00367

Bardos, S. (2008). Academy of economic studies. Bucharest Doctoral School of Finance and Banking, DOFIN, Bucharest.

Barkoulas, J. T., Baum, C. F., & Travlos, N. (2000). Long memory in the Greek stock market. *Applied Financial Economics*, *10*(2), 177–184. doi:10.1080/096031000331815

Beaver, W. H. (1966). Financial ratios as predictors of failures. *Journal of Accounting Research*, *4*, 71–111. doi:10.2307/2490171

Beck, T., Demirguc-Kunt, A., & Maksimovic, V. (2002). Financing patterns around the world: The role of institutions. *Proceedings of World Bank, 1*, 1–40.

Bekaert, G., & Harvey, C. (1995). Time-varing world market integration. *The Journal of Finance, 50*(2), 403–444. doi:10.1111/j.1540-6261.1995.tb04790.x

Bekaert, G., & Harvey, C. (1997). Emerging equity market volatility. *Journal of Financial Economics, 48*(1), 29–78. doi:10.1016/S0304-405X(96)00889-6

Bekaert, G., & Harvey, C. (2003). Emerging markets finance. *Journal of Empirical Finance, 10*(1-2), 3–55. doi:10.1016/S0927-5398(02)00054-3

Bekaert, G., Harvey, C., & Lundblad, C. (2007). Liquidity and expected returns: Lessons from emerging markets. *Journal of Financial Studies, 20*(6), 1783–1831. doi:10.1093/rfs/hhm030

Bekaert, G., & Hodrick, R. (1992). Characterizing predictable components in excess returns on equity and foreign exchange markets. *The Journal of Finance, 47*(2), 467–509. doi:10.1111/j.1540-6261.1992.tb04399.x

Bell, D. (1982). Regret in decision making under uncertainty. *Operations Research, 30*(5), 961–981. doi:10.1287/opre.30.5.961

Berg, A. (1999). *The Asia Crisis, Causes, Policy Responses and Outcomes.* NBER Working Paper.

Berger, A. N., & Udell, G. E. (2005). *A More Complete Conceptual Framework for Financing of Small and Medium Enterprises.* Policy Research Working Paper- 4481. The World Bank.

Berger, A. N., & Bouwman, C. (2013). How does capital affect bank performance during financial crises? *Journal of Financial Economics, 109*(1), 146–176. doi:10.1016/j.jfineco.2013.02.008

Berger, A. N., Frame, W. S., & Miller, N. H. (2005). Credit scoring and the availability, price, and risk of small business credit. *Journal of Money, Credit and Banking, 37*(2), 191–222. doi:10.1353/mcb.2005.0019

Berger, A. N., & Mester, L. J. (1997). Inside the Black Box: What explains differences in the efficiencies of financial institutions? *Journal of Banking & Finance, 21*(7), 895–947. doi:10.1016/S0378-4266(97)00010-1

Berger, A. N., & Udell, G. F. (1998). The economics of small business finance: The roles of private equity and debt markets in the financial growth cycle. *Journal of Banking & Finance, 22*(6-8), 613–673. doi:10.1016/S0378-4266(98)00038-7

Berger, A. N., & Udell, G. F. (2002). Small business credit availability and relationship lending: The importance of bank organisational structure. *Economic Journal (London), 112*(477), 32–53. doi:10.1111/1468-0297.00682

Berger, A. N., & Udell, G. F. (2006). A more complete conceptual framework for SME finance. *Journal of Banking & Finance, 30*(11), 2945–2966. doi:10.1016/j.jbankfin.2006.05.008

Berry, J. (2004). *Tangible Strategies for Intangible Assets.* New York, NY: McGraw-Hill Professional.

Bharat, S. T. (2013). *Investment Behaviour of Individual Investors in the Stock Market of Nepal.* Academic Press.

Bhattacharryya, M., & Ritolia, G. (2008). Conditional VaR using EVT-towards: A planned margin scheme. *International Review of Financial Analysis, 17*(2), 382–395. doi:10.1016/j.irfa.2006.08.004

Bhattacharya, S. N., & Bhattacharya, M. (2012). Long Memory in Stock Returns: A study of emerging markets. *Iranian Journal of Management Studies, 5*, 67–88.

Billio, M., & Pelizzon, L. (2000). Value-at-risk: A multivariate switching regime Approach. *Journal of Empirical Finance, 7*(5), 531–554. doi:10.1016/S0927-5398(00)00022-0

Black, F., & Scholes, M. (1973). The Pricing of Options and Corporate Liabilities. *Journal of Political Economy*, *81*, 637–654.

Black, F., & Scholes, M. S. (1973). the Pricing of Options and Corporate Liabilities. *Journal of Political Economy*, *81*(3), 637–659. doi:10.1086/260062

Blavy, R. (2002, Autumn). Changing volatility in emerging markets: A case study of two Middle Eastern stock exchanges. *Revue Entente Cordiale*, 1-35.

Bollerslev, T. (1986). Generalized autoregressive conditional heteroskedasticity. *Journal of Econometrics*, *31*(3), 307–327. doi:10.1016/0304-4076(86)90063-1

Bontisa, N., Dragonetti, N. C., Jacobsen, K., & Roosc, G. (1999). The knowledge toolbox: A review of the tools available to measure and manage intangible resources. *European Management Journal*, *17*(4), 391–402.

Boot, A. W. A. (2000). Relationship banking: What do we know? *Journal of Financial Intermediation*, *9*(1), 7–25. doi:10.1006/jfin.2000.0282

Boot, A. W. A., & Thakor, A. V. (2000). Can relationship banking survive competition. *The Journal of Finance*, *55*(2), 679–713. doi:10.1111/0022-1082.00223

Booth, L., Aivazian, V., Demirguc-Kunt, A., & Maksimovic, V. (2001). Capital structures in developing countries. *The Journal of Finance*, *56*(1), 87–130. doi:10.1111/0022-1082.00320

Bordo, M. D., & Haubrich, J. G. (2017). Deep recessions, fast recoveries, and financial crises: Evidence from the American record. *Economic Inquiry*, *55*(1), 527–541. doi:10.1111/ecin.12374

Bouchaud, J. P., & Potters, M. (2000). *Theory of Financial Risks. From Statistical Physics to Risk Management*. Cambridge, UK: Cambridge University Press.

Breidt, F. J., Crato, N., & de Lima, P. (1998). On the detection and estimation of long memory in stochastic volatility. *Journal of Econometrics*, *83*(1-2), 325–348. doi:10.1016/S0304-4076(97)00072-9

Brenner, Y. S., Kaelble, H., & Thomas, M. (1991). *Income Distribution in Historical Perspective*. Cambridge, UK: Cambridge University Press.

BRICS Economic Partnership Strategy. (2015). Retrieved April 15, 2017, from http://static.kremlin.ru/media/events/files/ru/KT0SBHnIZjOpIuAj2AOXCnszNQA8u7HL.pdf

Brown, E. (2009). *The Wall Street Ponzi Scheme called Fractional Reserve Banking, Borrowing from Peter to Pay Paul*. Retrieved from www.globalresearch.ca/index.php?context=va&aid=11600

Brunnermeier, M. K. (2016). *"Bubbles." Banking Crises*. Palgrave Macmillan.

Bry, G., & Boschan, C. (1971). *Cyclical Analysis of Time Series: Selected Procedures and Computer Programs*. Technical Paper 20. Retrieved from http://papers.nber.org/books/bry_71-1

Büberkökü, Ö. (2013). Kriz Döneminde Yükselen Piyasa Ekonomileri, Euro Bölgesi ve ABD piyasaları Arasındaki Volatilite Yayılmasının İncelenmesi: Varyansta-Granger-Nedensellik Testinden Kanıtlar. *EY International Congress on Economics I "Europe and Global Economic Rebalancing"*, Ankara, Turkey.

Buckmaster, N. (2000). The performance measurement panacea. *Accounting Forum*, *24*(3), 264–277. doi:10.1111/1467-6303.00041

Bugorskiy, V.N., & Sergienko, A.G. (2008). Ispolzovanie neyronnyh setey dlya modelirovaniya prognoza kotirovok tsennyh bumag. *Prikladnaya Informatika, 3*.

Bulot, N., Salamudin, N., & Abd Aziz, R. (2015). Size of indirect financial distress costs: Which factors are reliably important? *Jurnal Intelek, 12*(1), 12–20.

Burnside, C., Eichenbaum, M., & Rebello, S. (2001). Prospective Deficits and the Asian Currency Crisis. *Journal of Political Economy, 109*(6), 1155–1197. doi:10.1086/323271

Bushuev, K. (2000). Primeneniene yrosetevyh metodov pri prognozirovanii dinamiki rynka aktsiy. Modeli ekonomicheskih system I informatsionnye tehnologii. *Finansovay aakademiya pri Pravitelstve RF*, 30-35.

Bustelo, P., Garcia, C., & Olive, I. (1999). *Global and Domestic Factors of Financial Crises in Emerging Economies: Lessons from the East Asian Episodes (1997-1999)*. Retrieved from http://papers.ssrn.com/sol3/papers.cfm?abstract_id=201150

Byström, H. (2004). Managing extreme risks in tranquil and volatile markets using conditional extreme value theory. *International Review of Financial Analysis, 13*(2), 133–152. doi:10.1016/j.irfa.2004.02.003

Campello, M., & Giambona, E. (2011). *Capital structure and the redeployability of tangible assets*. Retrieved November 30, 2011, from https://papers.tinbergen.nl/11091.pdf

Carbone, A., Kaniadakis, G., & Scarfone, A. M. (2007). Where do we stand on econophysics? *Physica A, 382*(1), xi–xiv. doi:10.1016/j.physa.2007.05.054

Carter, D. A., & McNulty, J. E. (2005). Deregulation, technological change, and the business-lending performance of large and small banks. *Journal of Banking & Finance, 29*(5), 1113–1130. doi:10.1016/j.jbankfin.2004.05.033

Casey, E., & O'Toole, C. (2014). Bank lending constraints, trade credit and alternative financing during the financial crisis: Evidence from European SMEs. *Journal of Corporate Finance, 27*, 173–193. doi:10.1016/j.jcorpfin.2014.05.001

Castiglionesi, F., Feriozzi, F., & Lorenzoni, G. (2017). *Financial integration and liquidity crises. No. w23359*. National Bureau of Economic Research. doi:10.3386/w23359

Catasis, B., & Gröjer, J. E. (2003). Intangibles and credit decisions: Results from an experiment. *European Accounting Review, 12*(2), 327–356. doi:10.1080/0963818032000089418

Cavalcante, J., & Assaf, A. (2005). Long-range dependence in the returns and volatility of the Brazilian stock market. *European Review of Economics and Finance, 5*, 5–20.

Celasun, O. (1994). Macroeconomics and Growth Group Development Research Department. *The World Bank Review*.

Cesarini, F., & Ferri, G. (1997). *Credit and development: Local cooperative banks and small businesses*. Bologna, Italy: Il Mulino.

Chander, P. (1992). Corruption in tax administration. *Journal of Public Economics, 49*, 333-349.

Chapman, G. B., & Johnson, E. J. (2002). Incorporating the irrelevant: Anchors in judgments of belief and value. In T. Gilovich, D. Griffin, & D. Kahneman (Eds.), Heuristics and biases: The psychology of intuitive judgment (pp. 120–138). Cambridge, UK: Cambridge University Press.

Charest, G. (1978). Dividend Information, Stock Returns and Market Efficiency-II. *Journal of Financial Economics, 6*(2-3), 297–330. doi:10.1016/0304-405X(78)90033-8

Charvet S. (1995). *Words That Change Minds – Mastering the Language of Influence*. Kendall/Hunt Publishing Company.

Chegolin, A. P. (2011). - estimates for multideminsional integral operators with the spherical harmonic in oscillating symbol. *Izv. Vuz. Sev.-. Kav. Reg., 4*, 3–7.

Chegolin, A. P., & Nogin, V. A. (2001). On the L-characteristic of some multiplier operators with oscillating symbols in the non-elliptic case. *Fractional Calculus & Applied Analysis, 4*(1), 67–80.

Chegolin, A. P., & Nogin, V. A. (2002). Multidimensional potential integral transforms and description of L-characteristic of multiplier operators. *Integral Transforms and Special Functions, 13*(3), 193–197. doi:10.1080/10652460213512

Chegolin, A. P., & Nogin, V. A. (2013). The complex powers of some differentional operators in spaces of p-integrable functions. *Differential Equations, 34*(3), 402–409.

Cheung, Y. L., Cheung, Y. W., & Ng, K. C. (2003). *East Asian Equity Markets Financial Crises, and The Japanese Currency*. Retrieved from http://papers.ssrn.com/sol3/papers.cfm?abstract_id=1009369

Cheung, Y., & Lai, K. (1995). A search for long memory in international stock market returns. *Journal of International Money and Finance, 14*(4), 597–615. doi:10.1016/0261-5606(95)93616-U

Ching, C. L., & Yang, J. (2000). Knowledge value chain. *Journal of Management Development, 19*(9), 783–793. doi:10.1108/02621710010378228

Chirkova, E. (2014). *How to evaluate business by analogy*. Alpina Pablisher.

Chlon-Dominczak, A. (2016). Impact of changes in multi-pillar pension systems in CEE countries on individual pension wealth. *Journal of Pension Economics and Finance*, 1–12. doi:10.1017/S1474747216000238

Chomsky, N. (1975). *The Logical Structure of Linguistic Theory*. Springer.

Chybalski, F. (2011). The resilience of pension systems in the CEE countries to financial and economic crisis: the need for higher diversification. In *Proceedings of the 13th International Conference of Finance and Banking* (pp. 258–267). Ostrava, Czech Republic: Silesian University, School of Business Administration. Retrieved from http://www.opf.slu.cz/kfi/icfb/proc2011/pdf/22_Chybalski.pdf

Claessens, S., Honohan, P., & Rojas-Suarez, L. (2009). *Policy Principles for Expanding Financial Access*. CGD. Available at www.cgdev.org/files/1422882_file_Financial_Access_Task_Force_Report_FINAL.pdf

Cohen, J. A. (2005). *Intangible assets: valuation and economic benefit*. Hoboken, NJ: John Wiley & Sons.

Cohen, J. B., Black, F., & Scholes, M. (2012). The valuation of option contracts and a test of market efficiency. *The Journal of Finance, 27*(2), 399–417. doi:10.1111/j.1540-6261.1972.tb00969.x

Cont, R. (1997). *Beyond implied volatility: Extracting information from options prices. Econophysics*. Dordrecht: Kluwer.

Cootner, P. A. (1964). *The Random Character of Stock Market Prices*. Cambridge, MA: MIT Press.

Cowell, F. (1995). Auditing with "ghosts." In F. Cowell & G. F Gordon (Eds.), *The Economics of Organized Crime* (pp. 184-198). Academic Press.

Crafts, N. (1999). East Asian Growth Before and After the Crisis. *IMF Staff Papers, 46*(2), 139–166.

Crato, N. (1994). Some international evidence regarding the stochastic behavior of stock returns. *Applied Financial Economics, 4*(1), 33–39. doi:10.1080/758522123

Crato, N., & de Lima, P. (1994). Long range dependence in the conditional variance of stock returns. *Economics Letters, 45*(3), 281–285. doi:10.1016/0165-1765(94)90024-8

Cucculelli, M., & Bettinelli, C. (2015). Business models, intangibles and firm performance: Evidence on corporate entrepreneurship from Italian manufacturing SMEs. *Small Business Economics, 45*(2), 329–350. doi:10.100711187-015-9631-7

Czarnecki, L., Grech, D., & Pamula, G. (2008). Comparison study of global and local approaches describing critical phenomena on the polish stock exchange market. *Physica A, 387*(27), 6801–6811. doi:10.1016/j.physa.2008.08.019

Dacorogna, M. M., Gençay, R., & Olsen, R. B. (2001). An Introduction to High-Frequency Finance. London: Academic Press.

Dalio, R. (2015). How the Economic Machine Works - Leveragings and Deleveragings. *Economic Principles, How the Economic Machine Works.* Retrieved July 7, 2017, from http://www.economicprinciples.org/wp-content/uploads/ray_dalio__how_the_economic_machine_works__leveragings_and_deleveragings.pdf

Daniel, K., Hirshleifer, D., & Subrahmanyam, A. (2001). Overconfidence, arbitrage and equilibrium asset pricing. *The Journal of Finance, 56*(3), 921–965. doi:10.1111/0022-1082.00350

Daskalakis, N., & Psillaki, M. (2008). Do country or firm factors explain capital structure? Evidence from SMEs in France and Greece. *Applied Financial Economics, 18*(2), 87–97. doi:10.1080/09603100601018864

Dávid, L., & Baros, Z. (2007). A possible use of indicators for sustainable development in tourism Anatolia. *An International Journal of Tourism and Hospitality Research, 18*, 349–355.

Davies, R. B., & Harte, D. S. (1987). Tests for Hurst effect. *Biometrika, 74*(1), 95–101. doi:10.1093/biomet/74.1.95

De Bruyn, R. (2005). Le ragioni delle Banche Popolari: motivi teorici ed evidenze empiriche. Genova, Italy: Disefin.

De Laurentis, G. (2001). *Internal ratings and credit risk management.* Rome, Italy: Bancaria Editrice.

De Laurentis, G. (2011). *Business credit after the crisis.* Rome, Italy: Bancaria Editrice.

De Santis, G., & Imrohoroglu, S. (1997). Stock returns and volatility in emerging financial markets. *Journal of International Money and Finance, 16*(4), 561–579. doi:10.1016/S0261-5606(97)00020-X

DeAngelo, H., & Masulis, R. W. (1980). Optimal capital structure under corporate and personal taxation. *Journal of Financial Economics, 8*(3), 3–29. doi:10.1016/0304-405X(80)90019-7

Debreu, G. (1991). The Mathematization of Economic Theory. *The American Economic Review, 81*(1), 1–7.

Degryse, H., Goeij, P., & Kappert, P. (2010). The impact of firm and industry characteristics on small firms capital structure. *Small Business Economics, 38*(4), 431–447. doi:10.100711187-010-9281-8

Degryse, H., & Ongena, S. (2002). Bank-firm relationships and international banking markets. *International Journal of the Economics of Business, 9*(3), 401–417. doi:10.1080/13571510210000010000A

Delhi Declaration on the results of the 4th Summit on March 29, 2012. (n.d.). Retrieved March 27, 2017, from http: //www.brics.mid.ru/brics.nsf/WEBdocBric/1DB82FF3745BA8FE442579D5004778E6

Derman, E. (2012). *My life as a quant: Reflections on physics and finance.* Wiley.

DeYoung, R., Glennon, D., & Nigro, P. (2008). Borrower-lender distance, credit scoring, and loan performance: Evidence from informational-opaque small business borrowers. *Journal of Financial Intermediation, 17*(1), 113–143. doi:10.1016/j.jfi.2007.07.002

DeYoung, R., Hunter, W. C., & Udell, G. F. (2004). The past, present, and probable future for community banks. *Journal of Financial Services Research, 25*(2-3), 85–133. doi:10.1023/B:FINA.0000020656.65653.79

Diamond, D. (1984). Financial Intermediation and Delegated Monitoring. *The Review of Economic Studies, 51*(3), 393–414. doi:10.2307/2297430

Diamond, D. (1991). Monitoring and reputation: The choice between bank loans and directly placed debt. *Journal of Political Economy, 99*(4), 689–721. doi:10.1086/261775

Didier, S. (2004). A complex system view of why stock markets crash. *New Thesis, 1*(1), 5-18.

Didier, S., & Zhou, W.-X. (2004). Antibubble and prediction of China's stock market and real-estate. *Physica A, 337*(1), 243–268.

Diebold, F. X., & Yilmaz, K. (2008). *Measuring financial asset return and volatilty spillovers, with application to global equity markets.* CFS Working Paper, No.2008/26.

Dilts, R., Grinder, J., Delozier, J., & Bandler, R. (1980). Neuro-Linguistic Programming: Vol. 1. *The Study of the Structure of Subjective Experience.* Cupertino, CA: Meta Publications.

Ding, Z., Granger, C. W. J., & Engle, R. F. (1993). A long memory property of stock market returns and a new model. *Journal of Empirical Finance, 1*(1), 83–106. doi:10.1016/0927-5398(93)90006-D

Disario, R., Saraoglu, H., McCarthy, J., & Li, H. (2008). Long memory in the volatility of an emerging equity market: The case of Turkey. *Journal of International Financial Markets, Institutions and Money, 18*(4), 305–312. doi:10.1016/j.intfin.2007.01.001

Domino, K. (2011). The use of the Hurst exponent to predict changes in trends on the Warsaw Stock Exchange. *Physica A, 390*(1), 98–109. doi:10.1016/j.physa.2010.04.015

Domino, K. (2012). The use of the Hurst exponent to investigate the global maximum of the Warsaw Stock Exchange WIG20 index. *Physica A, 391*(1-2), 156–169. doi:10.1016/j.physa.2011.06.062

Dubovikov, M. M., Starchenko, N. S., & Dubovikov, M. S. (2004). Dimension of the minimal cover and fractal analysis of time series. *Physica A, 339*(3-4), 591–608. doi:10.1016/j.physa.2004.03.025

Dubrovski, D. (2001). The role of customer satisfaction in achieving business excellence. *Total Quality Management, 12*(7-8), 920–925. doi:10.1080/09544120100000016

Dzhafarov, A. S. (1966). Nailuchshee priblizhenie konechnymi sfericheskimi summami i nekotorye differencial'nye svojstva garmonicheskih v share funkcij. In L. Kudryavcev (Ed.), *Teoremy vlozhenija i ih prilozhenija. Trudy simpoziuma po teoremam vlozheniya. Baku, 1966 god* (pp. 75–81). Moscow, Russia: Nauka.

Dzhafarov, A. S. (1985). Konstruktivnoe opisanie obobshhennyh klassov Besova na mnogomernoj sfere. *DAN SSSR, 285*(3), 542–546.

Edison, H. J., Luangaram, P., & Miller, M. M. (1998). *Asset Bubbles, Domino Effects and 'Lifeboats': Elements of the East Asian Crisis.* Retrieved from http://papers.ssrn.com/sol3/papers.cfm?abstract_id=93688

Eichengreen, B. (2001). *Crisis Prevention and Management: Any New Lessons from Argentina and Turkey?* Background Paper for the World Bank's Global Development Finance.

Einstein, A. (1905). On the Movement of Small Particles Suspended in a Stationary Liquid Demanded by the Molecular-Kinetic Theory of Heat. *Annalen der. Physik, 17*, 549–560. doi:10.1002/andp.19053220806

Elder, A. (2007). *How to play and win on the stock exchange.* Alpina Pablisher.

Eldera, J., & Serletis, A. (2008). Long memory in energy futures prices. *Review of Financial Economics, 17*(2), 146–155. doi:10.1016/j.rfe.2006.10.002

Ellsberg, D. (1961). Risk, Ambiguity, and the Savage Axioms. *The Quarterly Journal of Economics, 75*(4), 643–669. doi:10.2307/1884324

Elman, J. L. (1990). Finding structure in time. *Cognitive Science, 14*(2), 179–211. doi:10.120715516709cog1402_1

Emel, A. B., Oral, M., Reisman, A., & Yolalan, R. (2003). A credit scoring approach for the commercial banking sector. *Socio-Economic Planning Sciences, 37*(2), 103–123. doi:10.1016/S0038-0121(02)00044-7

Enders, W. (2004). *Applied Econometric Time Series*. John Wiley & Sons Inc.

Engle, R. F. (1982). Autoregressive conditional heteroskedasticity with estimates of the variance of U.K. inflation. *Econometrica, 50*(4), 987–1008. doi:10.2307/1912773

Engle, R. F., & Bollerslev, T. (1986). Modeling the persistence of conditional variances. *Econometric Reviews, 5*(1), 1–50. doi:10.1080/07474938608800095

Engle, R. F., & Manganelli, S. (2004). CAViaR: Conditional autoregressive Value-at-Risk by regression quantiles. *Journal of Business & Economic Statistics, 22*(4), 367–381. doi:10.1198/073500104000000370

Erdonmez, A, P. & Tulay, B. (1999). Küresel Krizlere Yeni Yaklaşımlar. *Bankacılar Dergisi, S 31*, 64-76.

Evci, S., & Kandır, S. Y. (2015). Altın Piyasasında Piyasa Riskinin Ölçülmesi: Riske Maruz Değer (VAR) Yöntemi İle Bir Uygulama. *Bankacılar Dergisi, 92*, 53–70.

Evstigneev, I. V., Amir, R., Hens, T., & Schenk-Hoppé, K. R. (2013). *Mathematical Behavioral Finance*. Academic Press.

Falkenstein, E., Boral, A., & Carty, L. (2000). *RiskCalc for Private Companies: Moody's Default Model*. Academic Press.

Fama, E. (1970). Efficient capital markets: A review of theory and empirical work. *The Journal of Finance, 25*, 383–417.

Fama, E. (1985). What's different about banks? *Journal of Monetary Economics, 15*(1), 29–39. doi:10.1016/0304-3932(85)90051-0

Fama, E. (1998). Market efficiency, long-term returns, and behavioral finance. *Journal of Financial Economics, 49*(3), 283–306. doi:10.1016/S0304-405X(98)00026-9

Fama, E. F. (1965). The behavior of stock-market prices. *The Journal of Business, 38*(1), 34–105. doi:10.1086/294743

Fama, E. F. (1970). Efficient Capital Markets: A Review of Theory and Empirical Work. *The Journal of Finance*, 47.

Fama, E. F., & Malkiel, B. G. (2012). Efficient Capital Markets: A Review of Theory and Empirical Work. *The Journal of Finance, 25*(2), 383–417. doi:10.2307/2325486

Farjoun, M., & Levin, M. (2011). A Fractal Approach to Industry Dynamism. *Organization Studies, 32*(6), 825–851. doi:10.1177/0170840611410817

Federal Health Reporting Information System. (2016). *Entwicklung der beitragssatze in der sozialversicherung* [Development of contribution rates in social security]. Federal Health Reporting Information System of Germany. Retrieved from http://www.gbe-bund.de/gbe10/abrechnung.prc_abr_test_logon?p_uid=gast&p_aid=0&p_knoten=FID&p_sprache=D&p_suchstring=8862

Financial Times. (2004). *Book review: Memoirs of a physicist on Wall Street, By Stefan Stern*. Financial Times.

Finke, R., & Sabatini, M. (Eds.). (2016). *2016 Pension Sustainability Index. International Pension Papers 1*. Munich, Germany: Allianz SE. Retrieved from http://projectm-online.com/app/uploads/Allianz_2016_Pension_Sustainability_Index.pdf

Flegelskamp, G. (2008). Ubersicht der Rentenbeitragssutze ab 1949 [Contribution rates to social security] [Data file]. Available from http://www.flegel-g.de/beitragssatz.html

Fomicheva, E. V. (2015). Geoeconomic positioning of Russia in the current conditions of international cooperation: New challenges and protection of national interests. South Federal University.

Formisano, V. (2016). *Non-Knowledge Risk and Bank-Company Management. The Role of Intangibles in Rating Models*. Basingstoke, UK: Palgrave Macmillan.

Fox, C., & Tversky, A. (1995). Ambiguity aversion and comparative ignorance. *The Quarterly Journal of Economics*, *110*(3), 585–603. doi:10.2307/2946693

Fox, J. (2009). *Myth of the Rational Market*. Harper Business.

Fraedrich, J. P., & Ferrell, O. C. (1992). The Impact of Perceived Risk and Moral Philosophy Type on Ethical Decision Making in Organizations. *Journal of Business Research*, *24*(4), 283–295. doi:10.1016/0148-2963(92)90035-A

Frank, M., & Goyal, V. (2003). Testing the pecking order theory of capital structure. *Journal of Financial Economics*, *67*(2), 217–248. doi:10.1016/S0304-405X(02)00252-0

Frank, M., & Goyal, V. (2009). Capital structure decisions: Which factors are reliably important? *Financial Management*, *1*(38), 1–37. doi:10.1111/j.1755-053X.2009.01026.x

Friedman, J. H. (1991). Multivariate Adaptive Regression Splines. *Annals of Statistics*, *19*(1), 1–67. doi:10.1214/aos/1176347963

Fukao, K., Miyagawa, T., Mukai, K., Shinoda, Y., & Tonogi, K. (2009). Intangible investment in Japan: Measurement and contribution to economic growth. *Review of Income and Wealth*, *55*(3), 717–736. doi:10.1111/j.1475-4991.2009.00345.x

Galluccio, S., Caldarelli, G., Marsili, M., & Zhang, Y. C. (1997). Scaling in currency exchange. *Physica A*, *245*(3), 423–436. doi:10.1016/S0378-4371(97)00316-6

Gamba-Santamaria, S., Gomez-Gonzalez, J. E., Hurtado-Guarin, J. L., & Melo-Velandia, L. F. (2017). Stock market volatility spillovers: Evidence for Latin America. *Finance Research Letters*, *20*, 207–216. doi:10.1016/j.frl.2016.10.001

Gencay, R., Selcuk, F., & Whitcher, B. (2002). Scaling properties of foreign exchange volatility. *Physica A*, *289*(1-2), 249–266. doi:10.1016/S0378-4371(00)00456-8

Genesove, D., & Mayer, C. (2001). Loss aversion and seller behavior: Evidence from the housing market. *The Quarterly Journal of Economics*, *116*(4), 1233–1260. doi:10.1162/003355301753265561

German Federal Pension Insurance. (2016). *Rentenversicherung in Zahlen 2016* [Pension insurance in figures 2016]. Berlin, Germany: German Federal Pension Insurance. Retrieved from http://www.deutsche-rentenversicherung.de/cae/servlet/contentblob/238692/publicationFile/61815/01_rv_in_zahlen_2013.pdf

Giles, C.L., Lawrence, S., & Tsoi, A.C. (2006). *Rule inference for financial prediction using recurrent neural networks*. Academic Press.

Gilmore, C. G. (1993). A new test for chaos. *Journal of Economic Behavior and Organization, 22*, 209-237.

Gilovich, T., Griffin, D., & Kahneman, D. (Eds.). (2002). *Heuristics and Biases: The Psychology of Intuitive Judgment*. Cambridge University Press. doi:10.1017/CBO9780511808098

Ginsburg, A. I., & Karapetyants, N. K. (1994). Drobnoe integrodifferencirovanie v gel'derovskih klassah peremennogo porjadka. *Доклады российской академии наук*, *339*(4), 439–441.

Gligor, M., & Ignat, M. (2000). Dissipative Structures in the Stock Market Space. *Bulg. Journal of Physics*, *27*.

Glinkina, S. P. (Ed.). (2014). *New direction of the Russian foreign and foreign economic policy - interaction in BRICS*. Moscow: Institute of Economics, Russian Academy of Sciences.

Goddard, J., Molyneux, P., & Wilson, J. (2001). *European banking efficiency, technology and growth*. Chichester, UK: Wiley Finance.

Graham, J. R. (1999). Herding among Investment Newsletters: Theory and Evidence. *The Journal of Finance*, *54*(1), 237–268. doi:10.1111/0022-1082.00103

Granger, C. W., Ding, J. Z., & Spear, S. (1997). *Stylized facts of the temporal and distributional properties of daily speculative markets* (Unpublished manuscript). San Diego, CA.

Granger, C. W. J., & Joyeux, R. (1980). An introduction to long memory time series models and fractional differencing. *Journal of Time Series Analysis*, *1*(1), 15–39. doi:10.1111/j.1467-9892.1980.tb00297.x

Granger, C., & Newbold, N. (1986). *Forecasting economic time series* (2nd ed.). San Diego, CA: Academic Press.

Grech, D., & Mazur, Z. (2004). Can one make any crash prediction in finance using the local hurst exponent idea? *Physica A*, *336*(1-2), 133–145. doi:10.1016/j.physa.2004.01.018

Grech, D., & Pamula, G. (2008). The local hurst exponent of the financial time series in the vicinity of crashes on the polish stock exchange market. *Physica A*, *387*(16-17), 4299–4308. doi:10.1016/j.physa.2008.02.007

Greene, M. T., & Fielitz, B. D. (1977). Long-term dependence in common stock returns. *Journal of Financial Economics*, *4*(3), 339–349. doi:10.1016/0304-405X(77)90006-X

Greenfield, R. (2016). Chatbots Are Your Newest, Dumbest Co-Workers. *Bloomberg*. Retrieved July 11, 2017, from https://www.bloomberg.com/news/articles/2016-05-05/chatbots-are-your-newest-dumbest-co-workers

Greiner, L. E. (1998, May). Evolution and revolution as organizations grow. *Harvard Business Review*, 1–11. Retrieved from http://www.ceriumpartners.com/insights/Ev%20%20Rev%20as%20Orgs%20Grow-%20Greiner.pdf PMID:10179654

Grice, J., & Ingram, R. (2001). Tests of the generalizability of Altman s bankruptcy prediction model. *Journal of Business Research*, *54*(1), 53–61. doi:10.1016/S0148-2963(00)00126-0

Grinin, L. E., Devezas, T. C., & Korotayev, A. V. (2014). *Kondratieff Waves. Juglar – Kuznets – Kondratieff.* Volgograd, Russia: 'Uchitel' Publishing House.

Grinin, L., Korotayev, A., & Tausch, A. (2016). *Economic cycles, crises, and the global periphery*. Springer. doi:10.1007/978-3-319-41262-7

Guardiancich, I. (2010). *Germany. Current pension system: First assessment of reform outcomes and output.* Country Report. European Social Observatory. Retrieved from http://www.ose.be/files/publication/2010/country_reports_pension/OSE_2010_CRpension_Germany.pdf

Guillaume, D. M., Dacorogna, M. M., Davé, R. R., Müller, U. A., Olsen, R. B., & Pictet, O. V. (1997). From the bird's eye to the microscope: A survey of new stylized facts of the intra-daily foreign exchange markets. *Finance and Stochastics*, *1*(2), 95–129. doi:10.1007007800050018

Güloğlu, B., Kaya, P., & Aydemir, R. (2016). Volatility transmission among Latin American stock markets under structural breaks. *Physica A*, *462*, 330–340. doi:10.1016/j.physa.2016.06.093

Gupta, J., Gregoriou, A., & Ebrahimi, T. (2015). *Using hazard models correctly: a comparison employing different definitions of SMEs financial distress*. Retrieved September 10, 2015, from http://eprints.brighton.ac.uk/14274/

Gurgul, H., & Wojtowicz, T. (2006). Long memory on the German stock exchange. *Czeck Journal of Economics and Finance*, *56*, 447–468.

Hafner, C. M., & Herwartz, H. (2006). A Lagrange multiplier test for causality in variance. *Economics Letters*, *93*(1), 137–141. doi:10.1016/j.econlet.2006.04.008

Hall, M., & Bodenhamer, B. (1997). Figuring Out People - Design Engineering With Meta-Programs. Comer Press.

Hall, G. C., Hutchinson, P. J., & Michaelas, N. (2004). Determinants of the capital structures of European SMEs. *Journal of Business Finance & Accounting*, *31*(5-6), 711–728. doi:10.1111/j.0306-686X.2004.00554.x

Hanousek, J., & Shamshur, A. (2011). A stubborn persistence: Is the stability of leverage ratios determined by the stability of the economy? *Journal of Corporate Finance*, *17*(5), 1360–1376. doi:10.1016/j.jcorpfin.2011.07.004

Hargreaves, S., & Varoufakis, Y. (2003). *Game Theory – A Critical Introduction*. Taylor & Francis Group.

Harris, L. C., & Ogbonna, E. (2001). Leadership style and market orientation: An empirical study. *European Journal of Marketing*, *35*(5-6), 744–764. doi:10.1108/03090560110388196

Harris, M., & Raviv, A. (1990). Capital structure and the informational role of debt. *The Journal of Finance*, *45*(2), 321–349. doi:10.1111/j.1540-6261.1990.tb03693.x

Harris, M., & Raviv, A. (1991). The theory of capital structure. *The Journal of Finance*, *46*(1), 297–355. doi:10.1111/j.1540-6261.1991.tb03753.x

Harvey, A. C. (1993). Long memory in stochastic volatility. In J. Knight & S. Satchell (Eds.), *Forecasting Volatility in Financial Markets* (pp. 307–320). Oxford, UK: Butterworth-Haineman.

Harvey, C. (1995). Preditable risk and returns in emerging markets. *Journal of Financial Studies*, *8*(3), 773–816. doi:10.1093/rfs/8.3.773

Harvey, C., & Viskanta, T. (1996). Expected returns and volatility in 135 countries. *Journal of Portfolio Management*, *22*(3), 46–58. doi:10.3905/jpm.1996.409554

Hausdorff, F. (1919). Dimesion und Ausseres Mass. *Mathematische Annalen*, *79*(1-2), 157–179. doi:10.1007/BF01457179

Haykin, S. (2006). *Neural networks a comprehensive foundations*. Hamilton, Canada: McMaser University.

Heath, C., & Tversky, A. (1991). Preference and belief: Ambiguity and competence in choice under uncertainty. *Journal of Risk and Uncertainty*, *4*(1), 5–28. doi:10.1007/BF00057884

Heide, D. V. D. (Ed.). (2016). *Rentenversicherung in Zeitreihen* [Pension insurance in time series]. Berlin, Germany: German Federal Pension Insurance. Retrieved from http://www.deutsche-rentenversicherung.de/cae/servlet/contentblob/238700/publicationFile/50913/rv_in_zeitreihen.pdf

Heiens, R. A., Leach, R. T., & McGrath, L. C. (2007). The contribution of intangible assets and expenditures to shareholder value. *Journal of Strategic Marketing*, *15*(3), 149–159. doi:10.1080/09652540701319011

Hinz, R., Zviniene, A., Biletskyy, S., & Bogomolova, T. (2009, October). *The Long Term Impact of the Financial Crisis on Public Pension System Financing: Stress Testing Models of Mandatory Pension Systems in Middle Income and Developing Countries*. Discussion Paper provided for the Expert Group Meeting on Population Ageing, Intergenerational Transfers and Social Protection, Santiago, Chile. Retrieved from http://www.cepal.org/sites/default/files/events/files/hinz.pdf

Hiremath, G. S., & Bandi, K. (2010). Long memory in stock Market volatility: Evidence from India. *Artha Vijnana, 52*(4), 332–345. doi:10.21648/arthavij/2010/v52/i4/115316

Hirshleifer, D. (2001). Investor psychology and asset pricing. *The Journal of Finance, 56*(4), 1533–1597. doi:10.1111/0022-1082.00379

Hodrick, R. J., & Prescott, E. C. (1997). Postwar U.S. Business Cycles: An Empirical Investigation. *Journal of Money, Credit and Banking, 29*(1), 1–16. doi:10.2307/2953682

Holden, K., Peel, D. A., & Thomson, J. L. (1990). *Economic forecasting: an introduction*. Press Syndicate of the University of Cambridge.

Holmes, S., & Kent, P. (1991). An empirical analysis of the financial structure of small and large Australian manufacturing enterprises. *Journal of Entrepreneurial Finance, 1*(2), 141–154.

Hong, H., & Stein, J. (1999). A unified theory of underreaction, momentum trading and overreaction in asset markets. *The Journal of Finance, 54*(6), 2143–2184. doi:10.1111/0022-1082.00184

Hoque, Z., Mia, L., & Alam, M. (2001). Market competition, computer-aided manufacturing and use of multiple performance measures: An empirical study. *The British Accounting Review, 33*(1), 23–45. doi:10.1006/bare.2000.0149

Hosking, J. R. M. (1981). Fractional differencing. *Biometrika, 68*(1), 165–176. doi:10.1093/biomet/68.1.165

Hulten, C. R., & Hao, X. (2008). *What is a company really worth? Intangible capital and the market to book value puzzle*. Retrieved January 15, 2009, from https://papers.ssrn.com/sol3/papers.cfm?abstract_id=1327221

Hurst, H. E. (1951). Long-term Storage of Reservoirs: An Experimental Study. *Transactions of the American Society of Civil Engineers, 116*, 770–799.

Inaoka, H., Ninomiya, T., Taniguchi, K., Shimizu, T., & Takayasu, H. (2004). *Fractal Network derived from banking transaction – An analysis of network structures formed by financial institutions*. Bank of Japan Working Paper. Retrieved in January 06, 2018, from: http://www.boj.or.jp/en/ ronbun/04/data/wp04e04.pdf

International Labour Organization. (2014). *World Social Protection Report 2014/15: Building economic recovery, inclusive development and social justice*. Geneva, Switzerland: International Labour Office. Retrieved from http://www.ilo.org/wcmsp5/groups/public/---dgreports/---dcomm/documents/publication/wcms_245201.pdf

International Monetary Fund IMF. (1999). *Financial Sector Crises and Reconstruction, Lessons From Asia*. Washington, DC: World Economic Outlook.

Jayaratne, J., & Wolken, J. (1999). How important are small banks to small business lending? New evidence from a survey of small Firms. *Journal of Banking & Finance, 23*(2-4), 427–458. doi:10.1016/S0378-4266(98)00085-5

Jefferis, K., & Thupayagale, P. (2008). Long memory in Southern African stock markets. *The South African Journal of Economics, 75*(3), 384–398. doi:10.1111/j.1813-6982.2008.00200.x

Joia, L. A. (2000). Measuring intangible corporate assets: Linking business strategy with intellectual capital. *Journal of Intellectual Capital, 1*(1), 68–84. doi:10.1108/14691930010371636

Jomo, K. S. (2001). *Growth after the Asian Crisis What Remains of the East Asian Model?* Retrieved from http://ksghome.harvard.edu/~drodrik/g24-jomo.pdf

Jones, C. M., Kaul, G., & Lipson, M. L. (1994). Transactions, volume, and volatility. *Review of Financial Studies*, *7*(4), 631–651. doi:10.1093/rfs/7.4.631

Kahnemam, D., Slovic, P., & Tversky, A. (1982). *Judgment Under Uncertainty: Heuristics and Biases*. New York: Cambridge University Press. doi:10.1017/CBO9780511809477

Kalmykova, L. (2007). *Fundamental analysis of financial markets*. Peter.

Kaminsky, G., & Reinhart, C. M. (1999). *The Twin Crises: The Causes of Banking and Balance-of-Payments Problems*. Retrieved from http://home.gwu.edu/~graciela/HOME-PAGE/RESEARCH-WORK/WORKING-PAPERS/twin-crises.pdf

Kamışlı, M., Kamışlı, S., & Sevil, G. (2016). The Effects of Crises on Volatility Spillovers between Borsa Istanbul Sector Indexes. *Advances in Economics and Business*, *4*(7), 339–344. doi:10.13189/aeb.2016.040703

Kang, S. H., & Yoon, S. M. (2008). Long memory features in the high frequency data of the Korean stock market. *Physica A*, *5*, 189–196.

Karapetyants, N. K., & Ginsburg, A. I. (1995). Fractional integrodifferentiation in Hölder classes of arbitrary order. *Georgian Mathematical Journal*, *2*(2), 141–150. doi:10.1007/BF02257475

Katcher, B. L. (2003). Make more strategies use of customer satisfaction surveys. *The Journal of Business Strategy*, *24*(1), 34–37. doi:10.1108/02756660310734893

Kawata, R., & Kijima, M. (2007). Value-at-risk in a market subject to regime switching. *Quantitative Finance*, *7*(6), 609–619. doi:10.1080/14697680601161795

Keasey, K., Pindado, J., & Rodrigues, L. (2015). The determinants of the costs of financial distress in SMEs. *International Small Business Journal*, *33*(8), 862–881. doi:10.1177/0266242614529317

Keasey, K., & Watson, R. (1993). *Small Firm Management: Ownership, Finance and Performance*. Blackwell Business.

Kennard, E. H., & Kennard, E. H. (1938). *Kinetic theory of gases: with an introduction to statistical mechanics*. New York.

Kheifets, B. A. (Ed.). (2014). Russia and the BRICS. New opportunities for mutual investment. Moscow: Publishing and Trading Corporation "Dashkov and K".

Kindleberger, C., & Aliber, R. Z. (2005). *Manias, Panics and Crashes - A History of Financial Crises* (5th ed.). Palgrave Macmillan. doi:10.1057/9780230628045

Koksal, B., Orman, C., & Oduncu, A. (2013). *Determinants of capital structure: evidence from a major emerging market economy*. Retrieved July 19, 2013, from https://mpra.ub.uni-muenchen.de/48415/

Kondratev, N. D. (1922). *Mirovoye khozyaystvo i ego koniunktury vo vremia i posle voyny* [The world economy and its conjuncture during and after the war]. Vologda, Russia: Oblastnoe Otdelenie Gosudarstvennogo Izdatelstva.

Koong, C. S., Tsui, A. K., & Chan, W. S. (1997). On tests for long memory in Pacific Basin stock returns. *Mathematics and Computers in Simulation*, *43*(3-6), 445–449. doi:10.1016/S0378-4754(97)00030-X

Korkmaz, T., & Çevik, E. İ. (2009). Zımni Volatilite Endeksinden Gelişmekte Olan Piyasalara Yönelik Volatilite Yayılma Etkisi. *BDDK Bankacılık ve Finansal Piyasalar Dergisi*, *3*(2), 87–105.

Korkmaz, T., Cevik, E. I., & Ozatae, N. (2009). Testing for Long Memory in ISE using ARFIMA-FIGARCH Model and Structural Break Test. *International Research Journal of Finance and Economics*, *26*, 187–191.

Kovel, M. (2007). *Trend Following: How Great Traders Make Millions in Up or Down Markets*. FT Press.

Kristoufek, L. (2010). Local scaling properties and market turning points at prague stock exchange. *Acta Physica Polonica B*, *41*(6), 1223–1336.

Kristoufek, L. (2011). Multifractal height cross-correlation analysis: A new method for analysing long-range cross-correlations. *EPL*, *95*(6), 68001. doi:10.1209/0295-5075/95/68001

Kristoufek, L. (2013). Fractal Markets Hypothesis and the Global Financial Crisis: Scaling, Investment Horizons and Liquidity. *Scientific Reports*, *3*, 2857. doi:10.1038rep02857 PMID:24091386

Krugman, P. (1979). A Model of Balance-of-Payments Crises. *Journal of Money, Credit and Banking*, *11*(3), 311–325. doi:10.2307/1991793

Krugman, P. (2008). *The Return of Depression Economics* (1st ed.). New York: W.W. Norton & Company.

Kuester, K., Mittnik, S., & Paolella, M. (2006). Value-at-risk prediction: A comparison of alternative strategies. *Journal of Financial Econometrics*, *4*(1), 53–89. doi:10.1093/jjfinec/nbj002

Kupiec, P. (1995). Techniques for verifying the accuracy of risk management models. *Journal of Derivatives*, *3*(2), 73–84. doi:10.3905/jod.1995.407942

Kurosh, A. G., Markushevich, A. I., & Rashevskij, P. K. (Eds.). (1948). *Matematika v SSSR za 30 let (1917-1947)*. Moscow, Russia: OGIZ.

Kutlar, A., & Torun, P. (2014). *Gelişmiş Piyasa Ekonomileri İle Yükselen Piyasa Ekonomileri Arasindaki Volatilite Dinamiklerinin Ekonometrik Analizi*. The 24th International Conference of The International Trade and Finance Association, Kayseri, Turkey.

Laletina, A. (2015). Prognozirovanie finansovyh ryncov s ispol'zovaniem iskusstvennyh neyronnyh setey. Elektronnye sredstva I sistemy upravleniya. *Tomskiy gosudarstvennyj universitet system upravleniyai radioelektroniki*, 214-217.

Landkof, N. S. (1973). *Foundations of Modern Potential Theory*. Providence, RI: Springer.

Lanouar, C., & Ajmi, A. N. (2013). The Tunisian stock market index volatility: Long memory vs. switching regime. *Emerging Markets Review*, *16*, 170–182. doi:10.1016/j.ememar.2013.05.003

Lenz, R. T. (1981). Determinants' of organizational performance: An interdisciplinary review. *Strategic Management Journal*, *2*(2), 131–154. doi:10.1002mj.4250020204

Leonov, M. V. (2011). Osnovnyye problemy reformirovaniya pensionnykh sistem v stranakh byvshego SSSR [Main problems of pension systems reforming in former USSR countries]. *Ekonomika, predprinimatelstvo i pravo*, *1*(11), 37–54.

Lim, S. C., Macias, A. J., & Moeller, T. (2014). *Intangible assets and capital structure*. Retrieved October 25, 2014, from https://papers.ssrn.com/sol3/Papers.cfm?abstract_id=2514551

Li, M.-Y. L., & Lin, H.-W. W. (2004). Estimating value-at-risk via Markov-switching ARCH models-an empirical study on stock index returns. *Applied Economics Letters*, *11*(11), 679–691. doi:10.1080/1350485042000236539

Lindeman, D., Rutkowski, M., & Sluchynskyy, O. (2000). *The evolution of pension systems in Eastern Europe and Central Asia: opportunities, constraints, dilemmas and emerging best practices*. Washington, DC: The World Bank. Retrieved from www.oecd.org/daf/insurance-pensions/

Lin, X., & Fei, F. (2013). Long memory revisit in Chinese stock markets: Based on GARCH-class models and multiscale analysis. *Economic Modelling*, *31*, 265–275. doi:10.1016/j.econmod.2012.11.037

Liu, Y., Cizeau, P., Gopikrishnan, P., Meyer, M., Peng, C. K., & Stanley, H. E. (1998). Volatility studies of the S&P 500 index. Center for Polymer Studies and Department of Physics, Boston University.

Li, Y., & Giles, D. E. (2015). Modelling Volatility Spillover Effects between Developed Stock Markets and Asian Emerging Stock Markets. *International Journal of Finance & Economics*, *20*(2), 155–177. doi:10.1002/ijfe.1506

Lo, A. W. (1991). Long-term memory in stock market prices. *Econometrica*, *59*(5), 1279–1313. doi:10.2307/2938368

Los, C. A. (2003). *Financial Market Risk: Measurement & Analysis*. London, UK: Taylor & Francis Books Ltd. doi:10.4324/9780203987636

Lucas, D. J., & McDonald, R. L. (1990). Equity issues and stock price dynamics. *The Journal of Finance*, *45*(4), 1019–1043. doi:10.1111/j.1540-6261.1990.tb02425.x

Luriya, A. (1998). Language and Conscious. Publishing House of the Moscow State University.

Mac An Bhaird, C. (2010). *Resourcing Small and Medium Sized Enterprises: A Financial Growth Life Cycle Approach*. Berlin, Germany: Springer Science & Business Media. doi:10.1007/978-3-7908-2399-8

Maddison, A. (2010). Statistics on World Population, GDP and Per Capita GDP, 1-2008 AD (Vertical file, copyright Angus Maddison, university of Groningen) [Data file]. Retrieved from http://www.ggdc.net/maddison/oriindex.htm

Majorana, E. (1942). Il valore delle leggi statistiche nella fisica e nelle scienze sociali. *Scientia*, *71*, 58–66.

Maksishko, N. K., Perepelica, V. A., & Zahovalko, T. V. (2008). Ispol'zovanie metodov nelinejnoj dinamiki dlja ocenki stepeni dostovernosti informacii, predostavljaemoj nalogoplatel'shhikom. *Vestnik nacional'nogo Zaporozhskogo universiteta, 1*(3).

Mandelbrot, B. B. (2002). *Gaussian Self-affinity and Fractals.* Springer.

Mandelbrot, B. B. (1963). The variation of certain speculative prices. *The Journal of Business*, *36*(4), 394–419. doi:10.1086/294632

Mandelbrot, B. B. (1971). A fast fractional Gaussian noise generator. *Water Resources Research*, *7*(3), 543–553. doi:10.1029/WR007i003p00543

Mandelbrot, B. B. (1972). Possible refinement of the lognormal hypothesis concerning the distribution of energy dissipation in intermittent turbulence. In M. Rosenblatt & C. Van Atta (Eds.), *Statistical Models and Turbulence*. New York: Springer. doi:10.1007/3-540-05716-1_20

Mandelbrot, B. B. (1997). *Fractals and scaling in finance: Discontinuity, concentration, risk*. New York: Springer-Verlag. doi:10.1007/978-1-4757-2763-0

Mandelbrot, B. B., & Van Ness, J. W. (1968). Fractional Brownian motions, fractional noises and applications. *SIAM Review*, *10*(4), 422–437. doi:10.1137/1010093

Mandelbrot, B., & Hudson, R. L. (2006). *The Misbehavior of Markets: A fractal view of financial turbulence*. Basic Books.

Mantegna, R. N., & Stanley, H. E. (1999). *Introduction to Econophysics: Correlations and Complexity in Finance*. Academic Press.

Mantegna, R. N., & Stanley, H. E. (2000). An Introduction to Econophysics. Cambridge, UK: Cambridge University Press.

Mantegna, R. N., & Stanley, H. E. (1995). Scaling behavior in the dynamics of an economic index. *Nature*, *376*(6535), 46–49. doi:10.1038/376046a0

Marimoutou, V., Raggad, B., & Trabelsi, A. (2009). Extreme value theory and value-at-risk: Application to oil market. *Energy Economics*, *31*(4), 519–530. doi:10.1016/j.eneco.2009.02.005

Markowitz, H. (1952). Portfolio Selection. *The Journal of Finance*, *7*, 77–91.

Markowitz, H. (1959). *Portfolio Selection: Efficient Diversification of Investment*. New York: Wiley.

Marrocu, E., Paci, R., & Pontis, M. (2012). Intangible capital and firms' productivity. *Industrial and Corporate Change*, *21*(2), 377–402. doi:10.1093/icc/dtr042

MATLAB. (n.d.). Retrieved from http://www.mathworks.com/products/matlab/

Mayhew, L. (2001). *A Comparative Analysis of the UK Pension System Including the Views of Ten Pension Experts*. SEERL – Birkbeck College, University of London. Retrieved from https://www.researchgate.net/publication/255608035_A_Comparative_Analysis_of_the_UK_Pension_System_Including_the_Views_of_Ten_Pension_Experts

Mc Neil, A., & Frey, R. (2000). Estimation of tail related risk measures for heteroscedastic financial time series: An extreme value approach. *Journal of Empirical Finance*, *7*(3-4), 271–300. doi:10.1016/S0927-5398(00)00012-8

McMillan, D. G., & Thupayagale, P. (2009). Efficiency of the South African equity market. *Applied Financial Economics Letters*, *4*(5), 327–330. doi:10.1080/17446540701720717

McNair, C. J., Lynch, R. L., & Cross, K. F. (1990). Do financial and non financial performance measures have to agree? *Management Accounting Research*, *75*(5), 28.

Melihova, O. A., Veprintseva, O. V., & Chumichev, V. S. (2015). *Rezhimyobucheniya v iskusstvennyh neyronnyh setyah*. YUFU.

Mendes, B. V. de M., & Kolev, N. (2006). *How Long Memory in Volatility Affects True Dependence Structure* (Working Paper). Department of Statistics, Federal University, Brazil.

Mera, K. M., & Renaud, B. (2016). Asia's financial crisis and the role of real estate. New York: Routledge.

Merton, R. (1973). Theory of Rational Option Pricing. *The Bell Journal of Economics and Management Science*, *4*(1), 141–183. doi:10.2307/3003143

Merton, R., & Scholes, B. (1972). The Valuation of Options Contracts and a Test of Market Efficiency. *The Journal of Finance*, *27*(2).

Michaelas, N., Chittenden, F., & Poutziouris, P. (1999). Financial policy and capital structure choice in UK SMEs: Empirical evidence from company panel data. *Small Business Economics*, *12*(2), 113–130. doi:10.1023/A:1008010724051

Miller, G. (1956). The magical number seven, plus or minus two: Some limits on our capacity for processing information. *Psychological Review*, *63*(2), 81–97. doi:10.1037/h0043158 PMID:13310704

Mishkin, F. (1996). Understanding Financial Crises: A Developing Country Perspective. In *Annual World Bank Conference on Development Economics*. The World Bank. 10.3386/w5600

Mishkin, F. (2001). *Financial Policies and the Prevention of Financial Crises in Emerging Market Countries*. NBER Working Paper, 8087.

Mishkin, F. (2015). *The Economics of Money, Banking and Financial Markets* (11th ed.). London: Pearson Education.

Miyakoshi, T. (2003). Spillovers of stock return volatility to Asian equity markets from Japan and the US. *Journal of International Financial Markets, Institutions and Money*, *13*(4), 383–399. doi:10.1016/S1042-4431(03)00015-5

Mladen, L. (2012). A comparative review over the pension systems performance in Central and Eastern European countries. *Journal of Knowledge Management, Economics and Information Technology, 2*(5), 1–11. Retrieved from http://www.scientificpapers.org/wp-content/files/1311_Luise_Mladen__A_COMPARATIVE_REVIEW_OVER_THE_PENSION_SYSTEMS_.pdf

Modigliani, F., & Miller, M. H. (1958). The Cost of Capital, Corporation Finance and the Theory of Investment. *The American Economic Review*, *48*(3), 261–297.

Modina, M., Quintiliani, A., & Arnone, M. (2013). Il ruolo delle banche locali e del relationship lending in scenari economici in profondo cambiamento. *Arethuse Rivista di studi economico-gestionali, 2*, 91-106.

Modina, M. (2015). *Credit Rating and Bank-Firm Relationship: New Models to Better Evaluate SMES*. Basingstoke, UK: Palgrave Macmillan. doi:10.1057/9781137496225

Morales, R., Di Matteo, T., Gramatica, R., & Aste, T. (2012). Dynamical generalized Hurst exponent as a tool to monitor unstable periods in financial time series. *Physica A*, *391*(11), 3180–3189. doi:10.1016/j.physa.2012.01.004

Mukherjee, I., Sen, C., & Sarkar, A. (2011). Long memory in stock returns: Insights from the Indian market. *The International Journal of Applied Economics and Finance*, *5*(1), 62–74. doi:10.3923/ijaef.2011.62.74

Müller, U. A., Dacorogna, M. M., Olsen, R. B., Pictet, O. V., Schwarz, M., & Morgenegg, C. (1990). Statistical study of foreign exchange rates, empirical evidence of a price change scaling law, and intraday analysis. *Journal of Banking & Finance*, *14*(6), 1189–1208. doi:10.1016/0378-4266(90)90009-Q

Mundell, R. (2008). The international financial crisis and international monetary system. Paper presentation at Bahcesehir University, İstanbul, Turkey.

Murphy, J. (1999). *Technical Analysis of the Financial Markets*. Institute of Finance.

Murphy, J. (1999). *Technical Analysis of the Financial Markets: A Comprehensive Guide to Trading Methods and Applications*. New York Institute of Finance.

Myers, S. C. (1977). Determinants of corporate borrowing. *Journal of Financial Economics*, *5*(2), 147–175. doi:10.1016/0304-405X(77)90015-0

Myers, S. C. (1984). The capital structure puzzle. *The Journal of Finance*, *39*(3), 575–592. doi:10.2307/2327916

Myers, S. C., & Majluf, N. S. (1984). Corporate financing and investment decisions when firms have information that investors do not have. *Journal of Financial Economics*, *13*(2), 187–221. doi:10.1016/0304-405X(84)90023-0

Nance, D. R., Smith, C. W. Jr, & Smithson, C. W. (1993). On the Determination of Corporate Hedging. *The Journal of Finance*, *48*(1), 267–284. doi:10.1111/j.1540-6261.1993.tb04709.x

Natali, D. (2011). *Pensions after the financial and economic crisis: a comparative analysis of recent reforms in Europe*. ETUI Working Paper 2011.07. doi:10.2139/ssrn.2221827

Natarajan, V. K., Azariah, R. R. S., & Nagarajan, C. P. (2014). Examining mean-volatility spillovers across national stock markets. *Journal of Economics, Finance and Administrative Science*, *19*(36), 55–62. doi:10.1016/j.jefas.2014.01.001

Nath, G. C. (2001). Long memory and Indian stock market: An empirical evidence. *Proceedings of UTI Institute of Capital Markets Conference*.

National Bank of Ukraine. (2017). *Macroeconomic indicators*. Retrieved April 28, 2017, https://bank.gov.ua/control/en/publish/article?art_id=24038763&cat_id=47389

Neely, A. (1999). The performance measurement revolution: Why now and what next? *International Journal of Operations & Production Management, 19*(2), 205–228. doi:10.1108/01443579910247437

Nekrasova, I., & Karnaukhova, O. (2016). Plasticity and Memory in the Financial Markets. In Neuroeconomics and the Decision-Making Process. Hershey, PA: IGI Global.

Nesta. (2014). *Understanding Alternative Finance*. University of Cambridge. Available online at https://www.nesta.org.uk/sites/default/files/understanding-alternative-finance-2014.pdf

Nikolsky, S. M., & Lizorkin, P. I. (1984). Priblizhenie spericheskimi polinomami. *Tr. MIAN SSSR, 166*, 186–200.

Nison, S. (2003). *The Candlestick Course*. Wiley.

Odzijewicz, T., Malinowska, A. B., & Torres, D. F. M. (2013). Fractional variational calculus of variable order. In A. Almeida, L. Castro, & F.-O. Speck (Eds.), *Advances in Harmonic Analysis and Operator Theory: The Stefan Samko Anniversary Volume* (pp. 291–301). Basel, Switzerland: Birkhäuser. doi:10.1007/978-3-0348-0516-2_16

OECD. (2009). *OECD recommendation on core principles of occupational pension regulation*. Retrieved April 24, 2017, from http://www.oecd.org/finance/private-pensions/33619987.pdf

OECD. (2013). *Pensions at a Glance 2013: OECD and G20 Indicators*. Paris: OECD Publishing. doi:10.1787/pension_glance-2013-

OECD. (2013a). SME financing: Roadmap for future work – Addendum. Selected issues and overview of past and current work on SME finance at the OECD. OECD, DAF/CMF(2013)3/ADD1.

OECD. (2015). *Pensions at a Glance 2015: OECD and G20 Indicators*. Paris: OECD Publishing. doi:10.1787/pension_glance-2015-

OECD. (2017). Pension spending (indicator) [Data file]. Retrieved April 24, 2017, from https://data.oecd.org/socialexp/pension-spending.htm#indicator-chart

Ohlson, J. (1980). Financial ratios and the probabilistic prediction of bankruptcy. *Journal of Accounting Research, 18*(1), 109–131. doi:10.2307/2490395

Onali, E., & Goddard, J. (2009). Unifractality and multifractality in the Italian stock market. *International Review of Financial Analysis, 18*(4), 154–16. doi:10.1016/j.irfa.2009.05.001

Onali, E., & Goddard, J. (2011). Are European equity markets efficient? new evidence from fractal analysis. *International Review of Financial Analysis, 20*(2), 59–67. doi:10.1016/j.irfa.2011.02.004

Opler, T., & Titman, S. (1994). Financial Distress and Corporate Performance. *The Journal of Finance, 49*(3), 1015–1040. doi:10.1111/j.1540-6261.1994.tb00086.x

Orenstein, M. A. (2003). Mapping the diffusion of pension innovation. In R. Holzmann, M. A. Orenstein, & M. Rutkowski (Eds.), *Pension Reform in Europe: Process and Progress* (pp. 171–193). Washington, DC: The World Bank. Retrieved from http://www.tulane.edu/~dnelson/PolTransConv/Orenstein.pdf

Osborne, M. F. (1964). Brownian motion in the stock market. In P. H. Cootner (Ed.), *Random Character of Stock Market Prices*. Cambridge, MA: The MIT Press.

Özmen, A., Weber, G. W., & Batmaz, I. (2010). The new Robust CMARS (RCMARS) method. In *Proceedings of 24th Mini EURO Conference* (vol. 12, pp. 362-368). Izmir, Turkey: Izmir University of Economics Press.

Özmen, A., Weber, G. W., Çavuşoğlu, Z., & Defterli, O. (2012). The new robust conic GPLM method with an application to finance: Prediction of credit default. *Journal of Global Optimization*, *56*(2), 233–249. doi:10.100710898-012-9902-7

Parsons, C., & Titman, S. (2009). Empirical capital structure: A review. *Foundations and Trends in Finance*, *3*(1), 1–93. doi:10.1561/0500000018

Perepelica, V. A., Tebueva, F. B., & Temirova, L. G. (2005). *Strukturirovanie dannyh metodami nelinejnoj dinamiki dlja dvuhurovnevogo modelirovanija.* Stavropol: Stavropol'skoe knizhnoe izdatel'stvo.

Peters, E. E. (1994). *Fractal Market Analysis: Applying Chaos Theory to Investment and Economics.* New York: John Wiley and Sons.

Peters, E. (1994). *Fractal Market Analysis – Applying Chaos Theory to Investment and Analysis.* New York: John Wiley & Sons, Inc.

Peters, E. E. (1991). *Chaos and Order in the Capital Markets.* New York: John Wiley and Sons.

Peters, E. E. (1994). *Fractal Market Analysis.* Wiley.

Petersen, C., & Plenborg, T. (2006). Voluntary disclosure and information asymmetry in Denmark. *Journal of International Accounting, Auditing & Taxation*, *15*(2), 127–149. doi:10.1016/j.intaccaudtax.2006.08.004

Pettit, R. R., & Singer, R. F. (1985). Small business finance: A research agenda. *Financial Management*, *14*(3), 47–60. doi:10.2307/3665059

Pindado, J., Rodrigues, L., & Torre, C. (2008). Estimating financial distress likelihood. *Journal of Business Research*, *61*(9), 995–1003. doi:10.1016/j.jbusres.2007.10.006

Pitterle, I., Haufler, F., & Hong, P. (2015). Assessing emerging markets' vulnerability to financial crisis. *Journal of Policy Modeling*, 1–32.

Pownall, R. A. J., & Koedijk, K. G. (1999). Capturing downside risk in financial markets: The case of the Asian crisis. *Journal of International Money and Finance*, *18*(6), 853–870. doi:10.1016/S0261-5606(99)00040-6

Prechter, R. R. Jr, & Parker, W. D. (2007). The Financial/Economic Dichotomy in Social Behavioral Dynamics: The Socionomic Perspective. *Journal of Behavioral Finance*, *8*(2), 84–108. doi:10.1080/15427560701381028

Psillaki, M. (1995). Credit rationing and small and medium sized firms: A tentative of linkage. *Revue Internationale PME*, *8*(3-4), 67–90. doi:10.7202/1008359ar

Psillaki, M., & Daskalakis, N. (2009). Are the determinants of capital structure country or firm specific? *Small Business Economics*, *33*(3), 319–333. doi:10.100711187-008-9103-4

Quintiliani, A. (2017a). Financial Crisis and Local Banks: a Review of Bank-Firm Relationships in Italy. *Corporate Ownership & Control, 14*(3-1), 249-258.

Quintiliani, A. (2017b). Financial distress cost of Italian Small and Medium Enterprises: A predictive and interpretative model. *The Journal of Risk Finance.*

Raible, S. (2000). *Lévy processes in finance: Theory, numerics, and empirical facts* (PhD dissertation). Universität Freiburg.

Rajan, R. G., & Zingales, L. (1995). What do we know about capital structure? Some evidence from international data. *The Journal of Finance*, *50*(5), 1421–1460. doi:10.1111/j.1540-6261.1995.tb05184.x

Ravn, M., & Uhlig, H. (2002). On adjusting the Hodrick-Prescott filter for the frequency of observations. *The Review of Economics and Statistics*, *84*(2), 371–380. doi:10.1162/003465302317411604

Recht, G. (2007). *Fortentwicklung der gesetzlichen Rentenversicherung* [Further development of the statutory pension insurance] [PowerPoint slides]. Retrieved from: https://www.f04.th-koeln.de/imperia/md/content/hp-f04/ivw/forschungundevents/symposium/vortrag_recht.pdf

Robinson, P. M. (2003). Long memory time series. In M. R. Peter (Ed.), *Time series with long memory* (pp. 4–32). Oxford University Press.

Roch, A. (2011). Liquidity risk, price impacts and the replication problem. *Finance and Stochastics, 15*(3), 399–419. doi:10.100700780-011-0156-x

Rodrik, D. (1999). Where Did All the Growth Go? External Shocks, Social Conflict and Growth Collapses. *Journal of Economic Growth, 4*(4), 385–412. doi:10.1023/A:1009863208706

Ross, B., & Samko, S. G. (1995). Fractional integration operator of a variable order in the Holder spaces $H^{\lambda}(S_{n-1})$. *International Journal of Mathematics and Mathematical Sciences, 18*(4), 777–788. doi:10.1155/S0161171295001001

Ross, S. A. (1977). The determinants of financial structure: The incentive-signaling approach. *The Bell Journal of Economics, 8*(1), 23–40. doi:10.2307/3003485

Ross, S. A. (1989). Information and volatility: The no-arbitrage martingale approach to timing and resolution irrelevancy. *The Journal of Finance, 44*(1), 1–17. doi:10.1111/j.1540-6261.1989.tb02401.x

Roubini, N. (2008). 2008 US and Global Economic Outlook and Implications for Financial Markets. In *RGE Monitor*. Stern School of Business, NYU.

Roulstone, D. T. (2011). Discussion of intangible investment and the importance of firm-specific factors in the determination of earnings. *Review of Accounting Studies, 16*(3), 574–580. doi:10.100711142-011-9149-4

Sadique, S., & Silvapulle, P. (2001). Long-term memory in stock market returns: International evidence. *International Journal of Finance & Economics, 6*(1), 59–67. doi:10.1002/ijfe.143

Sakulrat, M., & Ramkishen, S. R. (2001). *Financial Crisis, Bank Restructuring and Foreign Bank Entry: An Analytic Case Study of Thailand.* CIES Working Paper No. 131. 10.2139/ssrn.294579

Samarukha, V. I. (n.d.). *On the role of the BRICS countries in the formation of a multipolar world economy*. Retrieved from https://library.ru/item.asp?id=26226239

Samko, N., & Vakulov, B. (2011). Spherical fractional and hypersingular integrals of variable order in generalized Holder spaces with variable characteristic. *Mathematische Nachrichten, 2-3*(284), 355–369. doi:10.1002/mana.200810113

Samko, S. G. (1995). Fractional integration and differentiation of variable order. *Analysis Mathematica, 21*(3), 213–236. doi:10.1007/BF01911126

Samko, S. G. (1998). Differentiation and integration of variable order and the spaces $L^{p(x)}$. *Contemporary Mathematics, 212*, 203–219. doi:10.1090/conm/212/02884

Samko, S. G., Kilbas, A. A., & Marichev, O. I. (1993). *Fractional integrals and derivatives: theory and applications*. Philadelphia, PA: Gordon and Breach Science Publishers.

Samko, S. G., & Ross, B. (1993). Integration and differentiation to a variable fractional order. *Integral Transforms and Special Functions, 1*(4), 277–300. doi:10.1080/10652469308819027

Samko, S. G., & Vakulov, B. G. (2000). On equivalent norms in fractional order function spaces of continuous functions on the unit sphere. *Fractional Calculus & Applied Analysis, 3*(4), 401–433.

Samuel, Z.-T. (2008). Value-at-risk and conditional extreme value theory via Markov regime switching models. *Journal of Futures Markets*, *28*(2), 155–181. doi:10.1002/fut.20293

Schumpeter, J. A. (1939). *Business cycles: A theoretical, historical and statistical analysis of the capitalist process*. New York: McGraw-Hill Book Company.

Schwager, J., & Turner, S. (1995). Futures: Fundamental Analysis. Wiley.

Schwager, J. (1997). *Technical Analysis*. Wiley.

Schwarz, A. M., Arias, O. S., Zviniene, A., Rudolph, H. P., Eckardt, S., Koettl, J., ... Abels, M. (2014). *The Inverting Pyramid: Pension Systems Facing Demographic Challenges in Europe and Central Asia*. Washington, DC: World Bank. doi:10.1596/978-0-8213-9908-8

Schwarz, J., Beloff, B., & Beaver, E. (2002). Use of sustainability metrics to guide decision-making. *Chemical Engineering Progress*, *98*(7), 58–63.

Scott, J. H. Jr. (1977). Bankruptcy, secured debt, and optimal capital structure. *The Journal of Finance*, *32*(1), 1–19. doi:10.1111/j.1540-6261.1977.tb03237.x

Sergeeva, L. N. (2002). Modelirovanie povedenija jekonomicheskih sistem metodami nelinejnoj dinamiki (teorii haosa). In *Ljudmila Nil'sovna Sergeeva*. Zaporozh'e: ZGU.

Sharp, W. F. (1970). *Portfolio Theory and Capital Markets*. McGraw-Hill.

Shavshukov, V. M. (2015). BRICS in the system of global finance: ten years later. World Economy, 6, 141-160.

Shaw, E., & Mckinnon, R. (1973). *Financial Deepening in Economic Development*. New York: Oxford University Press.

Shedlock, M. (2007). *Russia liquidity problems and other warning signs*. Retrieved from http://globaleconomicanalysis.blogspot.com/2007/10/russia-liquidity-problems-and-other.html

Smith, C. W., & Stulz, R. M. (1985). The Determinants of Firms' Hedging Policies. *Journal of Financial and Quantitative Analysis*, *20*(4), 391–404. doi:10.2307/2330757

Sogorb-Mira, F. (2005). How SME uniqueness affects capital structure: Evidence from A 1994–1998 Spanish data panel. *Small Business Economics*, *25*(5), 447–457. doi:10.100711187-004-6486-8

Sornette, Johansen, & Bouchaud. (1996). Stock market crashes, precursors and replicas. *Journal de Physique I France*, *6*(1), 167–175.

Sornette, D. (2004). *Critical phenomena in natural sciences: chaos, fractals, selforganization, and disorder: concepts and tools*. Springer Verlag.

Sornette, D., & Johansen, A. (1997). Large financial crashes. *Physica A*, *245*(3), 411–422. doi:10.1016/S0378-4371(97)00318-X

Soros, S. (2008). The New Paradigm for Financial Markets: The Credit Crisis of 2008 and What It Means. *Public Affairs*, 208.

Stanislavchik, E. N. (2002). Risk-menedzhment na predprijatii. *Teorija i praktika*, 89.

Stanley & Gabaix. (2006). *Economic Fluctuations and Statistical Physics: Quantifying extremely rare and less rare events in finance*. Academic Press.

Stanley, H. E., Afanasyev, V., Amaral, L. A. N., Buldyrev, S. V., Goldberger, A. L., Havlin, S., ... Viswanathan, G. M. (1996). Anomalous fluctuations in the dynamics of complex systems: From DNA and physiology to Econophysics. *Physica A, 224*(1-2), 302–321. doi:10.1016/0378-4371(95)00409-2

Stanley, H. E., Amaral, L. A. N., Canning, D., Gopikrishnan, P., Lee, Y., & Liu, Y. (1999). Econophysics: Can physicists contribute to the science of economics? *Physica A, 269*(1), 156–169. doi:10.1016/S0378-4371(99)00185-5

State Statistics Service of Ukraine. (2016). *Social protection: Statistical publication.* Retrieved April 24, 2017, from http://www.ukrstat.gov.ua/

StatSoft. (n.d.). Retrieved from http://www.statsoft.ru/products/STATISTICA_Base/

Stauffer, D. (2000). EconoPhysics-A New Area for Computational Statistical Physics? *International Journal of Modern Physics C, 11*(06), 1081–1087. doi:10.1142/S0129183100000948

Stein, J. C. (2002). Information production and capital allocation: Decentralized versus hierarchical firms. *The Journal of Finance, 57*(5), 1891–1921. doi:10.1111/0022-1082.00483

Stulz, R. M. (1984). Optimal hedging policies. *Journal of Financial and Quantitative Analysis, 19*(2), 127–140. doi:10.2307/2330894

Sun, S., & Hu, J. (2014). The Impact of Pension Systems on Financial Development: An Empirical Study. In M. Faure & N. Philipsen (Eds.), *The Role of Law and Regulation in Sustaining Financial Markets* (pp. 54–72). Oxon, UK: Routledge. Retrieved from https://papers.ssrn.com/sol3/papers.cfm?abstract_id=2481749

Takats, E. (2004). *Banking consolidation and small business lending.* Retrieved December 13, 2004, from https://papers.ssrn.com/sol3/papers.cfm?abstract_id=601027

Takayasu, H., Sato, A. H., & Takayasu, M. (1998). *Power law behaviours of dynamic numerical models of stock market prices.* Academic Press.

Takens, F. (1981). *Dynamical Systems and Turbulence.* Berlin: Springer.

Tarasov, V. E. (2010). *Fractional Dynamics: Applications of Fractional Calculus to Dynamics of Particles, Fields and Media.* Berlin, Germany: Springer. doi:10.1007/978-3-642-14003-7

Temin, P. (2016). *"Great Depression." Banking Crises.* Palgrave Macmillan.

Thaler, R., & Johnson, E. J. (1990). Gambling with the house money and trying to breakeven: The effects of prior outcomes or risky choice. *Management Science, 36*(6), 643–660. doi:10.1287/mnsc.36.6.643

The Evolution of Public Pension Plans: Past, Present and Future. (2008). Washington, DC: National Conference on Public Employee Retirement Systems. Retrieved from http://www.ncpers.org/files/evolution_of_public_pensions_2d.pdf

The World Bank. (n.d.). *Indicators, 2000-2016.* Retrieved April 11, 2017, from http: //data.worldbank.org/indicator

Titman, S., & Wessels, R. (1988). The determinants of capital structure choice. *The Journal of Finance, 43*(1), 1–19. doi:10.1111/j.1540-6261.1988.tb02585.x

Toloraya, G. D., & Yarygina, I. Z. (2016). Harmonization of financial relations as a factor in the development of the BRICS. Banking Services, 2-9.

Tosey, P., & Mathison, J. (2006). *Introducing Neuro-Linguistic Programming.* Centre for Management Learning & Development, School of Management, University of Surrey.

Tuesta, D. (2011). A review of the pension systems in Latin America. *BBVA Research, 11*(15), 1–27. Retrieved from https://www.bbvaresearch.com/KETD/fbin/mult/WP_1115_tcm348-255095.pdf

Turovskiy, Y. A. (2014). O pytmodelirovaniya alternativnyh cheloveko-mashinnyh interfeysov. *Aktualnye napravleniya nauchnyh issledovaniy XXI veka: teoriya I praktika, 5*, 249–252.

Turovskiy, Y. A. (2015). Otsenka skorosti raboty neyrokompyuternogo interfeysa, realizovannogo s ispolzovaniem gibridnogo intellekta. *Biomeditsinskaya radioelektronika, 3*, 61–70.

Tversky, A., & Kahneman, D. (1974). Judgment under uncertainty: Heuristics and biases. *Science, 185*(4157), 1124–1130. doi:10.1126cience.185.4157.1124 PMID:17835457

Udell, G. F. (2008). What's in a relationship? The case of commercial lending. *Business Horizons, 51*(2), 93–103. doi:10.1016/j.bushor.2007.10.005

UK Public Spending. (2017). Social Protection: Fiscal Years 1900 to 2020 [Data file]. Retrieved April 24, 2017, from http://www.ukpublicspending.co.uk/spending_chart_1900_2020UKp_16c1li011mcn_40t00t00t_Social_Protection

Unger, B. (2000). Innovation Systems and Innovative Performance: Voice Systems. *Organization Studies, 21*(5), 941–969. doi:10.1177/0170840600215005

Uzzi, B., & Gillespie, J. J. (1999). What small firms get capital and what cost: Notes on the role of social capital and banking networks. *Proceedings of Federal Reserve Bank of Chicago, 778*, 413–444.

Vacha, L., Janda, K., Kristoufek, L., & Zilberman, D. (2013). Time-frequency dynamics of biofuel-fuel-food system. *Energy Econ, 40*, 233–241. doi:10.1016/j.eneco.2013.06.015

Vakulov, B. G. (2001). O dejstvii operatora rissova potenciala kompleksnogo porjadka po Rn v vesovyh prostranstvah Gjol'dera. *Izvestija vuzov. Severo-Kavkazskij region. Estestvennye nauki, 4*, 47-49.

Vakulov, B. G. (2005). Sfericheskie operatory tipa potenciala v vesovyh prostranstvah Gjol'dera peremennogo porjadka. *Vladikavkazskij matematicheskij zhurnal, 7*(2), 26-40.

Vakulov, B. G., & Kochurov, E. S. (2010). Operatory drobnogo integrirovanija i differencirovanija peremennogo porjadka v prostranstvah Gjol'dera $H^{\omega(x,t)}$. *Vladikavkazskij matematicheskij zhurnal, 12*(4), 3-11.

Vakulov, B. G., & Kochurov, E. S. (2011). Ocenki tipa Zigmunda dlja operatorov drobnogo integrirovanija i di_erencirovanija peremennogo porjadka. *Izv. vuzov. Sev. Kavk. region. Estestv. nauki. Specvypusk*, 15-17.

Vakulov, B. G. (1986). Operator tipa potenciala na sfere v obobshhennyh klassah Gjol'dera. *Izv.vuzov. Matem., 11*, 66–69.

Vakulov, B. G., & Drobotov, Yu. E. (2017). Variable order Riesz potential over Ŕn on weighted generalized variable Hölder spaces [Russian, English abstract]. *Siberian Electronic Mathematical Reports, 14*, 647–656.

Vakulov, B. G., & Karapetjanc, N. K. (2003). Operatory tipa potenciala na sfere s osobennostjami na poljusah. *Doklady Akademii Nauk, 392*(2), 151–154.

Vakulov, B. G., Karapetjanc, N. K., & Shankishvili, L. D. (2003). Operatory sfericheskoj svertki so stepenno-logarifmicheskim jadrom v obobshhennyh prostranstvah Gjol'dera. *Izv. vuzov. Matem, 2*, 3–14.

Vakulov, B. G., Kochurov, E. S., & Samko, N. G. (2011). Ocenki tipa Zigmunda dlja operatorov drobnogo integrirovanija i differencirovanija peremennogo porjadka. *Izv. vuzov. Matematika, 6*, 25–34.

Vakulov, B. G., & Samko, S. G. (1987). Ob jekvivalentnyh normirovkah v prostranstvah funkcij drobnoj gladkosti na sfere tipa $C^{\lambda}(S_{n-1})$, $H^{\lambda}(S_{n-1})$. *Izv. vuzov. Matem, 12*, 68–71.

Van Der Wijst, D. (1989). *Financial structure in small business, theory, tests and applications*. Berlin, Germany: Springer-Verlag. doi:10.1007/978-3-642-45656-5

Vandewalle, N., & Ausloos, M. (1998). Multi-affine analysis of typical currency exchange rates. *J. Eur. Phys. B*, *4*(2), 257–261. doi:10.1007100510050376

Venkatraman, S. (1997). Value-at-risk for a mixture of normal distributions: The use of quasi-bayesian estimation techniques. *Economic Perspectives*, 2–13.

Vintizenko, I. G., & Yakovenko, V. S. (2008). Ekonomicheskaya tsiklomatika [Economic cyclomatics]. Moscow, Russia: Finance and Statistics. – Stavropol, Russia: AGRUS.

Vladimirova, D. (2014). Prognozirovanie finansovyh ryncov iskusstvennymi neiyronnymi setyami. Naukaibiznes: Putirazvitiya. *Fond razvitiya nauki I kultury*, 26-42.

Vladimirova, D. B., & Koksharova, A. A. (2014). Prognozirovanie finansovyh rynkov iskusstvennymi neyronnymi setyami. *FGBUVPO Permskiy natsionalnyy issledovatelskiy politehnicheskiy universitet, 3*, 45.

Wagner, H. (2007). *Liquidity problems and the stock market*. Retrieved from http://www.marketoracle.co.uk/Article1811.html

Wang, Z., Bovik, A., Sheikh, H., & Simoncelli, E. (2004). Image quality assessment: From error visibility to structural similarity. *Image Processing, IEEE Transactions on*, *13*(4), 600–612. doi:10.1109/TIP.2003.819861 PMID:15376593

Wasserman, F. (1992). *Neurocomputer Techniques: Theory and Practice*. Moscow: Mir Publ.

Watson, R., & Wilson, N. (2002). Small and medium size enterprise financing: A note on some of the empirical implications of a pecking order. *Journal of Business Finance & Accounting*, *29*(3-4), 557–578. doi:10.1111/1468-5957.00443

Weber, G. W., Batmaz, I., Koksal, G., Taylan, P., & Yerlikaya-Ozkurt, F. (2012). CMARS: A New Contribution to Nonparametric Regression with Multivariate Adaptive Regression Splines Supported by Continuous Optimisation. *Inverse Problems in Science and Engineering, 20*(3), 371–400. Retrieved from http://citeseerx.ist.psu.edu/viewdoc/download?doi=10.1.1.323.5864&rep=rep1&type=pdf

Weber, G. W., Batmaz, I., Köksal, G., Taylan, P., & Yerlikaya-Özkurt, F. (2012). CMARS: A new contribution to nonparametric regression with multivariate adaptive regression splines supported by continuous optimization. *Inverse Problems in Science and Engineering, 20*(3), 371–400. doi:10.1080/17415977.2011.624770

Weber, M., & Camerer, C. F. (1998). The disposition effect in securities trading: An experimental analysis. Journal of Economic Behavior and Organization. *Journal of Economic Behavior & Organization*, *33*(2), 167–184. doi:10.1016/S0167-2681(97)00089-9

Werens, M. (1981). Best approximation on the unit sphere in Rk. In P. Butzer, E. Gärlich, & B. Szökefalvi-Nagy (Eds.), *Proceedings of the Conference held at the Mathematical Research Institute at Oberwolfach Black Forest* (pp. 233-245). Basel, Switzerland: Birkhäuser.

Weron, A., & Weron, R. (2000). Fractal market hypothesis and two power-laws. *Chaos, Solitons, and Fractals*, *11*(1), 289–296. doi:10.1016/S0960-0779(98)00295-1

What is socialbot? (n.d.). In *WhatIs.com*. Retrieved from http://whatis.techtarget.com/definition/socialbot

Whitaker, R. (1999). The early stages of financial distress. *Journal of Economics and Finance*, *23*(2), 123–132. doi:10.1007/BF02745946

Wiener, N. (1923). Differential space. *Journal of Mathematical Physics*, *2*(1-4), 131–174. doi:10.1002apm192321131

World Bank. (2009). *Pensions in crisis: Regional report for Europe and Central Asia*. Washington, DC: The World Bank. Retrieved from http://siteresources.worldbank.org/FINANCIALSECTOR/Resources/Pensions_in_Crisis_Nov09.pdf

World Bank. (2017). World Development Indicators [Data file]. Available from http://databank.worldbank.org/data/reports.aspx?source=2&series=NY.GDP.MKTP.KD.ZG&country=

Yahyaeva, G. E. (2016). *Osnovy teorii neyronnyh setey*. Natsionalnyy Otkrytyy Universitet.

Yeager, J. (1985). *Thinking About Thinking with NLP*. Cupertino, CA: Meta Publications.

Yeldan, E. (2009). *Küresel Kriz ve Türkiye Ekonomisine Olası Etkileri*. Retrieved from http://www.bianet.org/bianet/kategori/biamag/110856/kuresel-kriz-ve-turkiye-ekonomisine-olasi-etkileri

Yerlikaya-Ozkurt, F. (2013). *Refinements, Extensions and Modern Applications of Conic Multivariate Adaptive Regression Splines*. (Doctoral dissertation). Retrieved from http://etd.lib.metu.edu.tr/upload/12616723/index.pdf

Yermo, J., & Severinson, C. (2010). The impact of the financial crisis on defined benefit plans and the need for counter-cyclical funding regulations. *OECD Working Papers on Finance, Insurance and Private Pensions, 3*. doi:10.1787/5km91p3jszxw-en

Yıldırım, S. (2010). 2008 Yılı Küresel Ekonomi Krizinin Dünya ve Türkiye Ekonomisine Etkileri. *KMÜ Sosyal ve Ekonomik Araştırmalar Dergisi*, *12*(18), 47–55.

Zangari, P. (1996). An improved methodology for measuring VaR. In *Risk Metrics Monitor*. Reuters/JP Morgan.

Zmijewski, M. (1984). Methodological issues related to the estimation of financial distress prediction models. *Journal of Accounting Research*, *22*, 59–82. doi:10.2307/2490859

Zumbach, O., & Lynch, P. (2001). *Heterogeneous Volatility Cascade in Financial Markets*. University of Manchester Institute of Science.

About the Contributors

Inna Nekrasova is Associate Professor in Economics Department of Economics, at Southern Federal University. Nekrasova I. defended her dissertation in 1992, at the Russian State Pedagogical University in St. Petersburg. She has been working in the Southern Federal University from 1993 to the present. Nekrasova I. lectures in the following disciplines - Investment, Financial Market and Financial Institutes, Financial Management, Mergers and Acquisitions in the Banking Sector Academic Positions South Federal University (Russia): 2009- Present Associate Professor of chair "Finance and Credit" Rostov Institute (Branch) of Russian State Trade and Economic University (RGTEU) (Russia): 2006-2009 Associate Professor, Head of chair "Finance and Credit" Rostov State University (Russia): 1998 -2006 Associate Professor of chair "Economics and entrepreneurship" Rostov State University (Russia): 1994-1998 Senior lecturer of chair "Economics and entrepreneurship" Ussuriysk State Pedagogical Institute (Russia): 1985-1988 Lecturer of chair "Economic theory."

Oxana Karnauhkova, PhD is an associate professor at the Institute of History and International Relations of Southern Federal University (Russia), the deputy director of the European Union Centre in South West Russia. Her academic background includes BA History and Social Psychology and PhD Cultural Studies with the specific interest in Economic Anthropology and Intercultural Communications. The publication list contains papers, chapters and a monograph highlighting the role of extraeconomic factors such as culture and social norms and patterns in the economic behavior at the macro- and mezzo-level. The achievements in this field are proven through participation and leadership in a number of projects, such as Communicative strategies in multicultural regions, Memory and Securitization in European Union and Neighbourhood, Super-Diversity and Citizenship in the European Union.

Bryan Christiansen is the Chief Executive Officer of Tactical Systems, LLC in Indiana, USA. A former business lecturer at universities in Turkey, Russia, and the USA, he has traveled to 41 countries where he has conducted international business since 1985 in multiple languages and various industries with Global 500 firms and smaller. Christiansen received his Bachelor of Science degree in Marketing at the University of the State of New York in 1996 and his MBA degree at Capella University in 2003. The author of 20 Reference books on business, economics, and psychology, he is currently working on his Doctor of Business Administration degree at Middlesex University in London, England and is expected to graduate in 2020.

* * *

Artem Aldyrev is a specialist of the Department of Calculations and Economic Analysis at the Southern Federal University since 2015. He holds a Master Degree in International Banking. His academic positions include assistant professorship at the Higher School of Economics of the Southern Federal University with focus on Finance and Credit and Management and Finance.

Dmitry Artemenko, PhD in Financial Security, focusing on mechanisms of financial security in banking activities. He served as a visiting professor in managerial accounting at the University of Kiel (Germany). His research topic is Tax Administration in Russia, Mechanisms and Directions of Modernization.

Galina Artemenko, PhD in Economics, associate professor at the Faculty of Economics of Southern Federal University (Russia).

Dmitry Averchenko is a financial analyst for technical analysis at Alpari since 2016. D. Averchenko holds Master Degree in Financial markets and financial instruments. His research focus is on Economics of Organizations and Mathematical Modeling.

Mamed Babaev, MA in Economics, the Faculty of Economics of Southern Federal University (Russia).

Azza Béjaoui is an Assistant Professor in accounting and finance at High Institute of Management of Tunis. She received his PhD in Finance in 2013 and Master in Finance in 2008 from Higher Institute of Management of Tunis, Tunisia. She has more than 8 years teaching experiences. Her research interest includes risk management, financial markets, behavioral finance, portfolio management and econometrics. She has published 3 papers related to her research major, with other papers as work in progress. Her two books entitled Adaptive Markets hypothesis: Reconciliation between Efficient Markets and Behavioral Finance as well as Market Risk- Attempt of Operationalization in the Tunisian Context: Traditional and Behavioral approaches published in French language. She is one of the authors that contribute in the book chapter title, Integrated Model of Perceived Risk and Risk-Reducing Strategies in the Tunisian Stock Market. Ethical and Social Perspectives on Global Business in Emerging Markets, IGI Global, E-Editorial Discovery.

Salim Ben Sassi is an Assistant Professor in Quantitative Methods and Economics at High Institute of Management of Tunis. He received his PhD in Economics in 2012 and Master in quantitative methods in 2006 from Faculty of Economics and Management of Tunis, Tunisia. He has more than 8 years teaching experiences. His research interest includes risk management, financial markets, behavioral finance, portfolio management and econometrics. He has published 4 articles related to his research major, with other papers as work in progress.

İbrahim Bozkurt was born on August 22, 1983, in Istanbul. He received his PhD from Gazi University in 2013, under the supervisor of Metin Kamil Ercan who is head of Finance Science Platform. He worked as an instructor in 2009-2013 and then he joined the faculty at Karatekin University in 2014 as an assistant professor. He also became an associate professor in 2017.

Cem Donmez is from Istanbul, Turkey. He holds degrees from Turkish University BSc in Physics Engineering and from UK Anglia Ruskin University a PhD on Econophysics. He is the leading expert in Quantitative approach analyser impact measurement and optimization. He has been leading teams that build such tools as well as perform exploratory research on Quantitative Econophysical burning questions. He has strong analysing skills also speaks on this topic extensively at statistical as well as conferences and seminar. Specialties: Stock Mechanics; Econophysics and stock market approaches, Financial Statistic analyser, Quantitative and Financial Market analyser, Financial Adviser, University Specialist and Education Counselor.

Yuri Drobotov is a PhD student at Institute of Mathematics, Mechanics and Computer Sciences named after I. I. Vorovich, lead engineer at Azov Sea Research Fisheries Institute, programmer & mathematician at SPE Vibrobit LLC.

Elena Fomicheva, PhD in the World Economy, specialized in Economic Security. She graduated from Rostov State University, Faculty of Economics with the focus on Economic theory. Her 18 years of professional and academic experience is linked with the research of Economics and Management of Russia and Eurasian Union. Since 2010she is an associate professor at the Department of World Economy and International Relations (Southern Federal University, Russia). The teaching experience includes delivering courses on International Economic Organizations, International Economic Security, WTO Regulation System.

Ellah Godwin is a Master's Student at the Nuertingen-Geislingen University.

Rıfat Karakuş received his PhD degree in Business Administration from Kırıkkale University in Turkey. In 2009 he was hired as the Instructor by the Çankırı Karatekin University Technical College and is currently Assistant Professor in Faculty of Economic and Administrative Sciences Banking and Finance Department. He writes and presents widely on firm value, investing and related financial topics.

Galina Kostetskaya is a candidate of Physical and Mathematical Science, Associate Professor, Department of differential and Integral Equations, Institute of Mathematics and Computer Science named after I. I. Vorovich, Southern Federal University.

Shikun Marina S. is a lecturer at the Faculty of Economics of Southern Federal University (Russia), having the background in Economics and Civil Law. She widely publishes her research on International Business, International and Applied Economics.

Andrea Quintiliani, PhD, Assistant Professor in the scientific field "Economics and Business Management". Pegaso Telematic University. Department of Law and Economic Sciences. Piazza Trieste e Trento, 48 – 80132, Naples (NA), Italy. Phone Direction: (+39) 081.19137500. Cell. Phone (+39) 347.6990431. Email address: andrea.quintiliani@unipegaso.it. Brief curriculum: I am currently Assistant Professor in the scientific field "Economics and Business Management" at the Pegaso Telematic University, where he holds courses in "Corporate Finance" and "Business Economics and Management". The research activity, carried out both on a personal basis and as a member of research groups, has developed in the following areas: a) Banking and finance; b) Intangible value; c) Internal rating banking; d) Corporate value; e) SMEs; f) The value creation in manufacturing and financial companies Basel 3, the effects on bank-firm relationship and the role of corporate finance; g) The management of the investment and financing techniques with a focus on the relationship between banks and firms, particularly small and medium-sized (SMEs), especially in light of the introduction of new assessment models credit (i.e. internal rating systems); h) The quality of financial and banking services for SMEs.

Dmitry Shevchenko is an Associate professor at the Southern Federal University, Rostov-on-Don, Russia.

Catherine Sutton-Brady is an Associate Professor of Marketing at the University of Sydney Business School, Australia. Dr Sutton-Brady holds a B.Comm (Hons) and Masters in Business Studies from University College Cork, National University of Ireland and a Ph.D. in International Marketing from the Western Sydney University. Her research concentrates on international marketing and relationships and networks in business markets. She has also published extensively in the area of marketing education. She is a member of the editorial board for Industrial Marketing Management.

Oleg Sviridov is a Professor of Finance and Credit Department of Economic Faculty Southern Federal University. O. Sviridov is the head of the master's program of the Southern Federal University "International Banking". Academic Positions Head of Finance and Credit Department of Economic Faculty Southern Federal University (Russia): 2016 - Present Head of Master's program of the Southern Federal University "International Banking": 2015- Present Head of Department of "State, Municipal Finance and Financial Engineering" of Higher School of Business Southern Federal University: 2011-2015 Deputy Director for Scientific Work of the North Caucasus Research Institute of Economic and Social Problems Southern Federal University: 2010-2011 Defended the doctoral thesis on the theme: "Strategy for the development of Russian commercial banks in the context of financial globalization crisis": 2009 Associate Professor of the Department of Finance and Credit, Rostov State University: 1998 - 2009 Teacher of the Department of Political Economy and Economic Policy of the Rostov State University: 1997-1998 Defended the PhD thesis in Rostov State University: 1995 Postgraduate Studies in Rostov State University: 1987 – 1991.

Boris Vakulov is an Associate Professor at Institute of Mathematics, Mechanics and Computer Sciences named after I. I. Vorovich.

Larisa Yakimova is a professor of Accounting and Taxation at Yuriy Fedkovych Chernivtsi National University, Ukraine. She is a graduate of Rostov State University (now known as Southern Federal University, Russia) where she majored in mathematics (1984). Prior to joining Chernivtsi National University (2014), she worked as an Associate Professor of the Department of Economic Cybernetics and Information Technologies at the Donbas State Technical University, Ukraine. Dr. Yakimova received the PhD degree in Economics in 2008 and the Doctor of Economic Sciences degree in 2014, both in mathematical methods, models, and information technologies in economics from the Simon Kuznets Kharkiv National University of Economics, Ukraine. Her research is in the general area of applied mathematics with particular focus on applications of mathematical techniques and computational tools to model pension systems development processes. She has authored 48 research papers.

Şeyma Yılmaz Küçük was born in 1987. She received her undergraduate degree in Business Administration department of Niğde University. She then completed her master's degree in banking education at Gazi University. She is continuing her finance PhD education at Kırıkkale University. Since 2012, she has been working as a lecturer in banking and insurance department at Çankırı Karatekin University.

Ülkü Yüksel is a Senior Lecturer at the University of Sydney Business School, Australia. Ülkü made the move from industry to an academic career after a 13-year-long professional business career in the international services sector which framed the basis of the context of her research. Her recent research focuses on consumer information processing and decision making in relation to corporate image of globals (corporate social responsibility, company crisis) and consumer (in)action (indecisiveness), consumer spending (overspending, withholding consumption) and branding (brand anthropomorphism, brand extensions). She has published several articles in leading academic journals, informing marketers, managers and social policy makers.

Asli Yuksel Mermod is a Professor of Finance in Marmara University Faculty of Business Administration and Economics. Prof Mermod holds Visiting Professor positions in Webster University in Geneva and Business School of Lausanne (BSL), Switzerland. Her research revolves around banking and corporate bank management, and sustainability in international and local banks and banking systems.

Index

I

J

K

L

M

N

O

P

Q

R

S

T

V

Z

www.ingramcontent.com/pod-product-compliance
Lightning Source LLC
LaVergne TN
LVHW062144300926
849161LV00010B/2622